COLLISIONS AND HALF-COLLISIONS WITH LASERS

COLLISIONS AND HALF-COLLISIONS WITH LASERS

Edited by
N. K. Rahman and C. Guidotti
University of Pisa, Italy

harwood academic publishers
chur - london - paris - utrecht - new york

Published under license by Harwood Academic Publishers GmbH

Harwood Academic Publishers :

Poststrasse 22
7000 Chur
Switzerland

42 William IV Street
London WC2N 4DE
England

58, rue Lhomond
75005 Paris
France

P.O. Box 15053
3501 BB Utrecht
The Netherlands

P.O. Box 786
Cooper Station
New York, NY 10276
United States of America

Library of Congress Cataloging in Publication Data
Main entry under title :

Collisions and half-collisions with lasers.

1. Collisions (Nuclear physics)—Congresses. 2. Photoionization—Congresses. 3. Lasers in chemistry—Congresses. I. Rahman, N. K. (Naseem K.). II. Guidotti, C.
QC794.6.C6C64 1983 539.7'54 83-22735
ISBN 3-7186-0192-3

CONTENTS

CONTENTS

CONTENTS

ACKNOWLEDGEMENT

As stated in the introduction, this book is an offshoot of a reunion of physicists and chemists, theoreticians and experimentalists, all very active in this particular area of research. They, as the authors of this volume, must take all the credit for this book. C.N.R. of Italy, and in particular I.C.Q.E.M. of Pisa, provided full support fpr this venture. Prof. O.Salvetti, Director of I.C.Q.E.M., must be singled out as the person without whose whole hearted encouragement this *tavola rotonda* could not be held. The technical assistance of Mr.O.Cosci and Ms. S.Grassini is gratefully acknowledged. Mr. A.Biagi, Ms. C.Grassini and Mr. P.Palla helped in running the meeting smoothly.

N.K.R.

C.G.

INTRODUCTION

Last year, a *tavola rotonda* (which is a cross between a workshop and a meeting and which when translated literally into English does not convey what it is) was held at Pisa, and this was a sequel to another one also held at Pisa a year before with the title "Photon-assisted Collisions and Related Topics". The scope of the second meeting was somewhat broader than the first, and the participants agreed that one could publish a book on what transpired at the meeting.

To cover the broadened nature of the meeting as well as to underscore the logical connection between some of the contributions, which at first sight might appear quite disparate, we chose the title "Collisions and Half-Collisions with Lasers". The heart of the matter is continuum (molecular, atomic, etc.) and how this is either probed gently to know what it is like or perturbed strongly to produce something new. This was to be done by photons, photons of a laser, which make some of the low probability processes become visible sometimes literally!

Continuum is normally the purview of two different branches of research. The first is the scattering where both the initial and the final states of a process are the continuum states. The second are the bound-free (or free-bound) transitions where the final (or the initial) state of the system is in the continuum. The experiments on photoionization and photodissociation are of this kind. With strong electromagnetic field both of these, i.e., collisional and half-collisional processes, take on a look which often is entirely different from the pre-laser days.

The book is not divided into sections, but the articles are arranged in a certain order. The first four articles concern theoretical treatment of atom-atom and atom-molecule collisions under strong radiations. A rather diverse group of researchers, using even more diverse theoretical methods, have investigated a variety of problems. This is followed by reports on current experiments in the area.

The next group of contributions is mostly theoretical and treats photon-electron atomic collisions (PEAC). This is a field of much concern for theorists because certain new effects are easily conceived, and the first theoretical treatment of these (unlike atom-atom or atom-molecule collisions) does not necessarily involve heavy numerical effort. It is hoped that the expectation of the theorists in this area would be matched soon by experimental programs, so that certain balance is restored, for this is surely not an abstract field of research.

The lack of experiments for PEAC is in contrast to the half-collision experiments with strong lasers on atoms. Two articles from the groups of Saclay and FOM-Institute bring out the most intriguing recent results of this area. A theoretical work on the other hand discusses how calculations are done for multiphoton ionization for real (i.e., non-hydrogenic) atoms. We come back now to atom-atom collisions but in an entirely different manner from the first. It is well known that line shapes are modified by collisions. This section is a typical example of how this basic phenomenon can appear in surprisingly rich and varied contexts. We see how collisions produce coherence in the four-wave mixing, how atom-atom interaction can produce a competitive third harmonic field with respect to ionization, and how resonance fluorescence in presence of collisions give us informations on atom-atom collisions.

The group of articles that bring up the last part of the book treats molecules excited, dissociated, or predissociated by single or multiple absorption of photons. This is today too vast a field to do justice to with a few articles. However, it is to be hoped that these contributions will convince readers that there is some significant new work in this area which is of a broader interest than the more routine work that one sometimes sees in the literature.

As editors, we hope the *tavola rotonda* of Pisa, which, was so successful as a meeting place for exchanges of ideas, leaves a somewhat permanent impression through this book. It is not a comprehensive monograph in any one field. On the other hand, we feel that it is worthwhile to point out that there is a common theme (and sometimes this may not be very apparent) among areas of research which tend not to communicate with each other. We believe at Pisa a theme was found and from what we hear and see, the theme will continue to flourish by the efforts of others, at least in the next few years. What more can we ask?

N.K. Rahman
C. Guidotti
Istituto di Chimica Fisica
dell'Università
Pisa, Italy

RADIATIVE COLLISION INDUCED ELECTRON CONTINUUM-CONTINUUM SCATTERING

MUNIR H. NAYFEH AND G. B. HILLARD
Department of Physics, 1110 West Green Street, University of Illinois at Urbana-Champaign Urbana, Illinois 61801

Abstract The two-photon radiative collision is examined when the continuum states of one of the atoms participate in the interaction. We find that the energy spectrum of the electrons produced consists of energies other than what is expected from direct photoionization. The additional energies are a result of the participation of the discrete levels of the two atoms that has been encountered previously (coherent excitation or deexcitation of superpositions of states of the two different types of the atoms). As a result of this coherent superposition, an intensity modulated C_6 is introduced in the potentials of the final scattering channel. This intensity-induced collisional shift which, when used to cancel out the collisional dephasing between the initial and final scattering states at wide range of internuclear separations, enhances the electron continuum scattering process.

Two photon laser induced radiative collisions have been observed recently[1] during the radiative collision between Ba and Tℓ ground state atoms. In the process two photons are absorbed which are not in resonance with any of the transitions in the atoms and which result in the simultaneous excitation of both atoms. We recently analyzed the theoretical aspect of this process using a semiclassical approach.[2-5] We found that there are two competing types of interactions. In the first, the laser interacts with one type of atom only resulting in a coherent superposition of two states of this type. A subsequent collision with an atom of the other type then produces the final scattering states. In the second type of interaction the laser excites a coherent superposition of two states of the two different types of atoms with the help of a collision. A subsequent collision between the two types then produces the final scattering states. In this paper we call the two types of interactions two photon-one collision, and two photon-two-collision cases. In the latter case we found a new intensity-induced collisional phase shift. This shift is used to control the overall phase between the initial and final scattering states. When the phase difference goes through zero (phase resonance) the cross section is enhanced and the two photon lineshape becomes both symmetric and highly sensitive to the intensity of the radiation.

In this paper we study the implications of the above two photon-two collision laser induced radiative collision process when it involves the continuum states of one of the colliding atoms.[6] We shall examine the photoionization of this atom during its collision with a foreign atom in the presence of an applied radiation field which is nonresonant with either of the atom's discrete transitions.

The transfer of excitation from an excited atom B to an acceptor atom A accompanied by the simultaneous absorption of a single photon sufficient to photoionize the initially excited state of atom B was previously analyzed.[7] The excited acceptor atom A then ionizes atom B via a Penning ionization process resulting in an increase in the photoionization efficiency. In contrast to the one photon-one collision process, we find here that a second order radiative process causes an intensity-induced collisional shift which can be used to cancel out the collisional dephasing shift between the initial and final scattering wavefunctions for a wide range of internuclear separations. As a result of this cancellation we find that the collision causes an enhancement in the cross section for electronic continuum-continuum scattering (i.e. the process which results in the production of electrons with different discrete energies) which is otherwise negligible.

The present process is quite different from the process involving only discrete states. In the discrete case, the frequency of radiation is taken not to coincide with any transition frequency in either atom. However, in the present case where continuum states are involved, the frequency of radiation coincides with some transitions to the continuum. Moreover, due to the nature of the continuum states, the present process involves an infinite level system.

We consider the collision of atoms A and B in the presence of a radiation field $\vec{E} = \vec{E}_o \cos \omega t$ via the process $A + B^* + 2\hbar\omega \rightarrow A + B(\nu) + \hbar\omega \rightarrow A^* + B + \hbar\omega \rightarrow A^{**} + B \rightarrow A^* + B^+ + e(\varepsilon)$, where $e(\varepsilon)$ represents the distribution of electron energies produced for both of the continuum states involved in the scattering process. In describing the process we treat the motion of the nuclei classically. Moreover, we assume that the dominant contribution to the collisional cross section comes from large internuclear separations where electronic overlap is negligible. Hence we represent the system with a product of atomic states, and write

$$\hat{H} = \hat{H}_A + \hat{H}_B + \hat{V}_{AB} - \vec{\mu}_A \cdot \vec{E} - \vec{\mu}_B \cdot \vec{E} \tag{1}$$

where $\hat{H}_A$ and $\hat{H}_B$ are the electronic Hamiltonians of isolated atoms A and B, $V_{AB}(t)$ is the atom-atom interaction, and the other terms are the laser field-atom interaction terms in the dipole-classical field approximation. We will treat the magnetic number degeneracy by treating the atom-atom interaction in the rotating atom approximation where $\hat{V}_{AB}$ matrix elements are evaluated by assuming the dipole transition moments are always aligned along the line joining the nuclei. The state vector of the system is taken to be of the form (see Fig. 1):

$$\begin{aligned}\psi = {} & a_o(t)\,|0a\rangle|1b\rangle \exp(-i\omega_o t) \\ & + \int a_\nu\,|0a\rangle|\nu b\rangle \exp\left[-i(\omega_o+\omega_\nu)t\right]\, d\nu \\ & + a_1(t)\,|1a\rangle|0b\rangle \exp(-i\omega_1 t) \\ & + a_2(t)\,|2a\rangle|0b\rangle \exp\left[-i(\omega_1+\omega_2)t\right] \\ & + \int a_{\nu'}\,|1a\rangle|\nu' b\rangle \exp\left[-i(\omega_{\nu'}+\omega_1)t\right]\, d\nu \qquad (2)\end{aligned}$$

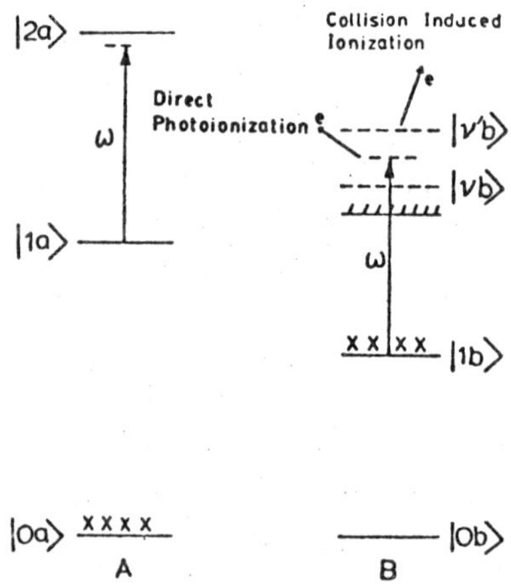

FIGURE 1. Restricted energy level diagram for the two-photon radiative collision involving continuum states, showing the ejection of electrons of different energies.

which is a coherent superposition of states of the two different types of atoms. In the process the initial state $|0a\rangle|1b\rangle$ is excited by the electromagnetic field to the continuum states $|0a\rangle|\nu b\rangle$. For continuum states nearly resonant with the applied field, the interaction results in real excitations and for those away from resonance the interaction results in virtual excitations. A virtual collision then transfers the excitation from $|0a\rangle|\nu b\rangle$ to the state $|1a\rangle|0b\rangle$ which in turn gets virtually excited by the electromagnetic field to $|2a\rangle|0b\rangle$. Finally, a collisional transfer from $|2a\rangle|0b\rangle$ to $|1a\rangle|\nu' b\rangle$ takes place.

During the process two sets of continuum states in atom B become excited, resulting in the emission of electrons with kinetic energies centered at two discrete values. One set of the continuum states becomes involved only as intermediate states in the overall process while the other set helps comprise the final state of our system. The cross section for this process will be obtained by substituting the above wavefunction and Hamiltonian into the time-dependent Schrödinger equation and solving for a_ν, the probability amplitude of the final state, $|1a\rangle|\nu' b\rangle$. Later in the paper we will estimate the cross section for the process using a high lying excited state of atom B for our initial state. The derivation which follows, however, is also applicable to the low lying excited states of atom B.

In this model we have neglected the direct electron scattering process $A + B^+ + e\ (\varepsilon) + \hbar\omega \rightarrow A + B^+ + e\ (\varepsilon')$ since this interaction is expected to be negligible in comparison to the radiative collision process when using low field intensities. The direct scattering process may become important, however, when using very intense fields.[8] In this case, a new set of high-lying continuum states become available to interact with the original continuum states. The interaction between these two overlapping electronic continua is predicted to become an important factor in the treatment of the collision dynamics. Furthermore, the participation of the high-lying continua due to the intense field is predicted to lead to interesting effects which include the emission of electrons having distributions in kinetic energies which are roughly shifted by $\hbar\omega$ on either side of the laser field-free emitted electrons.

Substituting Eqs. 1 and 2 in the time-dependent Schrödinger equation gives:

$$\frac{da_o}{dt} = i\int \mu_{\nu 1} E_o \exp(i\Delta_\nu t)\, a_\nu\, d\nu \tag{3}$$

$$\frac{da_\nu}{dt} = i\mu_{1\nu} E_o \exp(-i\Delta_\nu t)\, a_o - iV_1 \exp(i\bar{\Delta}_\nu t)\, a_1 \tag{4}$$

$$\frac{da_1}{dt} = -i\int V_1^* \exp(-i\bar{\Delta}_\nu t)\, a_\nu\, d\nu + i\mu_{2A} E_o \exp(i\Delta_2 t)\, a_2 \tag{5}$$

$$\frac{da_2}{dt} = i\mu_{2A}^* E_o \exp(-i\Delta_2 t)\, a_1 - i\int V_2 \exp(i\bar{\Delta}_{\nu'} t)\, a_{\nu'}\, d\nu' \tag{6}$$

$$\frac{da_{\nu'}}{dt} = iV_2^* \exp(-i\bar{\Delta}_{\nu'} t)\, a_2 \tag{7}$$

where (see Fig. 2):

$\Delta_\nu = \omega - \omega_\nu$, $\Delta_2 = \omega - \omega_2$, $\bar{\Delta}_\nu = \omega_o + \omega_\nu - \omega_1$,
$\bar{\Delta}_{\nu'} = \omega_1 + \omega_2 - (\omega_{\nu'} + \omega_1) = \omega_2 - \omega_{\nu'}$
$\mu_{1A} = \frac{1}{\hbar} \langle 1a|\hat{\mu}_{A_z}|0a\rangle$, $\mu_{2A} = \frac{1}{\hbar} \langle 2a|\hat{\mu}_{A_z}|1a\rangle$,
$\mu_{o\nu} = \frac{1}{\hbar} \langle \nu b|\hat{\mu}_{B_z}|0b\rangle$, $\mu_{o\nu'} = \frac{1}{\hbar} \langle \nu' b|\hat{\mu}_{B_z}|0b\rangle$,
$\mu_{\nu 1} = \frac{1}{\hbar} \langle \nu b|\hat{\mu}_{B_z}|1b\rangle$, $V_1 = \frac{1}{\hbar} \langle 0b|\langle 1a|\hat{V}_{AB_z}|0a\rangle|\nu b\rangle$
and $V_2 = \frac{1}{\hbar} \langle \nu' b|\langle 1a|\hat{V}_{AB_z}|2a\rangle|0b\rangle$.

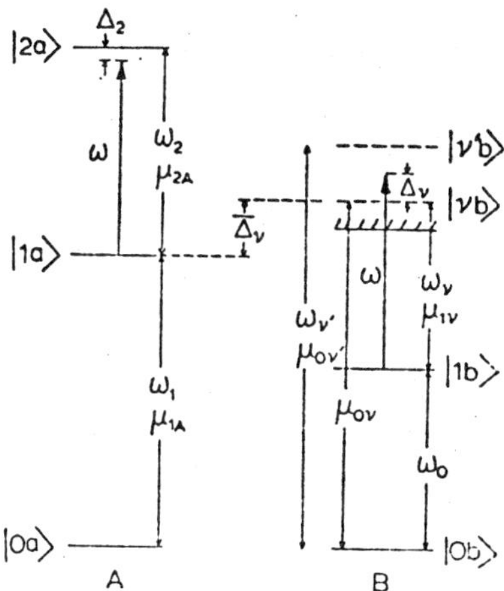

FIGURE 2. Detailed energy level diagram of the process, labelling the transition frequencies, oscillator strengths, and detunings.

We now proceed to reduce this system of infinite coupled levels to an effective two-level system involving only the initial and final states by sequentially eliminating a_ν, a_1 and a_2. This approach is justified if these intermediate states are chosen not to interact strongly with the electromagnetic field nor with the collisional field. In the elimination of the continuum states, however, the electromagnetic field will necessarily resonate with a portion of the continuum. We will show later that this radiatively resonant set of continuum states is responsible for a portion of the ionization current while the nonresonant states contribute primarily to the transfer of excitation.

The first step we carry out is the elimination of the amplitudes of the continuum states a_ν by considering them as intermediates to the transition between a_o and a_1. We integrate Eq. (4) in order to solve for a_ν and hence eliminate it from the rest of the equations. Note that if $|\Delta_\nu| = |\omega - \omega_\nu|$ were always large, we could integrate the first term of Eq. (4) by parts and keep only the zero order term. However, $|\Delta_\nu|$ can be small, and therefore this procedure is not accurate in evaluating the integral. The same argument holds in the contribution from the second integral of Eq. 4. Both cases must be considered in order to treat the problem accurately. In order to do this, we will use the following procedure:[9]

Since transitions from $|0a\rangle|1b\rangle$ to the continuum obey the Frank-Condon principle, we separate the continuum states into two sets; the first set is close to the resonance condition $[\,|\Delta_\nu| < |\Delta_\nu^o|\,]$ where $|\Delta_\nu^o|$ is a small quantity, and the second set is that which is sufficiently far away from resonance $[\,|\Delta_\nu| > |\Delta_\nu^o|\,]$.

When Δ_ν is large, the exponential in the first integral of Eq. 4 oscillates rapidly. If, in addition, the field amplitude E_o is a slowly varying function of time, we may approximate the integral by keeping the first order term in an integration by parts, i.e., $-(\mu_{\nu 1}\, E_o/\Delta_\nu)\, \exp\,(-i\Delta_\nu t) a_0$. However, when near resonant states are considered in the integral, Δ_ν is small and this part of the contribution to a_ν can only be represented by a full integral as $i \int_{-\infty}^{t} \mu_{\nu 1}\, E_o\, \exp\,(-i\Delta_\nu t')\, a_0\, dt'$. Similarly, we can write expressions for the second integral of Eq. 4 corresponding to the cases where $\bar{\Delta}_\nu$ is large and $\bar{\Delta}_\nu$ is small.

Thus substituting these expressions for a_ν in Eqs. 3 and 5 gives

$$\frac{da_o}{dt} + \gamma_o a_o = iG_1\, \exp\,(i\Delta_1 t)\, a_1 \qquad (8)$$

$$\frac{da_1}{dt} - iV_1'^2/\Delta_o' \; a_1 = iG_2 \exp(-i\Delta_1 t) \, a_o$$

$$+ \, i\mu_{2A} E_o \exp(i\Delta_2 t) \, a_2 \qquad (9)$$

where

$$\gamma_o = \pi\mu_R^2 E_o^2 \eta + i\frac{\mu_{1B}^2 E_o^2}{\Delta_1'} \, , \; \eta = \frac{d\nu}{dE} \, , \; \frac{\mu_{1B}^2}{\Delta_1'} = \int \frac{\mu_{\nu 1}^2}{\Delta_\nu} d\nu \, ,$$

$$G_1 = \alpha_1 + i\beta_1 \, , \quad G_2 = \alpha_2 + i\beta_2 \, , \quad \beta_1 = \pi E_o \mu_r V_r \eta \, ,$$

$$\beta_2 = \beta_1^* , \; \alpha_1 = - E_o \int \frac{\mu_{1\nu} V_1}{\bar{\Delta}_\nu} d\nu \, , \quad \alpha_2 = E_o \int \frac{\mu_{\nu 1} V_1^*}{\Delta_\nu} d\nu \, ,$$

$$\text{and } \frac{V_1'^2}{\Delta_o'} = \int \frac{V_1^2}{\bar{\Delta}_\nu} d\nu .$$

Here $\mu_R^2 = \int^r \mu_{1\nu}^2 \, d\nu$ and $\mu_r V_r = \int^r \mu_{1\nu} V_1 \, d\nu$ represent integrals over the radiatively resonant continuum states, $V_R^2 = \int^r V_1^2 \, d\nu$ represents an integral over the collisionally resonant states, $\Delta_1 = \omega_1 - \omega - \omega_o$ and $\eta = \frac{d\nu}{d\Delta}$ is the density of the continuum states. We also note that we used the quantities $\mu_{1B}^2 E_o^2/\Delta_1'$ and $V_1'^2/\Delta_o'$ as a compact way of respectively describing the Stark and collisional frequency shifts covered by all the nonresonant intermediate continuum states. Later in the paper we will, for some cases, show that the resonant states contribution to G_1 and G_2, $\pi E_o \mu_r V_r \eta$, is much smaller than the contribution of the nonresonant states.

We now derive an effective system coupling the initial and final amplitudes a_o and $a_{\nu'}$ respectively. We take $\Delta_1 \gg |G_1|$, $\Delta_2 \gg \mu_{2A} E_o$ and eliminate a_1 and a_2 sequentially from eqs. 8-9 and 6-7. We then integrate their equations by parts and keep the lowest terms. The resulting equations in the weak field limit are:

$$\frac{da_o}{dt} + iS_1 + i\mathrm{Re}\left(\frac{G_1G_2}{\Delta_1}\right) a_o = 0 \tag{10}$$

$$\frac{da_{\nu'}}{dt} + i\frac{V_2^2}{\bar{\Delta}_{\nu'}} a_{\nu'} = \frac{i\mu_{2A}V_2\ E_oG_2}{\Delta_1(\Delta_1 + \Delta_2)} \exp(-i\delta_\nu t)\ a_o \tag{11}$$

The phase S_1 is included phenomenologically to the phase of the ground state to account for the effect of all other nonresonant states which have not been included in the interaction. The quantity G_1G_2/Δ_1 depends on the intensity of the exciting field and on the collisional interaction, and as such constitutes an intensity induced collisional shift. Note that we kept the lowest order of the intensity-induced collisional shift since even in the weak field limit, this shift may be of the same order as the collisional shift of the final state.

The intensity-induced collisional shifts are interesting since their magnitude can be controlled by the intensity of the electromagnetic field. Hence the overall shift of the final state with respect to the ground state can be controlled by the intensity. In fact, it is possible to achieve complete cancellation of the relative phase shift between the initial and final transition probability amplitudes.

When E_o changes very little during the time of collision, Eq. (10) gives

$$a_o = \exp\left[-i\int_{-\infty}^{t}\left[S_1 + \mathrm{Re}\left(\frac{G_1G_2}{\Delta_1}\right)\right]dt\right] \tag{12}$$

and hence

$$|a_{\nu'}(\infty)|^2 = \frac{4\mu_{2A}^2\ E_o^{-4}}{\Delta_1^2\ (\Delta_1 + \Delta_2)^2} \left| \int_o^\infty V_2G_2\ e^{iS}dt\right|^2 \tag{13}$$

$$S = \int \left(\frac{V_2^2}{\bar{\Delta}_{\nu'}} + S_1 + \frac{\alpha_1 \alpha_2 - \beta_1^2}{\Delta_1} - \delta_\nu \right) dt \tag{14}$$

We now calculate the cross section for the continuum-continuum scattering process. Taking $\delta_\nu = 0$, and in the dipole-dipole interaction we find $S_1 = \int C' R^{-6} dt$ and $S = \int_0^t CR^{-6}\, dt$ where

$$C = \frac{\hbar^2 \mu_{o\nu'}^2 \mu_{2A}^2}{\bar{\Delta}_{\nu'}} + \frac{\mu_{1A} E_o^2}{\Delta_1} \left(\int \frac{\mu_{1\nu} \mu_{o\nu}}{\bar{\Delta}_\nu} d\nu \right) \left(\int \frac{\mu_{\nu 1} \mu_{\nu o}}{\Delta_\nu} d\nu \right)$$

$$+ C' - \frac{\pi^2 \hbar^2 \mu_{1A}^2 E_o^2 \left(\int^r \mu_{1\nu} \mu_{\nu o} d\nu \right)}{\Delta_1} \tag{15}$$

and

$$|a_{\nu'}(+\infty)|^2 = 4F_1^2 \; E_o^4 \; \left| \int_o^\infty R^{-6} \cos S \, dt \right|^2 \tag{16}$$

where

$$F_1^2 = \frac{\hbar^4 \mu_{2A}^4 \mu_{o\nu'}^2 \mu_{1A}^2}{\Delta_1^2 (\Delta_1 + \Delta_2)^2} \Big[\pi^2 \left| \int^r \mu_{\nu 1} \mu_{\nu o} d\nu \right|^2 \eta^2 + \left| \int \frac{\mu_{\nu 1} \mu_{\nu o}}{\Delta_\nu} d\nu \right|^2 \Big] \tag{17}$$

The ionization cross section σ is calculated from the integration of $|a_{\nu'}(\infty)|^2$ over the impact parameters. A thermal average of the cross section, $\bar{\sigma}$, then yields an ionization rate. For large C, all impact parameters can be integrated over because the frequency shift becomes large for R values $\lesssim$ 15 Å and there is no change in $a_{\nu'}$ at R values where overlap is important and deviations from straight line trajectory occur.

At some intensities, however, C can become very small even in the weak field limit. The situation where C is very small suggests a large coupling coefficient in the absence of any dephasing effect for all internuclear separations $R > 4Å$. This could lead to large cross sections for the process. However, because of the detuning at small R, orbiting phenomena play a significant role. An estimate of the magnitude of the cross section at the peak of the resonance can be determined from Eq. (13) by taking C = 0. In this case $|a_{\nu'}(\infty)|^2 = 9\pi^2 F_1^2 E_o^4/64\rho^{10}v^2)$. A lower limit on the estimate can be found by calculating the contribution from impact parameters where orbiting is not important; that is $\sigma > \int_{\rho_c}^{\infty} 2\pi\ \rho\, d\rho\, |a_{\nu'}(\infty)|^2 = 9\pi^3 F_1^2\ E_o^4/256V^2\rho_c^8)$ where F_1 is given by Eq. (17) along with the condition C = 0.

We now examine the cross section for a realistic case. The degree of contribution to the phase from the resonant and nonresonant continuum states depends on the atom in question (structure of the continuum), and the position of the resonant state with respect to the ionization limit as well as to the states of the atoms. Here we are interested in arriving at an order of magnitude estimate of the cross section. We take the case where the initial excited state of atom B is regarded as high-lying and hydrogen-like with a principle quantum number n because some of the expressions can be evaluated analytically. In this case the Kramers formulae, when applied to the absorption from bound states to free states, give the following expression for $\mu_{1\nu}^2\eta$ (Ref. 10):

$$\pi\mu_{1\nu}^2\eta = \frac{1}{\pi\sqrt{3}}\left(\frac{2R_y}{\hbar\omega}\right)^4 \frac{a_o^2 e^2}{\hbar R_y}\ \frac{1}{n^3} \qquad (18)$$

where R_y is the Rydberg energy, a_o is the Bohr radius, and e is the electronic charge. The couplings G_1 and G_2 can now be estimated by taking the effective nonresonant continuum state to be a high Rydberg state.[7] Thus

$$G_2 = - E_o \int \frac{\mu_{\nu 1} V_1^*}{\Delta_\nu} d\nu + i\pi\mu_{1\nu} E_o V_1^* \eta$$

$$= - \frac{B_1 E_o}{R^3} (1 - \frac{i}{2} \frac{B_2}{B_1}) \tag{19}$$

$$G_1 = - \frac{B_1' E_o}{R^3} (1 - \frac{i}{2} \frac{B_2'}{B_1'})$$

with

$$B_1 = \hbar(\mu_{1A}\, \mu_{on'}\, \mu_{nn'})/\omega \quad \text{and} \quad \frac{B_2}{B_1} = \frac{\omega_1}{\omega_o n^2}$$

$$B_1' = \hbar(\mu_{1A} \mu_{on'} \mu_{nn'})/\omega' \quad \text{and} \quad \frac{B_2'}{B_1'} = \frac{\omega_1}{\omega_o n^2} \tag{20}$$

where $\mu_{on'}$, $\mu_{nn'}$ are the matrix elements of the transition $0 \to n'$ and $n \to n'$. Taking $\mu_{1A}^2 \sim e^2 a_o^2$, $\mu_{on'}^2 \sim e^2 a_o^2/n^3$ and $\mu_{nn'}^2 \sim e^2 a_o^2/n^4$ gives $B_1 = e^6 a_o^6/(\omega n^7)$ and $B_1' = e^6 a_o^6/(\omega' n^7)$. The fact that $\frac{B_2}{B_1} << 1$ and $\frac{B_2'}{B_1'} << 1$ for $n > 5$ allows us to neglect the imaginary part of G_1 and G_2; hence

$$G_1 = - \frac{-B_1' E_o}{R^3} \quad , \quad G_2 = \frac{-B_1 E_o}{R^3} \tag{21}$$

and consequently

$$C = \frac{\hbar^2 \mu_{o\nu'}^2 \mu_{2A}^2}{\bar{\Delta}_{\nu'}} - \frac{B_1 B_1' E_o^2}{\Delta_1} \quad \text{and} \quad F_1^2 = \frac{\hbar^2 \mu_{2A}^4 \mu_{o\nu'}^2 B_1^2}{\Delta_1^2 (\Delta_1 + \Delta_2)^2} \tag{22}$$

These estimates show, in the case of a high-lying initial state, that the contribution of the resonant continuum states to the intensity induced shift and to the amplitude of $a_{\nu'}$ is negligible compared to that of the nonresonant states. However, these contributions are expected to become more important when the initial scattering states are low lying. We will examine the effect of these contributions in a later work.

We have numerically estimated the cross section for a typical case using the following values. We take $\mu_{1A} = 6.26$ a.u., $\mu_{2A} = .3$ a.u., $\mu_{gn'} = 1$ a.u., $\mu_{nn'} = .75$ a.u., $\mu_{02'} = 3.74 \times 10^{-2}$ a.u., $\Delta_1 = -1000\ \mathrm{cm}^{-1}$, $\Delta_2 = 3050\ \mathrm{cm}^{-1}$, $\bar{\Delta}\nu' = -4050\ \mathrm{cm}^{-1}$, $\omega = \omega' = 12000\ \mathrm{cm}^{-1}$,and $C' = 400$. Using these values the system will achieve phase resonance (C=0) for $E_o = 5\mathrm{x}10^6$ V/cm corresponding to a power density of 33 GW/cm^2. Despite the large power necessary, the quantity $\mu_{1A}^2\ E_o^{\ 2}/\omega^2$, which we require to be much less than 1 to establish the validity of the weak field limit approximation has a value of 2.2×10^{-2}. In obtaining our estimate we retain δ_ν in equation 14. We do this so that we may consider contributions from continuum states close to resonance. Eq. 16 is numerically integrated over time with a fixed value of δ_ν and then integrated over impact parameter. We chose the lower limit on impact parameter to be 3Å to avoid problems with overlap or orbiting. The resulting number represents the contribution to the total cross section from a particular continuum state labeled by the detuning δ_2, corresponding to the stars in Figs. 3 and 4. The lineshape built up in this way is then fitted with a cubic spline (the smooth curve) and the resulting spline numerically integrated. For our case, we find a peak contribution of 4×10^{-4} Å at $\delta_\nu = 0$ and a total integrated cross section of 2.2×10^{-2} Å^2. It is interesting to consider the behavior of the normalised cross section σ/E_o^4. We expect this quantity to have a peak value when E_o is chosen to achieve phase resonance since C = 0 results in the maximum contribution from the cosine in Eq. 16. For larger

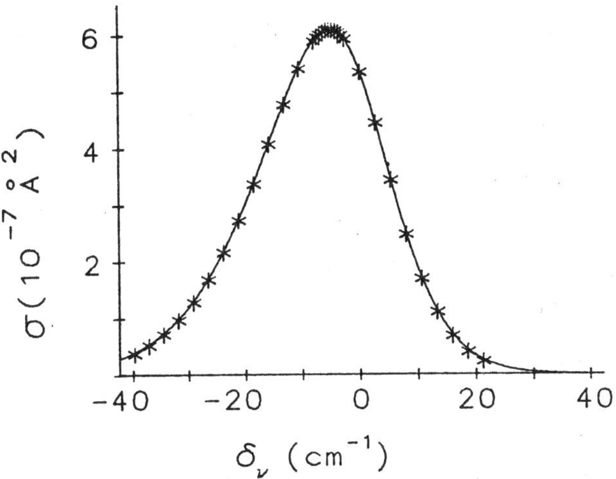

FIGURE 3 The absolute lineshape of the process at field intensity which does not achieve phase resonance, showing asymmetry and shift. Only one continuum final state has been included in the calculation.

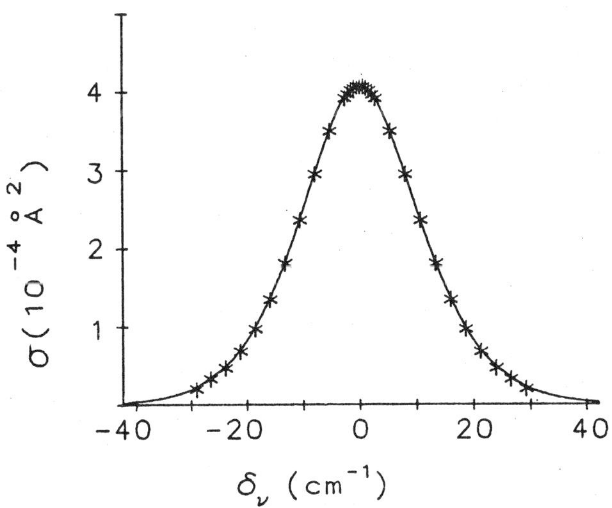

FIGURE 4. The absolute lineshape of the process when one continuum final state participate at phase resonance showing symmetry and no shift.

values of E_o, increasingly rapid oscillations of the cosine drive the normalised cross section to smaller values. For E_o less than the critical value, the parameter C approaches a constant. The lineshape in this case exhibits a (red or blue) wing and is shifted from the $\delta_\nu = 0$ center. An example is shown in Fig. 4 for $E_o = 10^6$ V/cm. Such phenomena were studied in detail for the case of discrete levels in Ref 2. In our case, the total normalised cross section has a value of 6.6×10^{-71} V^4/cm^2 at phase resonance and a limiting value of 6.1×10^{-71} V^4/cm^2 as $E_o \rightarrow 0$.

We note that when other discrete states of atom B are involved, other electrons will be ejected with appropriate energy to conserve the overall energy of the process. Hence due to this effect, a discrete spectrum of electrons is produced.

In conclusion, we have shown that the phase of the wavefunctions of the initial and final collisional states can be externally controlled, thereby allowing a large continuum-continuum scattering cross section. Due to the change in the energy of the electrons, this effect should be amenable to experimental investigations using energy analysis techniques.

This work was supported by NSF Grant PHY 81-09305

REFERENCES

1. J. C. White, Optics Lett. 6, 242 (1981).
2. M. H. Nayfeh and G. B. Hillard, Phys. Rev. A 24, 1409 (1981).
3. M. H. Nayfeh, G. . Hillard, and D. B. Geohegan, in Photon Assisted Collisions and Related Topics, N. Rahman and C. Guidotti Eds., Harwood Academic Publishers (London & New York) 1982, pp. 93-107
4. M. H. Nayfeh and G. B. Hillard, Phys. Rev. A (In press)

5. M. H. Nayfeh and G. B. Hillard, in the Proceedings of the 5th International Conference on Lasers "Lasers 82", New Orleans, Dec. 1982.
6. M. H. Nayfeh and D. B. Geohegan, Phys. Rev. A. (In press).
7. S. I. Yakovlenko, Zh. Eksp. Teor. Fiz. 64, 2020 (1973) [JETP, 37, 1019(1974)].
8. J. C. Bellum and T. F. George, J. Chem. Phys. 68, 134 (1978)
9. M. H. Nayfeh, Phys. Rev. A 16, 927 (1977); M. H. Nayfeh and M. G. Payne, Phys. Rev. A 17, 1695 (1978).
10. J. A. Gaunt, Proc. Roy. Soc., London A 128, 654 (1930).

LASER INDUCED COLLISIONAL EXCITATION TRANSFER: A DISCUSSION OF THE TWO AND THREE-LEVEL APPROXIMATIONS

ARTURO BAMBINI and ALBERTO STEFANEL
Istituto di Elettronica Quantistica CNR, Via Panciatichi 56/30, Firenze Italy

Abstract We discuss the effect of quasi-molecular levels close to the initial energy level in a laser induced collisional excitation transfer. Although those levels may be excited by virtual transitions only, their very presence may open a new channel of destructive interference which causes an overall reduction of the cross section in the wing of the LICET linesshape.

INTRODUCTION

In a Laser Induced Collisional Excitation Transfer (LICET) a laser field induces a transition between quasi-molecular energy levels (free-free transition) of the molecule that is formed during a collision between atoms of different species. Both atoms change their state during the collision, so that the reaction may be written as

$$A_i + B_i + \hbar\Omega \longrightarrow B_f + A_f \tag{1}$$

This reaction requires the simultaneous action of the laser field and the collision and may therefore be described as a single step, one photon + one collision process. No energy transfer occurs in the absence of either the laser field or the collision between the two atoms.

This process has been the focus of several experimental investigations[(1-4)] as well as of theoretical analyses (5-9).

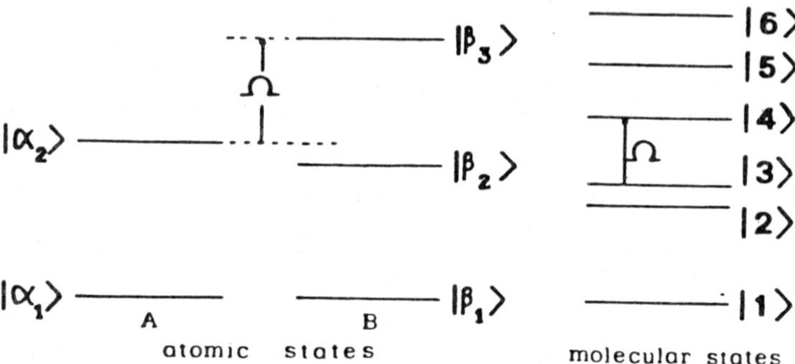

FIGURE 1. Typical umperturbed energy levels scheme for atoms A and B in a LICET process

A typical scheme for the reaction (1) is depicted in fig.(1). There atom A is prepared in an excited state α_2 by a pulse of radiation (the pump field). After the passage of the first pulse atom A interacts with atom B in its ground state, in the presence of a second pulse of radiation. Then the reaction (1) takes place, bringing atom A in its ground state and atom B in the excited state β_3.

Although level β_2 of B lies close to α_2 of A, their separation is such that there is no collisional energy transfer between A and B in the absence of the laser field. Furthermore, the two pulsed laser field must be well separated in time in order to avoid a two photon excitation of atom B.

When the laser field is tuned at the frequency $\hbar^{-1}(E_4 - E_3)$ the main contribution to the lineshape of the absorption process comes from collisions which occur at very large impact parameters. If all magnetic sublevels of 3 and 4 behave in much the same way when the two atoms come to a closer collision the line shape will be highly asymmetric around its line core: thus if levels 3 and 4 repel each other when the internuclear separation decreases, we expect the blue wing of the lineshape to be dominant over the red wing and vice versa.

This behaviour can be explained satisfactorily within the framework of the quasi static approximation. In this scheme, transitions occur in a time much shorter than the time during which the internuclear separation changes appreciably. Thus, using the Franck-Condon principle the transition will occur at the internuclear separation at which the energy levels spacing matches the laser field frequency (figure 2).

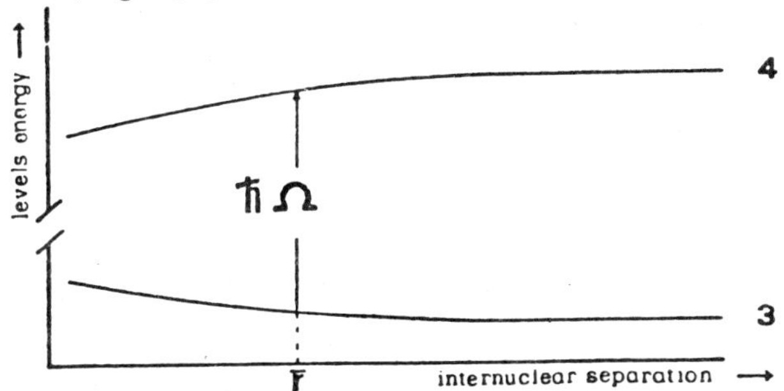

FIGURE 2. The Franck-Condon transition in the quasi-static approximation. At $r=r$ the laser frequency matches the displaced separation of levels 3-4.

However dynamical corrections may be important and one should make use of other approximation schemes such as for instance the impact approximation. All these methods, however, are based on perturbation expansions of energy levels and atomic/molecular states. These theories may therefore fail in predicting actual line shapes when the laser field becomes sufficiently large. Moreover, the use of a golden rule for the transition probability of the second order process may hide some important contributions which come from competing, simultaneous processes. On the other hand, one may approach the problem of finding the lineshape of the absorption profile by solving the Schrödinger equation for the amplitudes of the molecular levels involved in the reaction and then averaging the final level population over the impact parameter and the relative velocity distributions.

In what follows, we describe the application of this method to the determination of the absorption line shape for the case of the Eu + Sr reaction(2). We have performed numerical calculations for two and for three molecular energy levels, in view of the discrepancies reported in ref. 2 between experimental data and the theoretical predictions based on existing theories and calculations. We have found that inclusion of a third molecular level leads to a net reduction of the red wing of the lineshape, while the line-core is practically unchanged. This is a major effect which should be taken into account for a proper interpretation of the experimental data. A brief discussion follows of the assumptions made in our model (long-range dipole-dipole interaction, neglect of magnetic sublevels).

2. THE MODEL HAMILTONIAN

We focus our attention only on a few atomic states for atoms A and B. These include two states for atom A and three states for atom B (see fig.2). This scheme applies to the cases of Sr+Ca and Eu + Sr which have been investigated experimentally in the last years. All other energy states are out of reach for any first order and second order transitions and are expected to give only a small (if any) contribution to the lineshape.

Only long range, dipole-dipole interaction has been taken into account. This approximation is certainly not valid at short range interactions and should be lifted for a correct interpretation and for wing lineshape. We will discuss this point at the end of this article.

We also disregard the nearly degenerate magnetic sublevels, although in our numerical calculations their effect have been taken into account. Details of these calculations will be given elsewhere. In atom A, the two states are coupled by an electric dipole transition. In atom B, levels β_1 , β_2 and levels β_2 , β_3 are also coupled by an electric dipole transition. The two atoms interact with a laser electromagnetic field, whose amplitude is assumed to be constant over a time scale in which collision occurs. Furthermore, straight line, classical paths are assumed for the colliding atoms.

With these approximations, our model hamiltonian takes the form:

$$H = H_o + V_c + V_f \tag{2}$$

where H_o is the hamiltonian for the two atoms at infinite internuclear separation, V_c is their dipole-dipole coupling term

$$\begin{aligned} V_c = \frac{1}{r^3} \Big[& C_{32} \; |2\rangle\langle 3| + C_{51} \; |5\rangle\langle 1| + \\ & + C_{54} \; |5\rangle\langle 4| + C_{62} \; |6\rangle\langle 2| \Big] + \text{h.c.} \end{aligned} \tag{3}$$

and V_f is the atoms-field coupling

$$\begin{aligned} V_f = \; & g_A \Big[|3\rangle\langle 1| + |5\rangle\langle 2| + |6\rangle\langle 4| \Big] + \\ & + g_{B1} \Big[|2\rangle\langle 1| + |5\rangle\langle 3| \Big] + \\ & + g_{B2} \Big[|4\rangle\langle 2| + |6\rangle\langle 5| \Big] + \text{h.c.} \end{aligned} \tag{4}$$

In (3) and (4) use has been made of the molecular states $|n\rangle$ (with n=1,...6), given in the table below

$|1\rangle = |\alpha_1\rangle|\beta_1\rangle$	$|4\rangle = |\alpha_1\rangle|\beta_3\rangle$
$|2\rangle = |\alpha_1\rangle|\beta_2\rangle$	$|5\rangle = |\alpha_2\rangle|\beta_2\rangle$
$|3\rangle = |\alpha_2\rangle|\beta_1\rangle$	$|6\rangle = |\alpha_2\rangle|\beta_3\rangle$

TABLE 1. List of the relevant molecular states

With the adopted notation, the coupling between molecular states become self-explanatory. The coupling amplitudes have been evaluated for the Eu-Sr case taking into account the magnetic degeneracy of the levels involved in the process. Furthermore, the r^{-3} dependence of the dipole dipole interaction has been explicitly carried out of the coupling terms.

Although all these molecular states are involved in the LICET process, some of them are completely out of resonance and therefore out of reach for any real order transition induced by either the field or the collision or both. They just induce (slight) shifts in the resonances. On the assumption that their populations are negligibly small at any time during the interaction, they may be virtually eliminated from the Schrödinger equation. Their effects, however, are still retained and show up in the Schrödinger equation for the remaining states as shifts in the diagonal terms.

We have used this procedure to reduce the original Schrödinger equation to a two-level problem (elimination of states 1,2,5 and 6) or to a three level problem (elimination of states 1,5 and 6). In both cases we have set up a Runge Kutta method for numerical integration to evaluate the transfer of population form level 3 (Eu in the excited state, Sr in the ground state) to level 4 (the final level in the LICET process, where Eu is in the ground state and Sr in the upper excited state), in a single collisional event characterized by a relative velocity v and an impact parameter b. Thereafter, a Monte Carlo method takes care of the averaging over v and integrating over the range of b where energy transfer occurs. These calculations have been performed for several values of the detuning Δ between the frequency separation of levels 3 and 4 at infinite internuclear separation and the actual frequency carrier of the laser field.

3. NUMERICAL RESULTS AND DISCUSSION

In figure 3 we report the lineshapes obtained in the two cases. The two curves a and b refer to the two-level

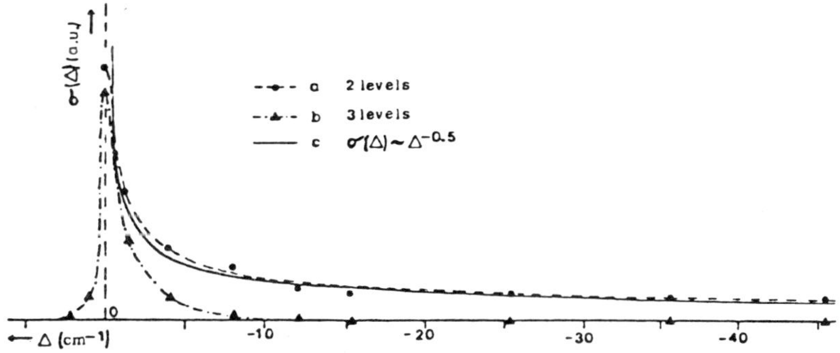

FIGURE 3.

and to the three level cases respectively. The surprising result is that the two curves differs appreciably in their red wings while curve A decreases slowly towards zero with a dependence on the detuning Δ of the form $\Delta^{-0.4}$, curve b shows a sharp cut off of energy transfer at $\Delta = -13\ cm^{-1}$. Curve c in fig.3 has been obtained by Gallagher and Holstein under the assumption that the quasi static approximation is valid at large detunings. As one would expect from their general analysis of a reaction in which total parity L changes by 1, curve c goes to zero as $\Delta^{-0.5}$ at large Δ. In this theory, a transition between levels 3 and 4 is assumed to occur at that internuclear separation where the frequency separation between the levels matches the applied laser field frequency. Curve c thus should be compared only with curve a obtained from our two level approximation scheme. Curve a shows a slower decrease to zero at large than curve c. This discrepancy may be due to the fact that when the quasi static approximation is lifted, transitions are allowed to occur over a finite range of internuclear separations. Numerical calculations were performed also by Harris and White for a two-level process in Ca + Sr reaction. Although their calculations cannot be compared directly with ours, they seem to confirm a slower decrease of the red wing than the $\Delta^{-0.5}$ decrease of the quasi static approximation.

However, both curve a and c predict a long, red-wing tail in the lineshape for the Eu + Sr reaction. This is manifestly not the case for curve b, obtained through a three level approximation scheme. In order to explain this discrepancy, we have investigated the dynamics of the system Eu + Sr in the absence of the external electromagnetic field, for several relative velocities and impact parameters. Calculations have shown evidence that at short range collisions, which originate the wing of the lineshape, energy transfer between levels 3 and 2 may occur. This collisional transfer can build up large populations(up to $\sim 10^{-1}$) in level 2, but only for a time interval of the order of b/v , when the two atoms are at their closest approach. Afterwards, population flows back to the original level 3, so that, when the two atoms are again at very large separations, level 2 is left almost unpopulated.

This population exchange is therefore only a virtual process, but it acts efficiently in depopulating level 3 just at the same time when the radiative transfer between level 3 and 4 (i.e. the LICET process) is expected to occur. Thus, level 2 opens a channel of deexcitation which interferes destructively with the LICET process itself, and a sharp reduction of energy transfer takes place. However this channel is very efficient only at short interaction ranges, so that it affects the wing of the lineshape but not its core.

The failure of a two level approximation scheme in the description of optical collisions was demonstrated by Light and Szöke[10] in a different situation: they found that the two level, Landau-Zener model of optical collision is inaccurate in as much it doesn't take into account the magnetic degeneracy of atomic states. Thus a four level scheme was proposed instead.

We have found that nearby levels, even if populated through short living, virtual processes may interfere heavily with the LICET process. These effects should be taken into account for a proper interpretation of the LICET lineshape.

However, even the three-level scheme does not adequately describe the LICET process. Curve d in figure 4 shows experimental results obtained by Brechignac, Cahuzac and Toschek[2]. The discrepancy between the experimental curve and curves A, B and C described above is evident. We think that the model described here is still inadequate because it doesn't take into proper account short range potentials in atomic interactions. Magnetic sublevels, too, may play an important role in the interaction process, since their degeneracy is lifted during the collision, thus making several channels available for energy transfer at different interatomic separations.

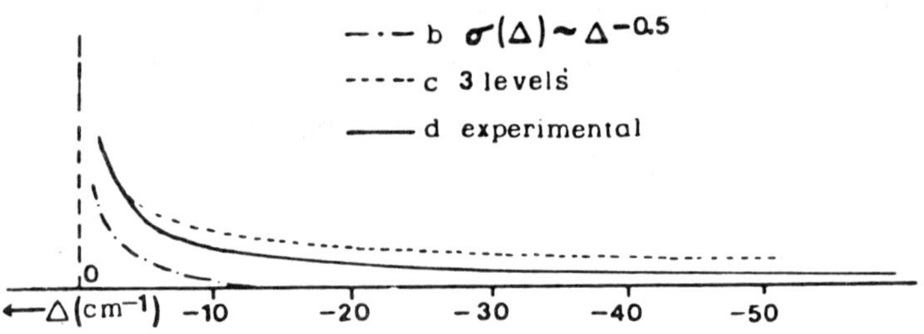

Fig.4 The line wing in the LICET experiment of Brechignac et al. (curve d). Curves b and c of Fig.3 have been reported here for comparison.

REFERENCES

1. R.W.Falcone, W.R.Green, J.C.White, J.F.Young and S.E. Harris, Phys.Rev., A15, 1333 (1977)
2. C.Brechignac, Ph.Cahuzac, P.E.Toschek, Phys.Rev., A21, 1969 (1980)
3. W.R.Green, J.Lukasik, J.R.Willison, M.D.Wright, J.F. Young and S.E.Harris, Phys.Rev.Lett., 42, 970 (1979)
4. B.Cheron and H.Lemery, Optics Comm. 42, 109 (1982)
5. L.I.Gudzenko and S.I.Yakovlenko, Sov.Phys.JETP, 35, 877, (1972)

6. V.S.Lisitsa, and S.I.Yakovlenko, Sov.Phys. JETP, 39, 759 (1974)
7. M.G.Payne and M.H.Nayfeh, Phys.Rev. A13, 595 (1976)
8. S.E.Harris and J.C.White, IEEE J.Quant.Elec., QE-13, 972 (1977)
9. A.Gallagher and T.Holstein, Phys.Rev., A16, 2413 (1977)
10. J.Light and A.Szöke, Phys.Rev.,A18, 1363 (1978)

THEORETICAL MODELS OF LASER-INDUCED CHEMICAL REACTIONS

KENNETH C. KULANDER*
Max-Planck-Institut für Quantenoptik, 8046 Garching West-Germany
and
ANN E. OREL
Lawrence Livermore National Laboratory, Livermore, California 94550

Abstract We have considered a model molecular system in which the reaction probability can be significantly enhanced by laser excitation to an excited electronic state during collision. Extensive calculations comparing exact quantum mechnical results with a completely classical treatment provide information about the effects on the dynamics of a strong laser field. An assessment of the accuracy of the classical model also is provided.

INTRODUCTION

Early enthusiasm for the promise of using lasers to influence the rates and directions of chemical reaction processes has dampened in recent years as the results of investigations both theoretical and experimental have shown the difficulty of obtaining practical results without requiring the use of immodest amounts of laser power.

In order to excite even a strongly allowed transition during a molecular collision, which for normal circumstances lasts for about 10^{-13} seconds, laser power on the order of 10^{12} W/cm^2 is necessary. Therefore, there is currently much interest in finding systems where the interaction time is significantly longer than the normal collision time either because of the possibility of the system being trapped in a collision complex or by substantially reducing the relative collision velocity by going to extremely low temperatures[1]. Detailed theoretical studies of the former of these possibilities involves calculations of the collisions dynamics on one or more molecular electronic potential energy surfaces. Even for a simple atom-diatom collision system a full quantum mechanical calculation is not feasible due to the large number of quantum states which is encountered even at thermal collision energies. Classical mechanics, however, has long been well established as the method of choice for studying molecular dynamics under these circumstances. Since there is a very large number of quantum states accessible, classical mechanics can be expected to be valid at least for the dynamics part of the calculation. There are two other degrees of freedom in this problem which must be considered: the photon field and the electronic state of the molecule. Semi-classical methods such as the trajectory surface-hopping model of Preston and Tully[2] treat the nuclear motion classically and the electronic and field degrees of freedom quantum mechanically. Similarly the complex trajectory methods of George and coworkers[3] can provide an accurate semi-classical way to treat the collision induced photoabsorption process. However, the complex trajectory method which is required if the transition region is not localized, can become difficult and expensive to apply. Therefore, recently Miller and coworkers[4] have developed a completely classical model which treats all degrees of freedom on the same footing. Orel and Miller[5] have applied this to a laser-induced chemically reactive system. The major advantage of this completely classical model is that it is computationally no more difficult than standard classical trajectory calculations. Therefore if the method can be relied upon to provide accurate predictions about the dependence of reaction probabilities on laser wave length and field strength and on the initial conditions of the molecular collision system, the collision energy and the vibrational and rotational states of the molecule, this would be an excellent tool for investigating possible candidate systems for laser-induced chemistry. The major question about the model is whether the electronic state and photon field "quantum numbers" can be treated as continuous variables. Calculations by Orel and Miller showed that this model generally produces physically reasonable results. It is the purpose of our work to establish whether these results are also quantitatively accurate. To do this we chose a model for which both classical and exact quantum mechanical results were obtainable. Our calculations[6] show that the classical model is useful to the extent that optimal laser wavelengths and collision energies for a particular molecular system can be determined. However, the magnitude of the effects of the laser on the collsion dynamics was found generally to be greatly exagerated in the classical model. We concluded that this was due to the possibility of absorbing a fraction of a photon, thus being partially effected by the excited state potential energy surface. Quantum mechanically, only the absorption or emission of a whole photon can occur. Therefore the classical model was found to be valuable for answering some questions but it appears that for quantitative answers a semiclassical treatment is necessary.

THE MODEL

We have chosen to study the model developed by Light and Altenberger-Siczek[7] of a laser-collision induced chemical reaction. The process involves the collinear collision of an H atom with an LiF molecule in the presence of a strong laser field. The laser couples the ground molecular ($^2\Sigma$) state to the excited ($^2\Pi$) state. The reaction on the two surfaces are

$$LiF(^1\Sigma) + H \rightarrow Li\ (^2S) + FH \qquad (1)$$

and

$$LiF(^1\Pi) + H \rightarrow Li(^2P) + FH \qquad (2)$$

Reaction (1) is slightly exothermic (0.103 eV) and has a barrier to reaction of ~0.4 eV. The reaction on the excited surface is very exothermic (5.85 eV). The separation of the ^2S and ^2P states of lithium is 1.85 eV. A correlation diagram for the system is shown in Fig. 1. The two surfaces are modelled by LEPS functions using parameters which are given in ref. 7.

The induced reaction occurs when the laser is tuned to a frequency such that a photon can be resonantly absorbed during the collision. This promotes the system to the upper surface which due to its high reactivity produces the desired albeit electronically excited products.

The difference between minima of the potential wells in the reactant arrangement is 6.57 eV. We chose laser frequencies from 6.2 to 6.4 eV which gives asymptotic resonance energy defects, Δ, of 0.37-0.17 eV. We assume a transition dipole independent of internuclear geometry.

The coupling strength between the scattering states on the two surfaces, which is given by $\mu \cdot E_o$ where E_o is the ave-

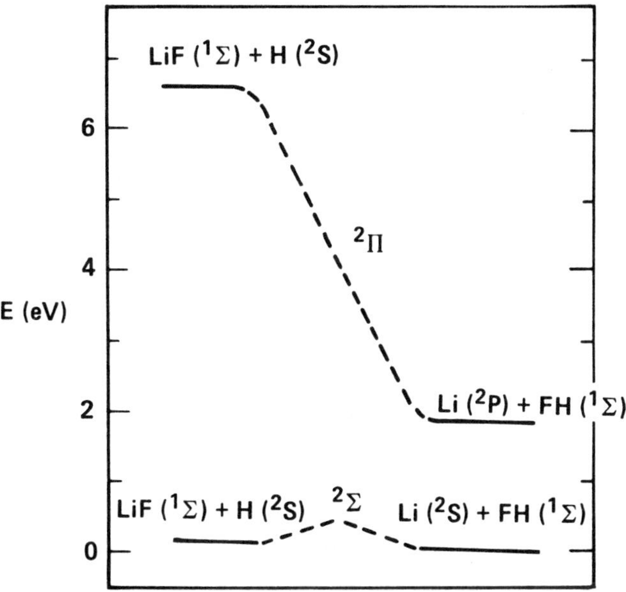

FIGURE 1. Correlation Diagram.

rage laser field strength, is chosen to lie between 0.005 and 0.01 eV. We obtain state-to-state transition probabilities between the diagonal (field-state) basis functions. In order to obtain a reasonable comparison between the quantum mechnaical and the classical results we consider only total reaction probabilities. Since we are quite far off resonance asymptotically, the initial states in both calculations can be assumed to be the field-free molecular states. Also by comparing only total probabilities we do not need to worry about binning the classical results to determine final quantum state populations.

CLASSICAL MODEL

The model developed by Miller and coworkers[4,5] treats all degrees of freedom classically. The electronic state of the system during the collision is taken to be a linear combination of the initial and final states. The photon field is unquantized and is equivalent to a classical oscillator. The Hamiltonian for the system is given by

$$H = \frac{P^2}{2\mu} + \frac{p^2}{2m} + (1-n)V_o(r,R) + nV_1(r,R) + N\hbar\omega \qquad (3)$$

$$- 2(8\pi\hbar\omega N/V)^{1/2}\mu_o \sin Q \cos q \, ((n+1/2)(3/2-n))^{1/2}$$

where r and R are the scattering coordinates and p and P then conjugate momenta, V_o and V_1 are the ground and excited potential energy surfaces, ω is the laser frequency, V the interaction volume, μ_o the transition dipole moment and (n,q) and (N,Q) are the action-angle variables for the electronic and field degrees of freedom. From this Hamiltonian we derive the following equations of motion

$$\dot{r} = p/m \qquad (4a)$$

$$\dot{p} = \begin{cases} -\dfrac{\partial V}{\partial r} & n < 0 \\ (n-1)\dfrac{\partial V}{\partial r} - n\dfrac{\partial V}{\partial r} & 0 < n < 1 \\ -\dfrac{\partial V}{\partial r} & n > 1 \end{cases} \qquad (4b)$$

$$\dot{R} = P/\mu \tag{4c}$$

$$\dot{P} = \begin{cases} -\dfrac{\partial V}{\partial R} & n < 0 \\[2ex] (n-1)\dfrac{\partial V}{\partial R} - n\dfrac{\partial V}{\partial R} & 0 < n < 1 \\[2ex] -\dfrac{\partial V}{\partial R} & n > 1 \end{cases} \tag{4d}$$

$$\dot{Q} = \hbar\omega - (8\pi\hbar\omega/NV)^{1/2}\mu_o \sin Q \cos q((n+1/2)(3/2-n))^{1/2} \tag{4e}$$

$$\dot{N} = 2(8\pi\hbar\omega N/V)^{1/2}\mu_o \cos Q \cos q((n+1/2)(3/2-n))^{1/2} \tag{4f}$$

$$\dot{q} = (V_1-V_o)f(n)$$

$$-(8\pi\hbar\omega N/V)^{1/2}\mu_o \sin Q \cos q \frac{(1-2n)}{((n+1/2)(3/2-n))^{1/2}} \tag{4g}$$

$$\dot{n} = -2(8\pi\hbar\omega N/V)^{1/2}\mu_o \sin Q \sin q((n+1/2)(3/2-n))^{1/2} \tag{4h}$$

with $f(n) = 1$ for $0< n < 1$ and $= 0$ otherwise. Slight modifications in Eqs. (4b),(4d) and (4g) from the obvious equations of motion are necessary due to the inconsistencies introduced when the Langer modification[4] is made for the variable n. The Langer modification is necessary because otherwise the coupling term is and remains zero for the initial state n=0 or 1.

The classical equations of motion are solved using a predictor-corrector integrator for a standard Monte Carlo random sampling of the initial conditions[8]. A sufficient number of trajectories is run that converged reaction probabilities can be determined.

QUANTUM MECHANICAL MODEL

Our quantal calculations were carried out using the approach developed by Light and Altenberger-Siczek[7]. We used the R-Matrix propagation method of Light and Walker[9] to perform the scattering calculations on the two coupled potential energy surfaces. This method involves dividing up configuration space into sectors, solving the coupled equations in each sector, then putting together the full scattering matrix by requiring the entire wave function and its derivative with respect to the scattering coordinate be continuous across the boundaries between sectors. In each sector a local adiabatic vibrational basis for both surfaces including the dipole coupling between the electronic states is constructed. The matching of the solutions at the sector boundaries provides the mixing between the locally adiabatic solutions. Further details of the calculations can be found in ref. 7. We found it necessary due to the strongly repulsive excited state potential in the interaction region to use a vibrational basis of up to 30 functions even though there were only about 12 open channels for the collision energies we considered. By using a basis that large of a basis we felt confident that we obtained converged results.

RESULTS AND CONCLUSIONS

We performed quite extensive calculations on this model system for a range of collision energies from well below the field-free reaction threshold to energies slightly above it[6]. We also considered different initial vibrational states for the target diatomic and different laser frequencies and field strengths. In addition we considered the reverse reaction, Li + FH, for which the effect of the laser is to reduce the reaction probability[10]. These calculations demonstrated the good qualitative agreement between the classical and quantum mechanical results. In Fig. 2 we show the typical results obtained in this comparative study for a particular photon energy, $\hbar\omega$ = 6.2 eV, field strength, $\mu \cdot E_0$=0.01 eV and initial vibrational state, v=0. The dashed curves in this figure are the field-free results. Above the minimum in the field-free results, > 0.45

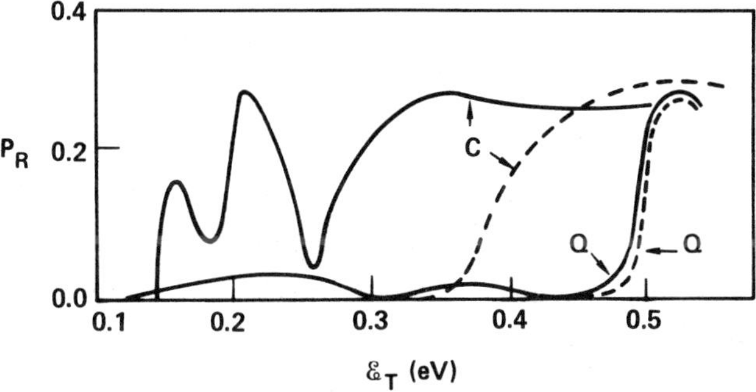

FIGURE 2. Reaction probabilites in the presence (solid line) and absence (dashed line) of the laser field as a function of the total collision energy. C denotes the classical results and Q the quantal.

eV classically and > 0.50 eV for the quantum case, the laser has no significant effect on the reaction probabilities. The classical model showed that the laser substantially reduced the reaction threshold with a number of peaks between this new threshold and the field-free threshold. Some of these peaks also appear in the quantum results but they are much broader and smaller. Note the difference between the classical and quantal reaction thresholds is due to the zero-point energy at the barrier in the latter calculation.

The origin of the several peaks in the laser enhanced reaction probabilities can be shown to be due to the turning points for the motion on the initial (or final) surface. The enhancement will be most pronounced if these points also coincide with points where the energy difference between the two surfaces matches the photon energy. Orel and Miller[5] showed that it was possible to predict the positions of the peaks by running trajectories on a single surface and plotting energy difference between the surfaces at the turning points as a function of the collision energies.

The great exageration of the effect of the laser on the reaction probabilities is due to the fact that in the classical calculations the absorption of only a fraction of a photon can lead to reaction whereas in reality only the absorption of a complete photon is allowed. It has been suggested[11] that this problem could be fixed if a standard Classical S-Matrix[8] approach were employed. This means that the trajectories which result in only a small fraction of a photon being absorbed would not be considered as reactive even though the final states otherwise look like products. The problem with this is that no trajectories resulted in a substantial fraction (>1/2) of a photon being absorbed so that no enhancement would be predicted at all. This is because the upper surface, once part of a photon is absorbed (n>0), forces the system quickly away from the resonance region. This way the fraction of the photon finally absorbed is smaller than it should be. The remedy for this problem might be to employ techniques similar to those proposed by Dahler and coworkers[12] in which the motion on the initial surface would continue with the photoabsorption causing only a loss of flux from the initial state but not effecting the dynamics unless an appreciable probability for reemission of the photon occurs.

In conclusion we have learned that the completely classical model of Orel and Miller is capable of reproducing most of the interesting features of the laser effects on chemical reactions. However, the results are only qualitatively correct. It can be used to predict the position of the peaks in the enhanced reaction probability with respect to the laser frequency and the collision energy. It also correctly predicted the laser inhibition of the reverse reaction, but again grossly overestimated the magnitude of the effect. Because of the great utility of a completely classical model it is hoped that some simple modification of the present prescription may lead to more quantitative results.

This work was performed partially under the auspices of the U.S. Department of Energy by the Lawrence Livermore National Laboratory under contract No. W-7405-ENG-48.

REFERENCES

*On leave from Lawrence Livermore National Laboratory, Livermore, California

1 M. Hutchinson, T.F. George: Mol. Phys. 46, 81 (1982)
2 R.K. Preston and J.C. Tully: J. Chem. Phys. 54, 4297 (1971); 55, 562 (1971)
3 Jian-Min Yuan, John R. Laing and T.F. George: J. Chem. Phys. 66, 1107 (1977)
4 W.H. Miller: J. Chem. Phys. 69, 2188 (1978); W.H. Miller and C.W. McCurdy: J. Chem. Phys. 69, 5163 (1978); H.D. Meyer and W. H. Miller: J. Chem. Phys. 71, 2156 (1979)
5 Ann E. Orel and W.H. Miller: J. Chem. Phys. 73, 241 (1980)
6 K.C. Kulander and A.E. Orel: J. Chem. Phys. 75, 675 (1981)
7 J.C. Light and A. Altenberger-Siczek: J. Chem. Phys. 70, 4108 (1979)
8 W.H. Miller: Adv. Chem. Phys. 25, 69 (1974)
9 J.C. Light and R.B. Walker: J. Chem. Phys. 65, 4272 (1976)
10 K.C. Kulander and A.E. Orel: J. Chem. Phys. 74, 6529 (1981)
11 Jon P. Davis: J. Chem. Phys. 77, 3775 (1982)
12 R.E. Turner and J.S. Dahler: J. Phys. B: Atom Molec. Phys. 13, 161 (1980); J.S.Dahler, R.E. Turner and S.E. Nielsen: J. Phys. Chem. 86, 1065 (1982); and H.P. Saha, J.S. Dahler and S.E. Nielsen "A Theory of laser-induced chemi-ionization" (preprint).

A ONE-DIMENSIONAL DIRAC DELTA FUNCTION MODEL FOR ATOM-ATOM COLLISIONS IN THE PRESENCE OF RADIATION FIELD

GIOVANNI PAOLO ARRIGHINI, NICOLA DURANTE
Istituto di Chimica Fisica dell'Università di Pisa, Via Risorgimento 35, 56100 Pisa, Italy

CARLA GUIDOTTI
Istituto di Chimica Quantistica ed Energetica Molecolare del CNR, Via Risorgimento 35, 56100 Pisa, Italy

Abstract An effective one-electron model of atom-atom collision in the presence of radiation is investigated. The quantal motion of the electron is assumed to be restricted to one dimension under the influence of a strongly screened internal core and the external field. Classical parameter impact approximation is utilized for the nuclear dynamics. Preliminary results for the dependence of the relevant cross section on radiation intensity and relative velocity of the colliding partners are obtained.

INTRODUCTION

The possibility of influencing a collisional process by intense laser radiation has attracted a lot of attention over the past decade[1,2]. As known, the role of the radiation is mainly that of making resonant (or near-resonant) a process which, in general, is not resonant in the absence of external field and the theory is entrusted with the task of predicting how the cross section for the given process depends on the typical parameters characterizing the radiation.

A rigorous treatment of the problem can but be founded on the formal apparatus of quantum mechanical scattering theory. A full quantal formulation of the atom-atom collision problem in a laser field has actually been given[3-5] (for a review of the parallel problem of electron-atom scattering in the presence of radiation, see ref. 6), but applications have invariably recourse to various kinds of approximations, ranging from the consideration of schematic models (most frequently of a two-state nature) to the introduction of simplifications either at the level of dealing with the matter-field interaction or in describing the nuclear motion dynamics.

Aim of the present paper is to start on investigating the response of a schematic collisional process to a strong electromagnetic external field, with the primary intention of assessing in terms of the same model the validity of simplifying approximations as well as the convenience of alternative approaches that can be put forward in the theoretical treatment of the problem. A list of points which are worthy of analysis includes: a) the comparative solution of the nuclear motion problem in terms of classical, semi-classical and quantum dynamics; b) the checking of the validity of both rotating wave and dipole approximations, as well as the use of different gauges for describing the matter-field interaction; c) the assessment of perturbation-like solutions concerning either the nuclear dynamics problem or that associated with the matter-radiation interaction; d) the extension of the model beyond the two-state approximation, so as to appreciate the role of the continuum spectrum mixing and the prominence of multiphoton effects.

In the present paper, the first in a series where we propose to shed light on some of the points described above, we start with investigating a simple two-state model system, which mimics an atom-atom collision in the presence of radiation, making use of "dressed" electron-field adiabatic states[7-9] and solving the nuclear motion dynamics in terms of classical trajectories approximated by the impact parameter method. In the next section we shall describe the model and the theoretical approach followed for its treatment.

THE MODEL

The type of collision we have in mind involves two different atomic partners, one of which neutral, the other negatively charged. The investigation of the model can therefore provide information about the charge exchange process

$$A^- + B + \hbar\omega \rightarrow A + B^- \tag{1}$$

induced by the presence of a (strong) radiation field. We shall assume that the quasi-molecule $(AB)^-$ instantaneously formed as the partners approach each other is characterized by a two-center core with a vanishing net electric charge, so that the problem comes down into one involving a single optical electron, which moves in the short-range field caused by the underlying core and an intense, external electromagnetic field. The internal potential field is then approximated by two attractive Dirac-delta wells, with different strengths.

A convenient specification of reference frames utilized in describing the collisional problem at issue is presented in Fig. 1[10]. The two colliding partners A and B, with masses m_A and m_B, have velocities $\underline{v}_A$ and $\underline{v}_B$, respectively, with respect to the center of mass (C.M.), so that

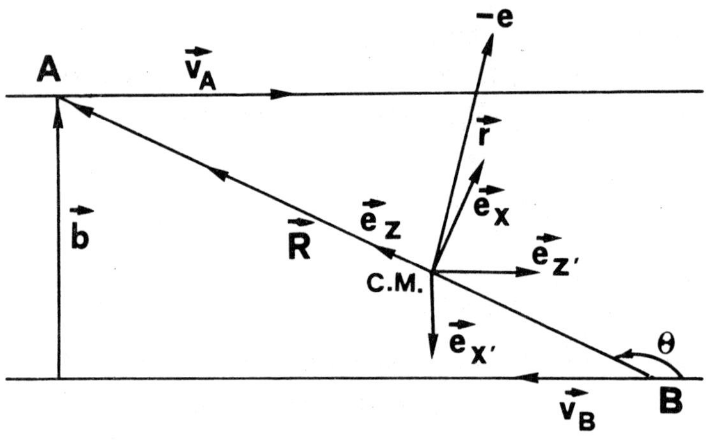

FIGURE 1. Coordinate frames used for describing the considered collision (see text).

the relative velocity of A to B is $\underline{v} = \underline{v}_A - \underline{v}_B$. The internuclear separation vector $\underline{R}$, which points from B to A, determines along with $\underline{v}$ the collision plane. The coordinate frame $(\underline{e}_{x'}, \underline{e}_{y'}, \underline{e}_{z'})$ is fixed at the laboratory, its origin being coincident with the center of mass. A second coordinate frame $(\underline{e}_x, \underline{e}_y, \underline{e}_z)$ attached to the origin, rotates during the motion of the (quasi) molecule in such a way that $\underline{e}_z$ is parallel to $\underline{R}$ and the plane $(\underline{e}_x, \underline{e}_z)$ is coplanar with the collision plane.

If the origin of time is chosen in coincidence with the perihelion point of the colliding partners, in the impact parameter approximation we have

$$\underline{R}(t) = \underline{b} + \underline{v}\, t \tag{2}$$

where b is the impact parameter (see Fig.1). The distance between the two centers is therefore $R(t) = (b^2 + v^2t^2)^{\frac{1}{2}}$.

After these short comments, which are useful for the next developments, we turn our attention to the electronic problem. On assuming that the optical electron is constrained to move one-dimensionally (along the axis $\underline{e}_z$) in the short-range field due to the internal core (see above),

$$\hat{V}(z) = -V_A\,\delta(z-R_A) - V_B\,\delta(z-R_B), \tag{3}$$

where $R_A = M_B\ R/(M_A+M_B)$, $R_B = M_A\ R/(M_A+M_B)$, the corresponding quantum-mechanical problem, in the Born-Oppenheimer approximation and in the absence of external fields, is exactly solvable[11-13]. In addition to the continuum of ionized states, the spectrum of $\hat{H}_{mol} \equiv \hat{T} + \hat{V}$ ($\hat{T} \equiv$ electron kinetic energy operator), displays two discrete bound states,

ground state:

$$\phi_1(z;R) = A_1\exp[-q_1|z-R_A|] + B_1\exp[-q_1|z-R_B|] \tag{4}$$

excited state:

$$\phi_2(z;R) = A_2\exp[-q_2|z-R_A|] + B_2\exp[-q_2|z-R_B|]$$

where

$$A_{1,2} = \left[\frac{V_A(V_B-q_{1,2})}{V_A+V_B-2q_{1,2}-2R(V_A-q_{1,2})(V_B-q_{1,2})}\right]^{\frac{1}{2}}$$

$$\frac{B_{1,2}}{A_{1,2}} = - \frac{V_A - q_{1,2}}{V_A} \exp(q_{1,2}R)$$

with energies $W_1(R) = -q_1^2(R)/2$, $W_2(R) = -q_2^2(R)/2$ a.u. (atomic units (a.u.) with $\hbar = e = m_e = 1$ are prevalently used throughout this paper), where q_1 and q_2 are the real solutions to the transcendental equations

$$2q_{1,2} = (V_A + V_B) + \left[(V_A - V_B)^2 + 4V_A V_B \exp(-2q_{1,2}R)\right]^{\frac{1}{2}} \tag{5}$$

An interesting feature of the model under study is that the excited bound state disappears below a treshold value R_o, R_o being determined by the equation $R_o = (V_A + V_B)/2V_A V_B$. In the presence of a sufficiently intense laser radiation, a treatment based on the use of perturbative techniques can be questionable. Under such circumstances a useful approach to the problem is offered by the utilization of "dressed" electron-field adiabatic states[7-9]. If we resort to a quantized description of the electromagnetic field in terms of Fock (number) states, the "dressed" electron-field adiabatic states are just the stationary states of the over-all Hamiltonian operator $\hat{H} \equiv \hat{H}_{mol} + \hat{H}_{rad} + \hat{V}_{rad}$ which includes, besides $\hat{H}_{mol}$, the free field Hamiltonian operator $\hat{H}_{rad}$ and the matter-field interaction term $\hat{V}_{rad}$. For a single- mode field of frequency ω and specified polarization

$$\hat{H}_{rad} |n\rangle = n\omega |n\rangle \tag{6}$$

where $|n\rangle$ represents a situation of the free field characterized by n photons (with given characteristics). The "dressed" adiabatic states $|\chi\rangle$ are then linear combinations of the non-interacting states $|\phi_j\rangle \otimes |n\rangle$, i.e.

$$|\chi\rangle = \sum_{jn} a_{jn} \, |\phi_j\rangle \otimes |n\rangle, \tag{7}$$

which come out, along with the eigenvalues E, on diagonalizing $\hat{H}$.

By a suitable choice of the gauge, in the electric dipole approximation the matter-field interaction operator $\hat{V}_{rad}$ can be cast into the form[14]

$$\hat{V}_{rad} = - \underline{\hat{\mu}} \cdot \underline{\hat{E}} \tag{8}$$

where $\underline{\hat{\mu}}$ is the electric dipole moment operator for the (quasi)-molecule and $\underline{\hat{E}}$ the electric field operator.

We now proceed to further simplify the model on assuming that: i) the continuum of ionized states supported by $\hat{H}_{mol}$ is not appreciably coupled to the two bound states, so that the sum over j of eq. (7) includes only the two states explictly given by eq.(4); ii) the diagonal matrix elements $\langle\phi_1|\underline{\hat{\mu}}|\phi_1\rangle$ and $\langle\phi_2|\underline{\hat{\mu}}|\phi_2\rangle$ are ignored; iii) the rotating wave approximation is valid[15,16]. None of these assumptions can actually be rigorously true for the model under study, but we enforce them because at this point our intention is to make the model as adherent as possible to those usually considered.

Among other things, the merit of the assumptions made is that of simplifying drastically the secular problem for the "dressed" states. Since only the basis states $|\phi_1\rangle \otimes |N\rangle$ $|\phi_2\rangle \otimes |N-1\rangle$ are now involved[9,16], we are led to the following two <u>"dressed" electron-field adiabatic potential</u>

curves

$$E_{\substack{1\\2}}(R) = N\omega + \tfrac{1}{2}\left[W_1(R)+W_2(R) - \omega\right] \pm \tfrac{1}{2}\left[\Delta^2(R)+4|d|^2\right]^{\frac{1}{2}} \tag{9}$$

Here $\Delta(R) \equiv W_2(R)-W_1(R)-\omega$ is the "detuning" at the separation R and

$$d(R) = i\left(\frac{2\pi I}{c}\right)^{\frac{1}{2}} \underline{\varepsilon}\cdot\langle\phi_2|\underline{\hat{\mu}}|\phi_1\rangle \tag{10}$$

the radiative coupling strength, which depends obviously on the radiation intensity I and the electric dipole moment transition matrix element between the two molecular states ϕ_1 and ϕ_2, as well as the polarization state of the radiation (unit vector $\underline{\varepsilon}$).

If we now assume that the polarization unit vector has components ($\sin\alpha\cos\beta$, $\sin\alpha\sin\beta$, $\cos\alpha$) with respect to the frame fixed in the laboratory, it is not difficult to verify that the scalar product $\underline{\varepsilon}\cdot\langle\phi_2|\underline{\hat{\mu}}|\phi_1\rangle$ appearing in eq. (10) can be expressed as follows[10]

$$\underline{\varepsilon}\cdot\langle\phi_2|\underline{\hat{\mu}}|\phi_1\rangle = \left[\frac{vt}{R(t)}\cos\alpha - \frac{b}{R(t)}\sin\alpha\cos\beta\right]\mu_{12}(R) \tag{11}$$

where

$$\mu_{12}(R) \equiv -\langle\phi_2|z|\phi_1\rangle = -A_1B_2\left\{\frac{e^{-q_1R}[R(q_1+q_2)+1]-e^{-q_2R}}{(q_1+q_2)^2} + \right.$$

$$\left. - \frac{e^{-q_1R}[R(q_1-q_2)+1]-e^{-q_2R}}{(q_1-q_2)^2}\right\} - \frac{2RB_1B_2}{q_1+q_2} + \tag{12}$$

$$-A_2B_1\{\frac{e^{-q_2R}[R(q_1+q_2)+1]-e^{-q_1R}}{(q_1+q_2)^2} - \frac{e^{-q_2R}[R(q_2-q_1)+1]-e^{-q_1R}}{(q_1-q_2)^2}\}$$

is the electric dipole moment transition matrix element between the adiabatic states ϕ_1, ϕ_2.

Since we agreed to approach the nuclear dynamics problem according to the classical impact parameter method, the time-dependent Schrödinger equation to solve is[1,7,9,10,16]

$$[\hat{H}_{mol}(R) + \hat{H}_{rad} + \hat{V}_{rad}(R)]|\Psi(t)\rangle = i\frac{\partial}{\partial t}|\Psi(t)\rangle \qquad (13)$$

where we have put in evidence the dependence on the internuclear separation $R = R(t)$. In terms of the "dressed" adiabatic eigenstates $|\chi_j(R)\rangle$ which diagonalize $\hat{H}$ at the given R value,

$$|\chi_j(R)\rangle = a_{j1}(R)|\phi_1(R)\rangle\otimes|N\rangle + a_{j2}(R)|\phi_2(R)\rangle\otimes|N-1\rangle \qquad (14)$$
$$j = 1,2$$

the state vector $|\Psi(t)\rangle$ can be expanded as

$$|\Psi(t)\rangle = c_1(t)|\chi_1(R)\rangle + c_2(t)|\chi_2(R)\rangle \qquad (15)$$

After straightforward manipulations, it is possible to show that the expansion coefficients $c_j(t)$ satisfy the coupled differential equation set

$$i\frac{d}{dt}\begin{pmatrix} c_1(t) \\ c_2(t) \end{pmatrix} = \begin{pmatrix} E_1(R) & \beta(t) \\ \beta(t) & E_2(R) \end{pmatrix}\begin{pmatrix} c_1(t) \\ c_2(t) \end{pmatrix} \qquad (16)$$

where the non-adiabatic coupling matrix element $\beta(t)$ between "dressed" states can be expressed in the form

$$\beta(t) = -i\, a_{11}^2(R) \frac{d}{dt} \left(\frac{a_{12}}{a_{11}}\right) \qquad (17)$$

a_{11}, a_{12} are known from the solution of the 2x2 secular problem leading to the "dressed" states $|\chi_j(R)\rangle$,

$$a_{11}(R) = \left\{1 + \frac{|d(R)|^2}{[W_2(R)+(N-1)\omega-E_1(R)]^2}\right\}^{-\frac{1}{2}}$$

$$a_{12}/a_{11} = -\frac{d(R)}{W_2(R)+(N-1)\omega-E_1(R)} \qquad (18)$$

so it is an easy task to calculate the function $\beta(t)$.

The quantities we are explicitly interested in this paper are the transition probability from the initial state $|\phi_1\rangle \otimes |N\rangle$ $(t=-\infty)$ to the final state $|\phi_2\rangle \otimes |N-1\rangle$ $(t=+\infty)$ and the corresponding cross section. If we assume a mode frequency ω non-resonant with the asymptotic separation $W_2(\infty)-W_1(\infty)$, and detuning $\Delta(\infty) < 0$, it is straightforward to verify that the searched transition probability is $|c_2(+\infty)|^2$, a function of the impact parameter and relative velocity, in addition to the intensity and polarization of the radiation.

After averaging over all possible orientations of the collision plane relative to the electric field direction $\underline{\varepsilon}$ and integrating with respect to the impact parameter, we are led to the desired cross section[17]

$$\sigma = 2\pi \int_0^\infty \overline{|c_2(+\infty)|^2}\; b\,db \qquad (19)$$

a function of the radiation intensity I and relative velocity v of the two colliding partners.

RESULTS AND COMMENTS

The collisional model we have considered requires first of all the specification of the two well strengths V_A, V_B (see eq. (2)). We have chosen $V_A = 0.235$, $V_B = 0.210$ a.u., which lead to an asymptotic energy separation $\Delta W(\infty) \simeq 0.006$ a.u., equal to the difference of electronic affinity between hydrogen atom and lithium atom in their ground states. So, if we are very confident in the model under study, we could hope to gain insight about the role of a suitable laser in promoting the charge exchange process $H^- + Li \rightarrow H + Li^-$. Since we are enabled to sum an arbitrary real constant to the Hamiltonian operator $\hat{H}_{mol}$, a function of the form $\exp(-kR)/R$, which mimics the repulsive interaction between screened nuclei, has been added to $W_1(R)$ and $W_2(R)$[11]. Accurate ab-initio calculations[18] for the molecule $(LiH)^-$ in its ground state lead to a minimum at $R = 3.15$ a.u., the electronic affinity being there 0.011 a.u. Choosing $k = 1.325$ a.u., the potential energy curve $W_1(R)$ exhibits a minimum just at $R = 3.15$ a.u., with $|W_1(3.15)| = 0.042$ a.u., a much higher prediction with respect to the accurate result.

In Fig. 2 we have reported the potential energy curves $W_1(R)$ and $W_2(R)$ (inclusive of the nuclear repulsion contribution), corresponding to the discrete bound states supported by the present model. One can note for $R<R_o \simeq 4.6$ a.u. the disappearance of the excited bound state; for R values below such a threshold we have taken $W_2(R) = 0$ and the

wavefunction $\phi_2(z;R)$ identically zero. In the same figure, the curve labeled $W_1 + \hbar\omega$ is simply the curve for $W_1(R)$ translated uniformly upward by $\hbar\omega$, with $\hbar\omega = 0.013$ a.u.: by

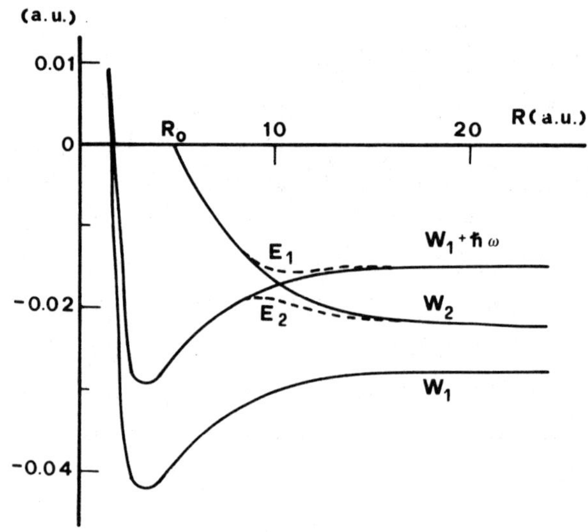

FIGURE 2. Behaviour of potential energy curves relevant for the model examined in this paper (see text).

this ω choice a curve crossing around $R \simeq 10$ a.u. occurs. The two "dressed" electron-field adiabatic curves $E_1(R)$, $E_2(R)$ (partially shown by dotted lines) correspond to a radiation intensity $I = 1$ MW/cm^2 (= $0.1554.10^{-9}$a.u.) and a polarization direction with angles $\alpha = \beta = 0$ (eq. (10)). Because of the radiative interaction, the diabatic curves $W_1(R) + \hbar\omega$ and $W_2(R)$ give rise to an avoided crossing around $R \simeq 10$ a.u., while at $R \simeq 1.9$ a.u., within the range where the radiative interaction vanishes, a crossing between the same curves stays on.

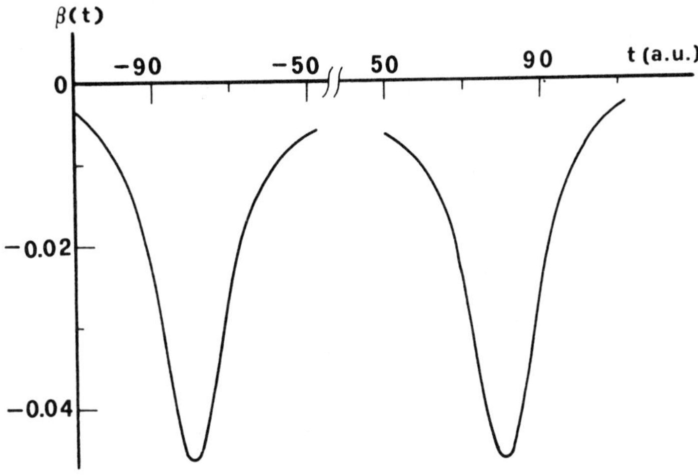

FIGURE 3. Behaviour of the non-adiabatic coupling term $\beta(t)$ |see eq. (17)| as a function of time t. The plot refers to b = 0., $\alpha = \beta = 0$, v = 0.12 a.u. and I = 1MW/cm .

The behaviour of the non-adiabatic coupling function $\beta(t)$ as a function of time t, eq. (17), can be appreciated by examining the Figs. 3,4. Fig. 3 refers to b = 0, $\alpha = \beta = 0$, v = 2.63×10^5 m/sec (=0.12 a.u.) and I = 1 MW/cm^2. The evident symmetry of this figure under $t \to -t$ is not at all a general feature, but is rather a consequence of the particular choice $\alpha = \beta = 0$ (see eq. (11)). This remark becomes particularly apparent from Fig. 4, which refers to b = 5. a.u., $\alpha = \pi/3$, $\beta = 0$, and values for all other parameters kept identical to those used in Fig. 3. In addition to the symmetry breaking between the -t and +t semiaxes, one may also observe the high strength as well as the strongly localized nature of the coupling in the zone around the avoided crossing on the way-in of the collision, characteristics which are therefore dependent upon the particular trajectory considered.

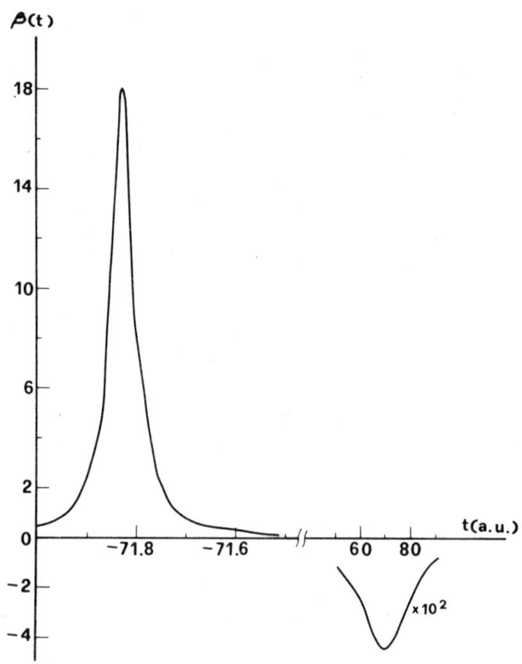

FIGURE 4. Behaviour of the non-adiabatic coupling term $\beta(t)$ as a function of time t. The plot refers to b = 5. a.u., $\alpha = \pi/3, \beta = 0$ and remaining parameters fixed at the same values as in Fig. 3.

In Figs. 5,6 we plot the behaviour of the transition probability $\overline{|c_2(+\infty)|^2}$, averaged over the orientations of the collision plane with respect to the external field, as a function of the impact parameter b, at given values of radiation intensity I and relative velocity v.

The integration of the 2x2 coupled differential equation set, eq. (16), has been carried out by resorting to a fourth-order Runge-Kutta algorithm, with variable step (its size being controlled by the variation rate of $\beta(t)$), starting at $t \simeq -200$ a.u. up to $t \simeq +200$ a.u., after preliminarily transforming the equation set into the form

$$\frac{d}{dt}\begin{pmatrix} d_1(t) \\ d_2(t) \end{pmatrix} = \begin{pmatrix} 0 & \beta(t)-\frac{i}{2}[E_1(R)-E_2(R)] \\ -\beta(t)-\frac{i}{2}[E_1(R)-E_2(R)] & 0 \end{pmatrix}\begin{pmatrix} d_1(t) \\ d_2(t) \end{pmatrix}$$

with

$$\begin{pmatrix} c_1(t) \\ ic_2(t) \end{pmatrix} = \frac{1}{2}\begin{pmatrix} d_1(t)+d_2(t) \\ d_1(t)-d_2(t) \end{pmatrix} \exp\{-\frac{i}{2}\int^t dt'[E_1(R)+E_2(R)]\}$$

Averaging with respect to all possible orientations of the collision plane with respect to the external field direction requires to iterate the computations, for a prefixed set of values of the impact parameter, in correspondence with a sufficient number of orientations (α_j,β_j). Since $|c_2(+\infty)|^2$ displays a rather regular behaviour as (α, β) vary and some symmetry properties can moreover be exploited, it is possible to perform the average in terms of a reasonably modest grid of (α_j,β_j) values.

In Fig. 6, the dotted curve, which represents the behaviour of $|c_2(+\infty)|^2$ as a function of the impact parameter,

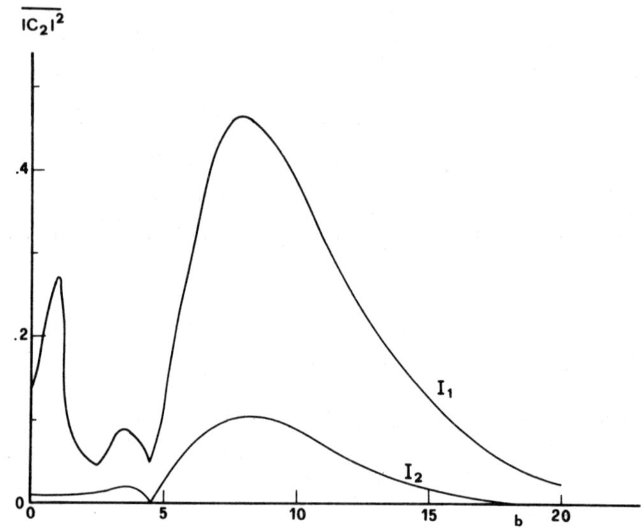

FIGURE 5. Behaviour of the orientationally averaged transition probability $\overline{|c_2(+\infty)|^2}$ as a function of the impact parameter b. I_1 = 10MW/cm^2, I_2 = 1MW/cm^2. Both curves refer to the relative velocity v = 0.12 a.u. ($\equiv 2.63\text{x}10^5$ m/sec).

for the particular orientation $\alpha = \beta = 0$, should provide us with some feeling about the order of magnitude of the fluctuations in the transition probability with respect to the mean value as the impact parameter changes.

The general behaviour of the curves in Figs. 5,6 is similar to that put in evidence elsewhere[17,19-22], i.e. after a more or less extended interval where undulations are evident, the probability transition $\overline{|c_2(+\infty)|^2}$ drops to negligible values as the impact parameter becomes larger and larger.

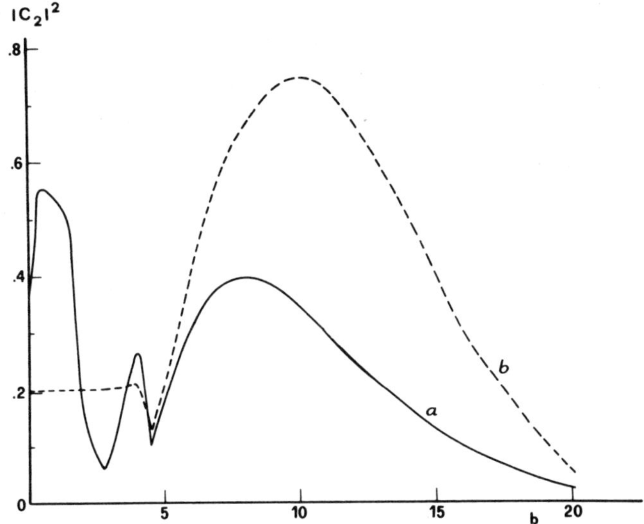

FIGURE 6. Behaviour of the transition probability as a function of the impact parameter b. (——): orientationally averaged; (----): $\alpha = \beta = 0$. Both curves refer to v = 0.18 a.u. and I = $10 MW/cm^2$.

TABLE I Cross section values (a_o units) for some radiation intensities (MW/cm^2) and relative velocities (m/sec).

I(MW/cm^2) / v(m/sec)	1.	10.
2.63×10^5	44.3	223.
3.94×10^5		220.

Some preliminary σ cross section values (a_o units), obtained from eq. (19), are collected in Table I; they refer to radiative intensities I = 1 MW/cm^2 and 10 MW/cm^2, and collision relative velocities v = $2.63x10^5$ m/sec and $3.94x10^5$ m/sec (hyperthermal regime). Sizeable σ values are clearly suggested by the model investigated. Further developments of the calculations here presented are in course of elaboration which will allow a quantitatively more complete picture of the cross section dependence on the relevant parameters, particularly the relative velocity (thermal regime).

REFERENCES

1. N.K.Rahman and C.Guidotti (eds.), Photon-Assisted Collisions and Related Topics, Harwood A. P., N.Y., 1982.
2. M.H.Mittleman, Introduction to the Theory of Laser-Atom Interactions, Plenum Press, N.Y., 1982.
3. P.L.De Vries and T.F.George, Mol.Phys. 36, 151 (1978).
4. P.L.De Vries and T.F.George, Mol.Phys. 38, 561 (1979).
5. F.H.Mies in D.Henderson (ed.), Theoretical Chemistry: Advances and Perspectives, vol. 6B, p.127, Academic Press, N.Y., 1981.
6. L.Rosenberg, Adv.At.Mol.Phys. 18, 1 (1982).
7. N.M.Kroll and K.M.Watson, Phys.Rev. A 13, 1018 (1976).
8. T.F.George, I.H.Zimmerman, P.L.De Vries, J-M.Yuan, K-S.Lam, D.K.Bhattacharyya and M.Hutchinson, J.Phys. Chem. 86, 1075 (1982).
9. G.P.Arrighini and C.Guidotti, J.Molec.Struct.(THEOCHEM) 93, 49 (1983).
10. D.A.Copeland and C.L.Tang, J.Chem.Phys. 65, 3161 (1976)
11. A.A.Frost, J.Chem.Phys. 25, 1150 (1956).
12. P.D.Robinson, Proc.Phys.Soc. 78, 537 (1961).
13. S. Gasoriowicz, Quantum Physics, J. Wiley, N.Y., 1974.
14. R.Loudon, The Quantum Theory of Light, Clarendon Press, Oxford, 1973.
15. P.L.De Vries, K-S.Lam and T.F.George, Int.J.Quantum Chem. 13, 541 (1979).

16. K-S.Lam and T.F.George in M.S.Child (ed.), Semiclassical Methods in Molecular Scattering and Spectroscopy, D.Reidel Publ. Co., Dordrecht, 1980.
17. S.E.Harris, J.C.White, IEEE J.Quantum Electron., QE-13, 972 (1977).
18. B.Liu, K.O-Ohata, K.Kirby-Docken, J.Chem.Phys. 67, 1850 (1977).
19. D.R.Bates, H.C.Johnston and I.Stewart, Proc.Phys.Soc. 48, 517 (1964).
20. E.J.Robinson, J.Phys.B 13, 2359 (1980).
21. J.F.Seely, J.Chem.Phys. 75, 3321 (1981).
22. H.A.Hyman, J.Phys.B 15, L773 (1982).

EXPERIMENTAL STUDIES of LASER-INDUCED IONIZING COLLISIONS

JOHN WEINER
Department of Chemistry, University of Maryland, College Park, Maryland 20742

I. INTRODUCTION:

It has now been slightly more than a decade since Gudzenko and Yakovlenko[1] first suggested the radiative collision process -- a seminal idea from which germinated a bumper crop of research projects aimed at the investigation of diverse laser-controlled collisions. Among the rich variety of possible interactions, long-range electrostatic forces such as dipole-dipole[2,3] or dipole-quadrupole coupling[4] have played a crucial role in establishing laser-induced energy transfer. In an early discussion of the processes Yakovlenko[5] drew the distinction between radiative and optical collisions. The purpose of the present article is to review how laser-induced ionizing collisions illustrate this classification and to discuss how line-shape studies may lead to determination of the dominant interaction mechanism.

An inelastic process which alters inital states of both colliding partners and in which the maximum cross section occurs at a laser frequency about equal to the energy difference between atomic states in both partners is termed a radiative collision:

$$A(2)+B(1)+\hbar w \rightarrow A(1)+B(2)$$
$$\hbar w \simeq B(2)-A(2). \qquad (1)$$

By contrast an optical collision occurs when only one partner changes state and the laser frequency is close to an allowed atomic transition of that partner:

$$A(1)+B(1)+\hbar w \rightarrow A(2)+B(1)$$
$$\hbar w \simeq A(2)-A(1). \qquad (2)$$

Gallagher et al.[6-8] have shown that the essential nature of the two processes is identical and a unified picture may be drawn from the quasistatic theory of collision broadening. From an experimental viewpoint, however, process (1) creates a new feature in the absorption spectrum while process (2) simply modifies the shape of the isolated absorption profile. The more dramatic appearance of an entirely new peak explains the emphasis on radiative collisions in recent experiments.

II. OPTICAL COLLISIONS

Optical collisions possess, however, a dramatic property of their own when they take place between identical particles. Long-range dipole dispersion forces may give rise to an interaction potential which falls off as the inverse cube of the internuclear distance rather than the more familiar inverse sixth power. This special dipole-dipole "resonance" force obtains in collisions between two identical atoms when one of them is in an excited state whose symmetry permits dipole energy transfer to its partner. This electrostatic interaction must perforce be the longest-range coupling between colliding neutral particles with its physical origin discussed long ago by Van Vleck[9] and Margenau.[10] The optical collision line profile is a direct probe of the dispersion force dominant at long range.

To illustrate the optical collision we discuss the observation of resonance dipole forces between Li*(2p) and Li(2s) atoms by means of a two-step, two-photon ionizing collision. In the first step Li*(3d) is populated via an optical collision

$$\mathrm{Li}^*(2p)+\mathrm{Li}(2s)+\hbar w \rightarrow \mathrm{Li}^*(3d)+\mathrm{Li}(2s) \tag{3}$$

followed in the second step by photoionization from Li*(3d:

$$\mathrm{Li}^*(3d)+\hbar w \rightarrow \mathrm{Li}^+ + e. \tag{4}$$

The optical collision line shape is reflected in the observed rate of Li^+ production as a function of laser frequency. As pointed out by Nayfeh,[11-13] in a series of pioneering articles, exploitation of two- and three-photon ionizing collisions is attractive because of inherent efficiency and sensitivity of charged-particle detection.

III. THEORETICAL DESCRIPTION

The development proceeds in two steps: First, we write down a simple formula for the optical collision cross section by applying standard time-dependent perturbation theory in second order. Secondly, using previously calculated photoionization cross sections for Li*(3d), we determine the overall two-photon ionizing rate from coupled rate equations which are then integrated over the time of a laser pulse to find the total ion yield per pulse. The rate equation approach is justified because the two steps, optical collision followed by photoionization, occur with random relative phase in the time development of the overall system wave function. Photoionization takes place essentially in the time between optical collisions not simultaneously with them.

A. Optical collision cross section for a three-level system

We begin by forming appropriate linear combinations of Li atomic-state wave functions to reflect nuclear exchange symmetry,

$$\psi_1^{(g)} = \frac{1}{\sqrt{2}}[\phi_s(1)\phi_p(2)-\phi_p(1)\phi_s(2)], \qquad (5)$$

$$\psi_2^{(u)} = \frac{1}{\sqrt{2}}[\phi_s(1)\phi_p(2)+\phi_p(1)\phi_s(2)], \qquad (6)$$

where the superscripts g,u have their customary meaning, and subscripts s,p denote the orbital angular momentum of the active electron in atoms (1) or (2). At internuclear separation R the two quasimolecular wave functions interact through the dipole-dipole coupling operator (mks units)

$$H'_{12}=-\left(\frac{e^2}{4\pi\varepsilon_0}\right)\left(\frac{2z_1z_2}{R^3}\right), \qquad (7)$$

where z_1z_2 are the space-fixed coordinates of the active electron on atoms 1 and 2. In fact, the collisional perturbation should include two other terms for the x and y coordinates. Including them gives rise to four quasimolecular states ($\Sigma_{u,g};\Pi_{u,g}$) which elaborates the discussion unnecessarily for the results reported here. Using (5) and (6) to take matrix elements of (7) we find the expression for the energy shifts of states (1) and (2),

$$\Delta E_{12}\binom{g}{u}=(\pm)\frac{2(\mu_{12})^2}{4\pi\varepsilon_0 R^3}\ , \tag{8}$$

where the gerade state is repulsive, the ungerade state attractive, and μ_{12} is the transition dipole matrix element between Li(2s) and Li(2p). Next we make the "fixed atom" assumption in which the initial orientation of the dipole is considered random with respect to the internuclear axis but remains unchanged during the course of the collision. Averaging over initial orientations,[14] one finds

$$\Delta E_{12}=\pm\frac{2}{\sqrt{3}}\frac{(\mu_{12})^2}{4\pi\varepsilon_0 R^3}\ , \tag{9}$$

The energy splitting between states (1) and (2) in the case of lithium is depicted in Fig. 1.

The second perturbation comes from the laser field which couples one of the collision states (1) and (2) to the final state (3) of the system

$$\psi_3=\phi_s(1)\phi_d(2), \tag{10}$$

where $\phi_d(2)$ is the atomic wave function of the Li*(3d) level. Taking, for example, an optical transition from state (2) to state (3), we write the appropriate coupling expression as

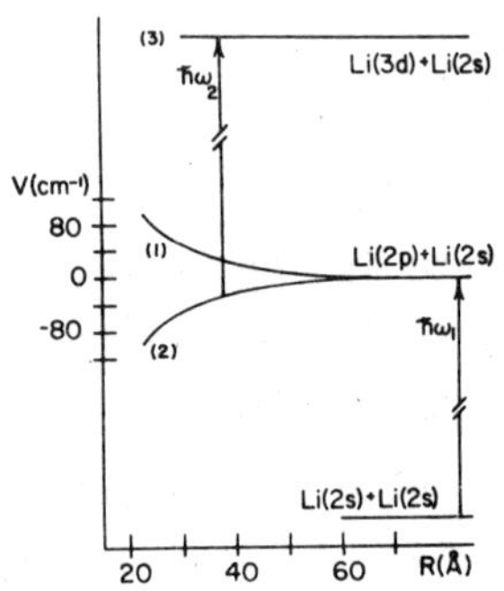

FIG. 1. Three-level-model system showing C_3/R^3 van der Waals' collisional interaction.

$$V'_{23}=\mu_{23}E\ \cos wt, \tag{11}$$

where μ_{23} is the transition dipole between Li(2p) and Li(3d), and E,w are the optical field amplitude and frequency. The time evolution of the system wave function is

$$\psi(t)=a_1(t)\phi_1 e^{-i\int w_1 dt}+a_2(t)\phi_2 e^{-i\int w_2 dt} +a_3\phi_3 e^{-e\int w_3 dt} \tag{12}$$

Substituting (12) into Schrödinger's time-dependent wave equation (where H'_{12} and V'_{23} are perturbative terms of the Hamiltonian) and solving the resultant set of time-dependent coupled differential equations for the coefficients a_n, we write the expression for the probability of finding the system in state (3),

$$|a_3|^2 = \frac{\pi C_3 E^2 (\mu_{23})^2}{6h^3 v R_X^2 (\delta w)^2}, \tag{13}$$

where

$$|C_3| = \frac{2}{\sqrt{3}} \frac{(\mu_{12})^2}{4\pi\varepsilon_0},$$

and R_X is the point of stationary phase where the laser field of frequency w is resonant with the energy difference between states (2) and (3). The term δw in the denominator is the laser detuning from the Li(2p,3d) atomic transition, and v is the relative velocity of collision. In writing (13) we have assumed that the region around R_X is traversed twice and the transition probability to state (3) remains the same for both the incoming and outgoing branches of the trajectory. Finally we calculate the optical collision cross section as

$$\sigma_C = 2\pi\int_0^\infty |a_3|^2 R dR = \frac{\pi^2 C_3 E^2 (\mu_{23})^2}{3h^3 v (\delta w)^2}. \tag{14}$$

For crossed-beam lithium collisions with relative velocity corresponding to source temperature of ~1000 K,

$$\sigma_C(cm^2) = 1.42\times10^{-17}\frac{I}{(\delta w)^2}, \tag{15}$$

where I is laser power density in W cm^{-2} and δw is the detuning in cm^{-1}. Equation (14) agrees with an earlier result of Nayfeh[12] who analyzed the problem in terms of potential curve crossings. Equation (15) is for the particular case of Li-Li collisions, where C_3 and μ_{23} are calculated from a compilation[15] of critically evaluated oscillator strengths.

B. Photoionization rate

The next step is to calculate the rate of photoionization from the Li*(3d) populated by the optical collision. The rate equations are

$$\frac{d}{dt}[Li^*(3d)] = [Li(2s)][Li^*(2p)]\sigma_C v - A[Li^*(3d)]\sigma_p, \tag{16}$$

$$\frac{d}{dt}[Li^+]=I[Li(3d)]\sigma p, \qquad (17)$$

where the square-bracketed quantities are number densities, A is the spontaneous radiation rate $Li^*(3d)\rightarrow Li^*(2p)$, and σp is the photoionization cross section from the $Li^*(3d)$ excited level. Taking the spontaneous radiation rate A from Wiese et al.,[15] the photoionization cross section from Manson and Lahiri,[16] and the laser power density from measurement,

$A=7.44x10^7 cm^{-1}$,
$\sigma p=6.7x10^{-18} cm^2$,
$I=7.98x10^{25}$ photons $cm^{-2} sec^{-1}$,

we find that

$$\frac{[Li^+]}{[Li(2s)]^2} = \frac{(8.28x10^{-35}) I^2 \sigma_p}{(A+I\sigma p)(\delta w)^2}, \qquad (18)$$

which expresses the ion yield per pulse normalized to atomic beam intensities as a function of laser intensity and detuning from line center. Details of the development of equation (18) may be found in reference (17). The most important feature of (18) is the inverse square dependence of the detuning. This is a direct consequence of Eq.(9) and is characteristic of collisions between identical atoms. The familiar van der Waals' inverse sixth power law yields a detuning function decreasing as $(\delta w)^{-3/2}$. Equation (18) also shows that laser power dependence should increase quadratically for low power but gradually change to a linear form at high power.

IV. DESCRIPTION OF THE EXPERIMENT

The apparatus has been described in previous laser-collision studies,[18,19] but for convenience we summarize the essential points here.

Two atomic lithium beams cross at right angles in the horizontal plane of a cylindrical high-vacuum system. One laser beam enters through an adjacent port between the atomic beam sources. A second laser beam enters from a port opposite the first. Thus the two lasers propagate in opposite directions and are very nearly collinear. All four beams overlap in a well-defined interaction region at the center of the apparatus. A quadrupole mass spectrometer, situated above the collision plane on a vertical axis passing through the interaction region, filters the ion masses produced in the experiment. Signal pulses of mass-selected ions are detected and amplified by a channeltron electron multiplier, integrated over the time of the laser pulse by a charge-sensitive amplifier, and fed to a boxcar averager gated synchronously with the lasers. The laser sources are flash-lamp pumped tunable dye lasers operated at a 10-Hz repetition rate with a pulse duration of about 1 μsec. Laser line width is about 3 cm^{-1}. Hot-wire detectors monitor the lithium beam flux. During a typical run atomic beam density in the collision region is about 10^9 cm^{-3}. Focused laser power density is 2.6×10^7 W/cm^2 in a spot size of about 0.015 mm^2. In the experiments reported here one unfocused laser is tuned to populate the Li*(2p) level with power sufficient only to saturate the transition. The second, focused laser is tuned in the vicinity the Li*(2p,3d) transition frequency. The quadrupole mass filter is set to pass Li^+ and the ion intensity, produced by photoionization of Li*(d) in the field of the strong laser, is recorded as a function frequency.

V. EXPERIMENTAL RESULTS AND DISCUSSION

Figure 2 shows the principal result of the experiment. The ion profile maximum is peaked at the Li*(2p,3d) transition frequency. Note that ion intensity at the peak does not increase with laser power due to saturation of the atomic transition, but that broadening in the wings, due to quasimolecular absorption, is dramatic. The effect is more than an order of magnitude above that power broadening of the atomic line. For example at maximum laser power ($I=2.60\times10^7 Wcm^{-2}$) the Li(2p,3d) transition profile is power broadened to about 2 cm^{-1} while the optical collision intensity is still detectable at 125 cm^{-1} into the wing (see Fig.3). As atomic beam fluxes are reduced the broadening effect diminishes and finally disappears. The observed residual spectral profile centered on the atomic line reflects

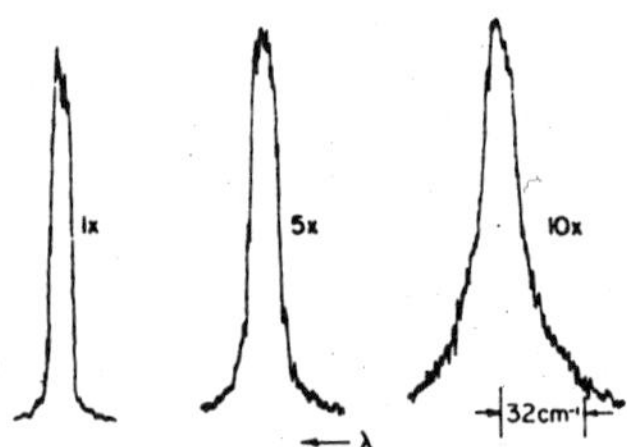

FIG. 2. Ion line shape vs laser power. Note that the peak intensity, corresponding to the Li(2p,3d) atomic transition, is saturated at all three laser powers but that in the wings, corresponding to the quasimolecular absorption, increases dramatically. Maximum laser intensity is $\approx1.3\times10^7 W/cm^2$.

only the linewidth of the laser and a small contribution from power broadening.

Another discernable feature in Fig. 2 is the slight asymmetry of the line shape toward the blue wing. This may be due to a C_6/R^6 dispersion interaction between the manifold of states correlating to Li*(3d)+Li(2s) and those correlating to Li*(2p)+Li*(2p). The energy level of the excited pair of atoms in the 2p level at infinite internuclear separation is only 0.18 eV below Li*(3d)+Li(2s). As internuclear separation decreases, the two manifolds will "repel" each other, thus effectively shifting state (3) in Fig. 1 to higher energy. Transitions from state (1) and (2) will therefore be shifted to the blue. A detailed analysis of the line-shape asymmetry using the quasistatic theory of line broadening is clearly possible.

Figure 3 shows in detail a portion of the blue wing. To verify a $(\delta w)^{-2}$ dependence, the intensity profile was fit to a function of the form

$$F(\delta w)=a(\delta w)^{-3/2}+b(\delta w)^{-2},$$

and the coefficients a,b were varied to obtain best agreement by the method of least squares. The best-fit ratio of b/a is 8/1 clearly demonstrating that the wing intensity reflects a C_3/R^3 interaction rather than the more usual C_6/R^6 van der Waals law.

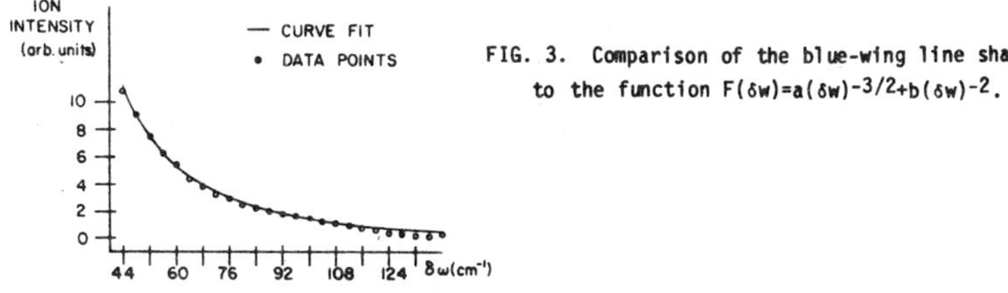

FIG. 3. Comparison of the blue-wing line shape to the function $F(\delta w)=a(\delta w)^{-3/2}+b(\delta w)^{-2}$.

The magnitude of the cross section, given by Eq.(15), shows that the resonance interaction gives rise to very large optical collision cross sections. At a detuning of 100 cm^{-1} and a laser intensity of $7x10^{7}$ W cm^{-2}, for example, the cross section is about $10^{-13}cm^{2}$. A recent study[20] of resonance stark tuning in dipole-dipole collisions of alkali Rydberg states has also found remarkably large cross sections.

VI. RADIATIVE COLLISIONS

We continue to discuss here results on two-photon collisional ionisation. Firstly, the ionising event takes place in a crossed-beam environment under single-collision conditions. Secondly, two lasers are utilised: initially a weak resonant field populates the excited Ba level

$$Ba(^1S_0)+hw_1 \rightarrow Ba^*(^1P_1)$$

then a strong non-resonant field tuned close to the frequency difference between Na*(4d) and $Ba^*(^1P_1)$ effects the energy transfer. The strong laser switches excitation energy from $Ba^*(^1P_1)$ to Na*(4d) by means of a <u>radiative collision</u>[21]

$$Ba^*(^1P_1)+Na(3s)+hw_2 \rightarrow Na^*(4d)+Ba(^1S_0).$$

Thirdly, subsequent photoionisation of Na*(4d) in the strong laser field is detected directly by a quadrupole mass filter and charged particle multiplier unit mounted above the collision plane.

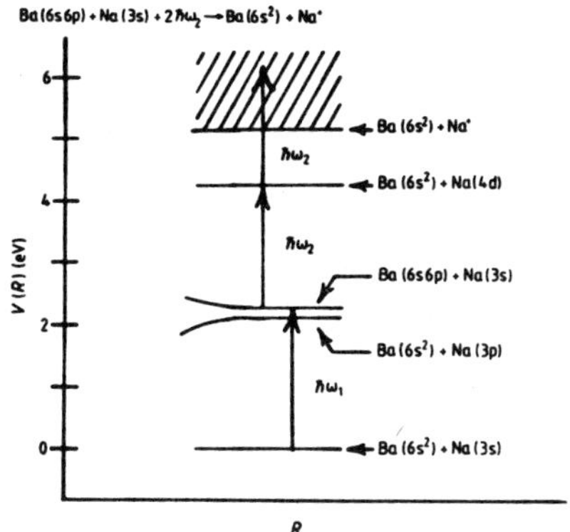

Fig. 4. Potential energy levels and transitions which play a role in the radiative ionising collision. Note that $\hbar\omega_1$ is resonant with the first allowed excited state of barium but that $\hbar\omega_2$ corresponds to the energy difference between Na((4d) and Ba*($^1P_1$). The laser field of $\hbar\omega_1$ is several orders of magnitude less intense than that of $\hbar\omega_2$ and therefore makes an insignificant contribution to the ionisation.

In order to write a formula for the total ion yield, we begin by first writing an expression for the peak in the radiative collision cross section in terms of optical field intensity. Then using previously calculated photoionisation cross sections for Na*(4d), we determine the overall two-photon ionising rate from coupled rate equations. The expression for the rate of ionisation is finally integrated over the laser pulse duration to calculate the total ion yield. The rate equation approach is justified because the two steps (excitation and ionisation) are incoherent. Photoionisation occurs essentially in the time between collisions, not simultaneously with them.

The off-resonance condition of the strong laser frequency and the long-range nature of the dipole-dipole collision interaction allow us to develop a formula for the radiative energy transfer from time-dependent perturbation theory in second order. The essential result,

$$\sigma_c = \frac{\pi}{\hbar^4}\left[\frac{\mu_a^{12}\,\mu_b^{12}\,\mu_b^{23}}{\upsilon \rho_0 \Delta w}\right]^2 E^2$$

where

$$\mu_a^{12} = \langle Ba(6s^{2\,1}S_0)| \, E \cdot r \, | \, Ba^*(626p^1P_1)\rangle$$

$$\mu_b^{12} = \langle Na(3s^2S_{1/2})| \, E \cdot r \, | \, Na^*(3p^2P_{3/2})\rangle$$

$$\mu_b^{23} = \langle Na^*(3p^2P_{3/2})| \, E \cdot r \, | \, Na^*(4d^2D_{5/2,3/2})\rangle.$$

υ is the average relative collision speed, ρ_0 the Weisskopf radius, Δw the energy defect between $Ba^*(6s6p^1P_1)$ and $Na^*(3p^2P_{3/2})$, and E the amplitude of the electric component of the strong laser optical field with polarisation vector E. for the Na, Ba case studied here

$$\sigma_c(cm^2)=7.13\times10^{-19} I(W\ cm^{-2}). \tag{19}$$

The rate equations governing Na^+ production are

$$d[Na^*(4d)]/dt=[Na(3s)][Ba^*(6s6p^1P_1)]\sigma_c\upsilon - A[Na^*(4d)-I[Na^*(4d)\sigma_p \tag{20}$$

$$d[Na^+]/dt=[Na^*(4d)]I\sigma_p \tag{21}$$

where σ_c, σ_p are the radiative and photoionisation cross sections respectively. I is the laser flux density and A is the $Na^*(4d\rightarrow3p)$ spontaneous decay rate. Bracketed quantities are particle number densities.

Solving (20), substituting the result in (21), and integrating over the laser pulse duration we obtain

$$[Na^+] = \left(\frac{[Na(3s)][Ba^*(^1P_1)]\sigma_c\sigma_p I\bar{\upsilon}}{A+I\sigma_p}\right)\left(\frac{\tau-\exp[-(A+I\sigma_p)\tau]}{A+I\sigma_p}\right) \quad (22)$$

with τ, the time of the light pulse, close to one microsecond. From known values of A, I, σ_p, τ one finds that the exponential term on the right-hand side of (22) is negligible. We then write the expression for the ion yield per pulse, normalised to the crossed-beam flux, as

$$\frac{[Na^+]}{[Na(3s)][Ba^*(^1P_1)]} \simeq \frac{\sigma_c\sigma_p I\bar{\upsilon}\tau}{A+I\sigma_p}. \quad (23)$$

Converting the numerical constant in (19) so that I is in units of photons $cm^{-2}\ s^{-1}$ and substituting (19) in (23) gives finally

$$\frac{[Na^+]}{[Na(3s)][Ba^*(^1P_1)]} = \frac{(2.33\times10^{-37})I^2\sigma_p\bar{\upsilon}\tau}{A+I\sigma_p}. \quad (24)$$

The rate A for spontaneous radiative decay is taken from reference (15) and the photoionisation cross section σ_p is taken from reference (16).

Figure 5 shows the result of the experiment. The very large peak to the right is the two-photon absorption Na*(5s) from the ground state followed by photoionisation. This signal provides a convenient internal frequency reference from which the absolute frequency of the two-photon collisional ionisation signal may be determined. The peak intensity occurs, as expected, at the difference frequency between Na*(4d) and Ba*(626^1P_1). The left-hand side of (23) or (24) is measured in the experiment. From known quantities on the right-hand side we verify that the radiative collision cross section σ_c is in reasonable agreement (within a factor of three) with that calculated from equation (24). Note also that (24) implies a quadratic growth of ion yield with laser intensity at low power that smoothly changes to a linear dependence as power increases.

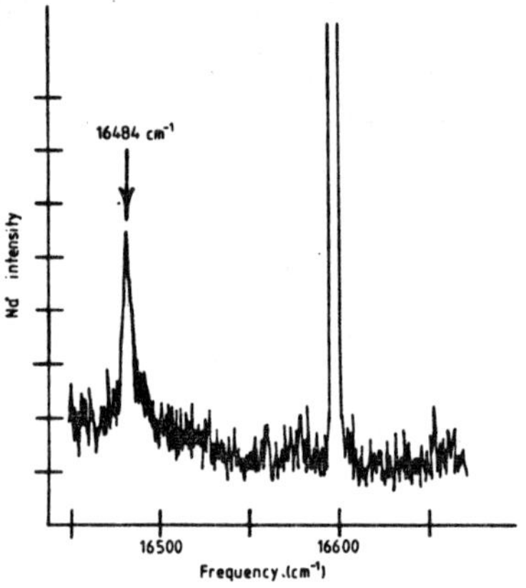

FIG. 5. The small peak to the left corresponds to the two-step ionising process of figure 1. The large peak to the right is due to two-photon absorption from Na(3s) to Na*(5s) followed by photoionisation. The abscissa is a frequency scan of the second laser.

An obvious complementary experiment is the resonant excitation of Na followed by a radiative collision with ground state Ba to produce Ba* which is subsequently photoionized.

$$Na(^2S_{1/2})+\hbar w_1 \rightarrow Na^*(^2P_{3/2})$$

$$Na^*(^2P_{3/2})+Ba(6s^2\ ^1S_0)+\hbar w_2 \rightarrow Ba^*(6p^2\ ^1S_0)$$

$$Ba^*(6p^2\ ^1S_0)+\hbar w_2 \rightarrow Ba^+ + e$$

For this ionization pathway the predicted transfer frequency ($\hbar w_2$) for the radiative collision step is 5748.6 Å. Figure (6) shows the Ba^+ peak appearing very close to this frequency in a scan of ion intensity vs. laser frequency. Note that the transfer frequencies in the complementary pair of experiments are different. Equation (24) implies that ion signal intensity should be quadratic with laser intensity in the low regime. Figure (7) confirms this behavior.

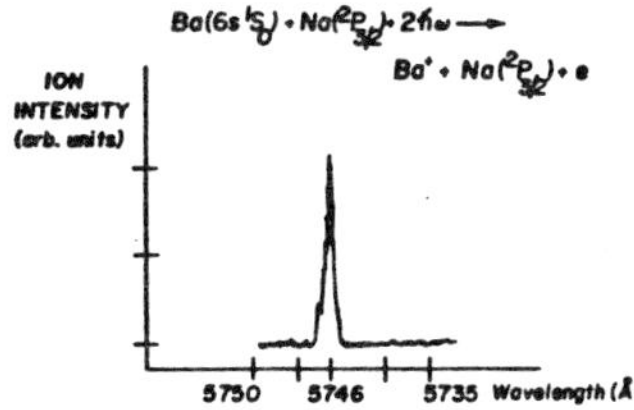

FIG. 6. Ba^+ ion signal vs. scanning laser wavelength.

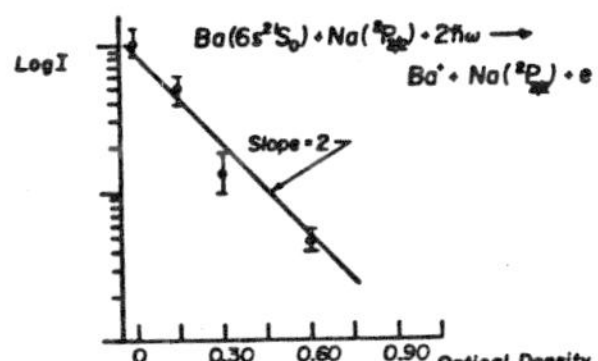

FIG. 7. Variation of Ba^+ intensity with laser power. Laser frequency fixed at 5746 A.

Summary:

This article has discussed two-photon collisional ionization within the context of optical and radiative collisions. New results from this laboratory on Na-Na collisions indicate that an entirely different mechanism may be responsible for laser-induced production of Na_2^+. To report of that work, however, will have to await a later conference in Pisa.

REFERENCES

1. L.I. Gudzenko and S.I. Yakovlenko, Zh. Eksp. Teor. Fiz. 62, 1686 (1972) [Sov. Phys.-JETP 35, 877 (1972)].

2. R.W. Falcone, W.R. Green, J.C. White, J.F. Young and S.E. Harris, Phys. Rev. A15, 1333 (1977).

3. S. Geltman, J. Phys. B10, 3057 (1977).

4. J.C. White, Opt. Lett. 5, 239 (1980).

5. V.S. Lisitsa and S.I. Yakovlenko, Zh. Eksp. Teor. Fig. 66, 1550 (1974) [Sov. Phys.-JETP 39, 759 (1974)].

6. R. Hedges, D. Drummond, and A. Gallagher, Phys. Rev. A6, 1519 (1972).

7. D.L. Drummond and A. Gallagher, J. Chem. Phys. 60, 3426 (1974).

8. A. Gallagher and T. Holstein, Phys. Rev. A16, 2413 (1977).

9. G.W. King and J.H. Van Vleck, Phys. Rev. 55, 1165 (1939).

10. H. Margenau, Rev. Mod. Phys. 11, 1 (1939).

11. M.H. Nayfeh, G.S. Hurst, M.G. Payne, and J.P. Young, Phys. Rev. Lett. 39, 609 (1977).

12. M.H. Nayfeh, Phys. Rev. A16, 927 (1977).

13. M.H. Nayfeh, Phys. Rev. A20, 1927 (1979).

14. S.E. Harris and J.C. White, IEEE J. Quantum Electron. QE-13, 972 (1977).

15. W.L. Wiese, M.W. Smith, B.M. Glennon, Atomic Transition Probabilities, Nat'l. Bur. Stand. (U.S.) Publ. No. NSRDS-NBS4 (U.S. GPO, Washington, D.C., 1966), pp.16-17.

16. J. Lahiri, and S.T. Manson, Abstracts of Contributed Papers, XII International Conference on the Physics of Electronic and Atomic Collisions, edited by S. Datz (Gatlinburg, Tennessee, 1981) p.39, and private communication.

17. P. Polak-Dingels, R. Bonanno, J. Keller, John Weiner, and J.C. Gauthier, Phys. Rev. A25, 2539 (1982).

18. J. Weiner and P. Polak-Dingels, J. Chem. Phys. 74, 508 (1981).

19. P. Polak-Dingels, J.F. Delpech, and J. Weiner, Phys. Rev. Lett. 44, 1663 (1980).

20. K.A. Safinya, J.F. Delpech, F. Gounand, W. Sander, and T.F. Gallagher, Phys. Rev. Lett. 47, 405 (1981).

21. P. Polak-Dingels, R. Bonanno, J. Keller, and J. Weiner, J. Phys. B: At. Mol. Phys. 15, L41 (1982).

EXPERIMENTAL METHOD FOR MEASURING ELECTRONIC ENERGY TRANSFER CROSS SECTIONS IN LASER EXCITED ATOMIC VAPORS.

M.ALLEGRINI, P.BICCHI* and L. MOI
Istituto di Fisica Atomica e Molecolare del C.N.R. - Pisa, Italy
* Also at Istituto di Fisica dell'Università-Siena, Italy.

Abstract The electronic energy transfer in collisions between two laser excited atoms is now a widely studied process. Here we present a novel method for the determination of its cross section, which is based on a specially built cell and on an absolute calibration procedure. Experimentally such a method simply requires a broadband, low power, cw dye laser in resonance with an atomic transition. It can be applied to any atomic vapor which may be contained in the sealed cell and it is reliable to ∿30%. Its application to the process Na (3P)+ Na (3P) → Na(5S,4D)+Na(3S) is presented.

INTRODUCTION

The transfer of electronic energy in collisions between two atoms, i.e.

$$A(n\ell) + B(ground) \rightarrow A(ground) + B(n'\ell') \qquad (1)$$

is a well known process, widely studied in the past by using spectral lamps to excite the atoms[1]. Its cross section σ has been measured for many pairs of atoms, especially for the alkali atoms. Solution of the rate equations of process (1) is straightforward and, in terms of the intensity of the fluorescence lines I, which is the quantity usually measured in the experiment, σ results

$$\sigma = \frac{\alpha}{\bar{v}} \quad \frac{1}{N_B \ \tau_A} \quad \frac{I_B}{I_A} \tag{2}$$

where α is a factor which takes into account the response of the detection apparatus and the characteristics of the transitions (such as frequency factors and braching ratios), $\bar{v} = \sqrt{\frac{8\pi k}{\pi\mu}}$ is the mean velocity of the colliding partners, N_B the density of the B species and τ_A the radiative lifetime of the excited level of the A atoms. The accuracy on σ depends on the accuracy in measuring the density of the atomic species B and the lifetime of the excited state $A(n\ell)$, which may be complicated in presence of radiation trapping.

Recently the laser, owing to the power and frequency selectivity available compared to a spectral lamp, has allowed the observation of many kinds of collisions involving more than one excited atom. In analogy to process (1), known as sensitized fluorescence, also the process[2]

$$A(n\ell)+A(n\ell)\rightarrow A(n'\ell')+A(\text{ground}) \tag{3}$$

involving collisions between two atoms excited

to the same quantum level, has a cross section which depends on the population $N_{n\ell}$ and the lifetime $\tau_{n\ell}$ as

$$\sigma = \frac{\beta}{\bar{v}} \quad \frac{1}{N_{n\ell}\tau_{n\ell}} \quad \frac{I_{n'\ell'}}{I_{n\ell}} \tag{4}$$

where the notations have the same meaning of those used in equation (2).

The volume in which the fluorescence appears does not enter specifically in equation (2) because it cancels out in the ratio of the intensities $\frac{I_B}{I_A}$, provided the single fluorescence lines I_A and I_B come out from the same volume and provided the volume remanins constant during the measurements. Also for process (3) the absolute measurement of the laser-vapor interaction volume may be avoided if the population density of the excited state $|n\ell>$ is independently determined (see Equ.(4)). The absolute value of the volume from which $I_{n\ell}$ and $I_{n'\ell'}$ are observed is not important, however it must be well defined and remain constant during the measurements. Unfortunately this in not usually the case because in a vapor the optical absorption length varies considerably with changes in the atom density and/or in the laser intensity, resulting in a variation of the volume of the laser-vapor interaction. Moreover, because of the diffusion, the excited atoms are not confined to the focused laser column and great care must be taken in the evaluation of the resulting volume of excitation[3].

In spite of these difficulties great effort has been paid to get reliable quantitative results of σ for process (3), at least in the alkali vapors[4-9]. In effect process (3) followed by the photoionization of the highly-excited species $A(n'\ell')$ has proven to be important for the production of the seed electrons in the experiments where total ionization of the vapor is achieved[10]. Lasing action has recently been observed from one of the $A(n'\ell')$ level[11], populated through process (3). The interest in collisions between excited atoms is more general since they are certainly involved in stellar processes and in chemical reactions induced by laser radiation. From a theoretical point of view the cross sections of process (3) give informations on the long-range interactions between atoms, in the intersting region where the atoms are not trapped to form the stable neutral dimers but cannot be considered isolated either.

Here we present a novel experimental procedure to measure σ with about 30% accuracy, which avoids most of the error sources responsible for the discrepancies between previous measurements.

"Energy transfer in laser excited atomic vapors"

EXPERIMENTAL METHOD

The construction of a special capillary cell, sketched in figure 1, has solved the main part the problem mentioned before.

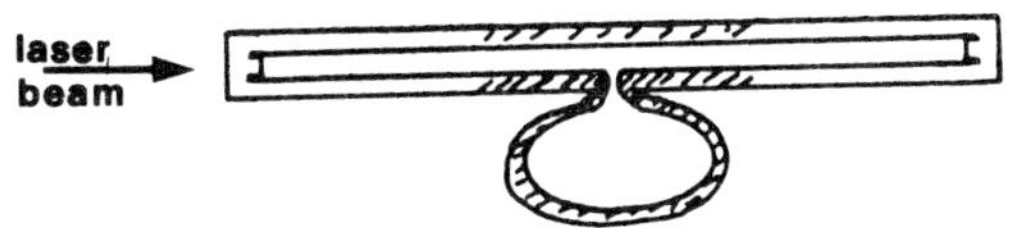

FIGURE 1. Capillary cell (actual dimensions).

The cell body is a pyrex capillary tube of internal radius 0.9 mm and the windows are slides of pyrex rods optically flattened on the internal side before welding to the capillary tube. After welding the ends of the cell are also flattened. The alkali metal is contained in a side arm of the cell and to prevent window contamination there is a small temperature gradient between the cell body and the side arm. The laser is focused to the diameter of the cell and illuminates the entire volume of the cell. To further define the volume from which fluorescence is detected a slit of 150 µm is placed near the input window. The resulting volume is 4×10^{-4} cm^3; however its absolute value is not crucial. The important point is that it is well defined, insensitive to variations in the optical absorption path and uniformly illuminated. Moreover, since the cell is sealed and the vapor is saturated, the atomic density N_{at}, whose value is necessary to obtain $N_{n\ell}$ can be determined for each temperature from the Nesmeyanov tables[12], without any sophisticated measurement. The accuracy of the Nesmeyanov tables, tested in sodium at ∿100°C higher than the temperature typical of our experiment, has been proved to be ∿2-13%[4].

The density of the excited state $N_{n\ell}$ is determined by an original calibration method based on the saturation of the resonance fluorescence. At low temperature, when the atomic density is also low, it is possible to achieve saturation of the resonance fluorescence transition $A(n\ell)\rightarrow A(\text{ground})$, as it is shown in figure 2, even if the excitation is made with a low power broadband dye laser.

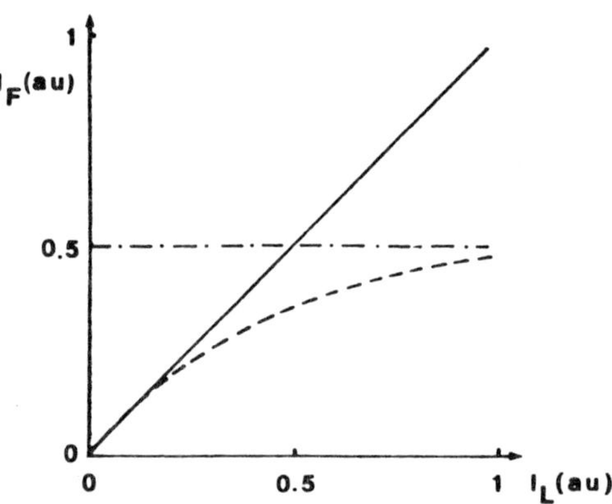

FIGURE 2. Resonance fluorescence signal versus the laser intensity, in absence of saturation (dashed line) and in presence of saturation (full line).

At saturation the ratio between $N_{n\ell}$ and N_{at} is known from the degeneracies of the ground and excited state and in first approximation $N_{n\ell} \simeq \frac{1}{2} N_{at}$ More details on this particular point, including the treatment of the line shape of the saturated fluorescence, are in ref.5. Once the detection apparatus is calibrated ,the resonance fluorescence signals at different atomic densities are compared and the density $N_{n\ell}$ is obtained from the expression

$$N_{n\ell}(T_1) = \frac{I\ (T_1)}{I\ (T_o)}\ N_{n\ell}(T_o)\ \frac{\tau^*}{\tau_o} \qquad (5)$$

where τ^* is the trapped lifetime of the state $|n\ell\rangle$ at T_1 and τ_o is the natural lifetime at T_o, when there is not radiation trapping. From equation(5) it is clear that the measurement of τ^* is of primary importance. Once again the system which has been studied in more detail involves the 3P-level of sodium. The experiment of ref. 3 has proved that the complicated mechanism of radiation trapping can be well understood and also controlled; however this is done at the expenses of the simplicity of the experimental set-up. In the application of our method to the process

$$Na(3P) + Na(3P) \rightarrow Na(5S,4D) + Na(3S) \qquad (6)$$

we have avoided the direct measurement of $\tau^*(3P)$

APPLICATION TO SODIUM

The atomic levels of interest in this experiment are sketched in figure 3.

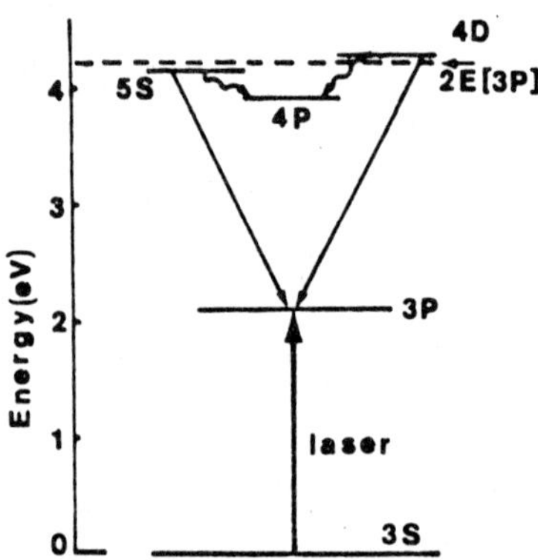

FIGURE 3. Sketch of the atomic sodium levels involved.

The rate constant $k_{n\ell}$ (with $n\ell$ = 5S or 4D) is related to the cross-section so far considered simply through the mean interatomic velocity $k_{n\ell} = \sigma_{n\ell}\bar{v}$. In the experiment we measure the density N_{3P} of the excited 3P-atoms, the intensity of the fluorescence transitions $I_{5S\to 3P}$, $I_{4D\to 3P}$ and $I_{3P\to 3S}$, corrected for the response of the apparatus. These quantities are related to the product $k_{n\ell}\,\frac{\tau^{*2}}{\tau_o}$. To minimize effects present in high vapor pressure, which may obscure process (6), the experiment is performed in a range of relatively low densities. Although in this situation the intrinsic signal-to--noise ratio is reduced we take advantage of this fact by minimizing the self trapping corrections and the contribution of other possible processes enhanched by the high vapour density. At these temperatures the colliding 3P-atoms have thermal energy so that their relative energy remains constant as does the collision frequency. As a consequence $k_{n\ell}$ has to remain constant and the temperature dependence of $k_{n\ell} \frac{\tau^{*2}}{\tau_0}$ is due only to τ^*. We have checked that in our experiment the Milne theory of radiation trapping[13] satisfies this condition (as shown in figure 4 for the 5S level) and provides the correct value for τ^*.

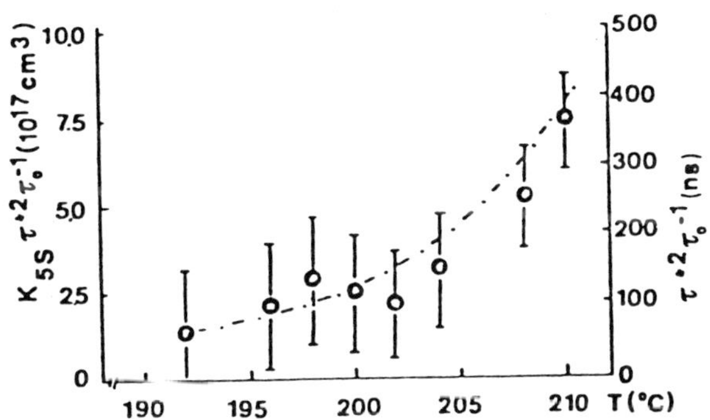

FIGURE 4. The dashed line gives the values of τ^{*2}/τ_0 from the Milne theory. The dots represent the experimental values of $k_{5S}\ \tau^{*2}/\tau_0$.

As an example, the resulting values for k_{5S} is

plotted in figure 5.

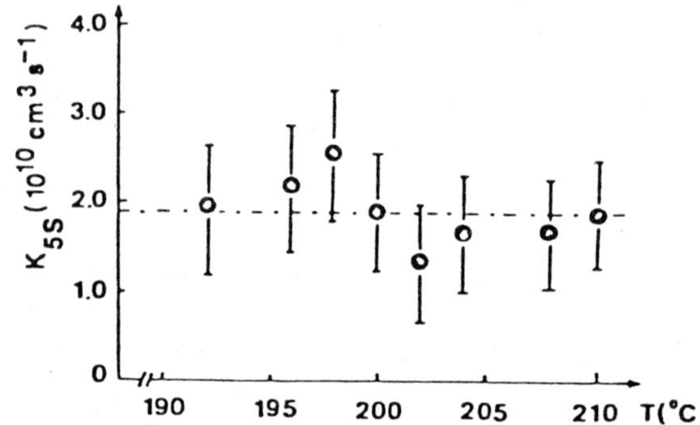

FIGURE 5. Experimental results of k_{5S} versus the temperature.

CONCLUSION

The previous disagreement between the experimental determinations of the rate constants for process (6), as shown in table 1, makes clear the need for reliable methods of measurement. The approach is obviously not unique. Huennekens and Gallagher[4], for example, have demonstrated that using a pulsed dye laser for the excitation and a single mode cw dye laser tuned to the wings of the resonance transition it is possible to measure both the spatial and temporal distribution of the excited atoms. Thus they can control the radiation diffusion and get N(3P) and τ^*(3P) in the changing volume. The special cell designed for our experiment instead ,allows us to take constant the excitation volume and to minimize the radiation trapping problems. It is satisfying that such different approaches give very close values (see table I) for the cross section of process (6).

Table I Rate coefficients k and cross sections σ for the 5S and 4D levels of sodium.

Level	T(°C)	k($cm^3 sec^{-1}$)	σ(cm^2)	Ref.
5S	227	7.0×10^{-15}	7.3×10^{-20}	8,(1982)
	∿324	1.63×10^{-10}	1.6×10^{-15}	4,(1983)
	210	1.9×10^{-10}	2.0×10^{-15}	5,(1983)
4D	277	2.2×10^{-11}	2.3×10^{-16}	14,(1979)
	397	1×10^{-12}	9.0×10^{-18}	7,(1981)
	227	4.5×10^{-15}	4.7×10^{-20}	8,(1982)
	307*		$\sim 10^{-14}$*	9,(1982)
	∿324	2.46×10^{-10}	2.3×10^{-15}	4,(1983)
	210	3.0×10^{-10}	3.2×10^{-15}	5,(1983)

* This experiment used a sodium beam: T=307°C is the temperature of the oven and $\sigma \simeq 10^{-14} cm^2$ is an indirect measurement.

REFERENCES

1. For a review see M. Elbel "Energy and Polarization Transfer" in Progress in Atomic Spectroscopy, W.Hanle and H.Kleinhoppen Ed. (Plenum Press, 1978) Part.B pag.1299.

2. M.Allegrini, G. Alzetta, A. Kopystyńska, L.Moi and G.Orriols, Opt.Commun.19,96 (1976); 22, 329 (1977). For a review see : A. Kopystyńska and L.Moi; Phys.Rep.92, 135 (1982).

3. J.Huennekens and A.Gallagher, Phys.Rev.A ..., ... (1983) and references there in.

4. J.Huennekens and A.Gallagher, Phys. Rev. A ..., ... (1983)

5. M. Allegrini, P.Bicchi and L. Moi, Phys. Rev. A ..., ... (1983)

6. L.Barbier and M. Chèret in this book

7. D.J. Krebs and L.D. Schearer, J.Chem.Phys. 75, 3340 (1981)

8. V.S.Kushawaha and J.J. Leventhal, Phys.Rev.A 25, 570 (1982)

9. J.L. Le Gouet, J.L.Piqué, F.Wuilleumier, J.M.Bizeau, P.Dhez, P.Kock and D.L.Ederer, Phys. Rev. Lett. 48,600 (1982)

10. T.B.Lucatorto and T.J.Mc.Ilrath, Phys. Rev. Lett. 37, 428 (1976)

11. W. Müller and I.V.Hertel, Appl.Phys. 24,33 (1981);

12. A.N. Nesmeyanov "Vapor Pressure of the Chemical Elements" (Elsevier, Amsterdam 1963)

13. E.A.Milne, J.London Math. Soc.1, 40 (1926)

14. P. Kowalczyk, Chem. Phys. Lett. 68, 203 (1979).

LASER PHOTODISSOCIATION OF Na_2^+ IONS PRODUCED BY LASER RESONANT IONISATION OF Na VAPOR

B. CARRE, F. ROUSSEL, P. BREGER and G. SPIESS
Service de Physique des Atomes et des Surfaces
Centre d'Etudes Nucléaires de Saclay
91191 Gif-sur-Yvette, Cedex, FRANCE

Abstract : We report direct measurements of the photodissociation of Na_2^+ ions created when a Na vapor is irradiated by a pulsed laser with intensity in the 10^5-10^6 $W.cm^{-2}$ range. The Na_2^+ ions, accelerated by a weak electric field, are photodissociated in a second laser beam tuned to 6040 Å.

INTRODUCTION

Ionisation of Na vapor irradiated by a pulsed laser light tuned to the resonances $3S_{1/2}$-$3P_{1/2,3/2}$ has been the subject of growing interest in the past few years [1-7].

For high density vapors, it has been shown that the free electrons produced in a particular seeding process, were efficiently heated in superelastic collisions, with excited states and played a determinant role in further ionization of the vapor. The seeding processes have been studied by several groups for various laser intensities. Associative ionization of two 3p states, Penning ionization or photoionization of highly excited nl states of collisional origin, have been proposed as the dominant mechanisms. The Na_2^+ ions coming from associative ionization of two 3p states, are believed to be produced in high vibrational levels.

This hypothesis is grounded on the first available calculations of energy potential curves for Na_2 excited states correlated to 3p + 3p atomic system. Very recently, electron spectroscopy studies of the Na excited sodium have concluded to such a distribution of v-levels in the final state of Na_2^+. It is attractive to investigate the photodissociation of these Na_2^+ ions and to compare the measured photodissociation cross-section with the accurate calculation of Kirby-Döcken et al[10] and of Uzer and Dalgarno[23]. From this comparison, the dis -

tribution of rovibrational levels, in the Na_2^+ final state of associative ionization may be discussed.

EXPERIMENTAL SET-UP

The experimental set-up consists basically of a Na cell and has been described in details elsewhere [6a-22]. For the present measurements two synchronous pulsed dye laser beams are focussed in the cell in front of an extraction diaphragm (fig. 1). The first laser (called the ionisation laser later on) is tuned to one of the $3S_{1/2}$-$3P_{1/2,3/2}$ resonances of sodium in order to create the Na_2^+ ions to be studied. The ions are accelerated in the

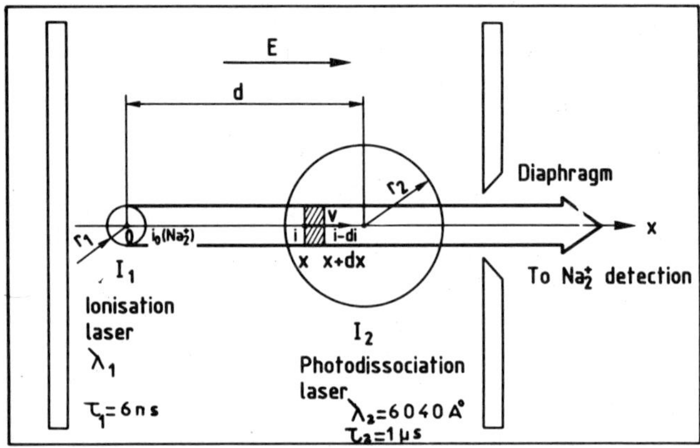

FIGURE 1. Schematic view of the laser beam waists in a plane containing the Ox axis of the extraction diaphragm

cell by a weak transverse electric field applied between two parallel plates. They are extracted through the diaphragm and travel along a drift tube where a time-of-flight mass analysis is carried out. On their path to the extraction diaphragm between the 2 plates the created ions are forced to cross the light beam of the second dye laser (called the photodissociation laser later on) where photodissociation occurs.

The photodissociation dye laser (Chromatix CMX4) works with Rhodamine 6G and is pumped by a flashlamp. For all the photodissociation measurements this laser is tuned to the wavelength λ_2 = 6040 Å . The light pulse energy is of the order of 1 mJ for a duration of 1 µs. The beam waist is adjusted to r_2 = 0.27 mm (at field amplitude 1/e) by suitable lenses in such a way that all the extracted ions have crossed the photodissociation light beam before being extracted through the diaphragm.

The ionisation dye laser also works with Rhodamine 6G and is pumped by a XeF exciplex laser (Sopra exciplex). The pulse energy is typically 10 µJ for a duration of 6 ns. The beam waist is adjusted to r_1 = 0.1 mm, which is smaller than the extraction diaphragm radius (0.25 mm).

A fast photodiode acts as a monitor of the photodissociation light pulse and as a trigger of the ionisation laser whose pulse duration is much shorter. The delay between the start of the two laser pulses is adjusted in such a way that the photodissociation intensity has reached its peak value I_{2p} when the flowing Na_2^+ ions, created during the 6 ns of the ionisation light pulse, are crossing the second light beam. During the total time the Na_2^+ ions are in the second light beam (∿ 50 ns) the variation of the photodissociation intensity, in the vicinity of its peak value, may be neglected.

During the measurements the Na_2^+ ion signal, i_o, is first recorded without the photodissociation laser beam. When the photodissociation laser is turned on, with peak intensity I_{2p}, an attenuated Na_2^+ ion signal $i(I_{2p})$ is then recorded and a survival Na_2^+ fraction :

$$F\ (I_{2p}) = i/i_o \qquad (1)$$

is determined. The survival fraction F has been measured for different values of I_{2p} and under various experimental conditions.

The survival fraction F is related to a photodissociation cross-section σ through the geometry of the extraction region and the photodissociation intensity I_{2p}.[22] It takes the form

$$F = \exp\ (-\sigma I_{2p} t_E) \qquad (2)$$

where t_E is an effective crossing time equal to 43 ns under the present experimental conditions.

B. CARRE, F. ROUSSEL, P. BREGER, G. SPIESS

PRELIMINARY EXPERIMENT : PHOTODISSOCIATION OF Na_2^+ PRODUCED BY IONISATION OF THE Na_2 DIMERS

In a preliminary experiment the photodissociation of the lowest vibrational levels of the Na_2^+ ions, formed from the Na_2 dimers in the vapour, has been studied. For the ground state $X^2 \Sigma_g$ (v = 0, J = 0) of Na_2^+ an accurate calculation has been performed by Kirby-Docken et al [10] for photodissociation light wavelengths in the range 4700-6200 Å including our photodissociation wavelength (λ_2 = 6040 Å).

We have used the same technics as Nitz et al [11] to form the Na_2^+ ions in the lowest vibrational levels and we have measured their photodissociation cross section at the wavelength 6040 Å.

The ground state of Na_2^+ is formed from the Na_2 dimers present in the vapour by a two-step process at the wavelength 4926 Å in the ionisation laser beam waist. The first step is a resonant excitation of Na_2 via the transition $X^1\Sigma_g$ (0,0) $\rightarrow$ $B^1 \Pi u$(0,0). The excited Na2 dimers are photoionised at the same wavelength in the second step and molecular ions are predominantly formed in the three lowest vibrational levels.[23]

The measured survival Na_2^+ fraction is represented in figures 2 . It is noted that 83 % of the created Na_2^+ ions may be photodissociated at the maximum intensity. When the ionisation laser intensity is strongly reduced the Na_2^+ signal is also reduced but the survival fraction, which is the ratio of the two ion signals, does not vary significantly within the uncertainty of the measurements.

A least-squares straight line has been plotted through the experimental points in figure2 for photodissociation intensities lower than 8 x 10^5 W cm^{-2}. The corresponding photodissociation cross section is 9 x 10^{-18} cm^2 which is close to the value calculated by Kirby-Docken et al[10] and Uzer and Dalgarno[23] for the lowest vibrational levels of Na_2^+ . Within the present experimental uncertainty, which is estimated to be 20%, the agreement may be considered as satisfactory, thus eliminating any serious systematic error in the determination of the photodissociation cross section.

For photodissociation intensities larger that 8 x 10^5W. cm^{-2} a reproducible and significant deviation from the straight line is observed in the survival Na_2^+ fraction.

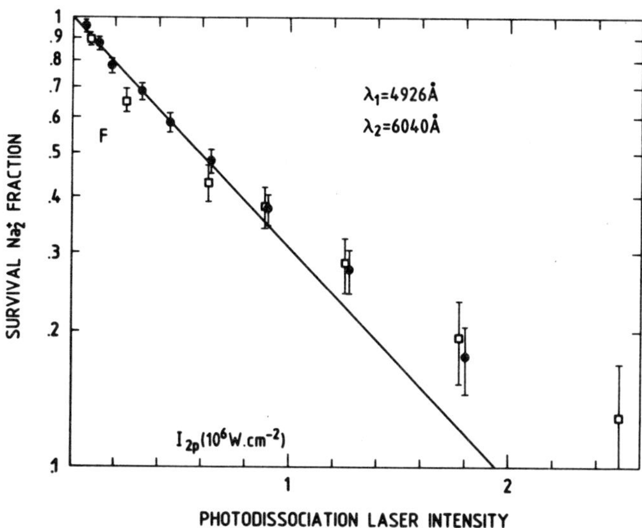

FIGURE 2. Survival fraction in photodissociation (at 6040Å) of Na_2^+ ions produced by two-photon ionisation (at 4926Å) of the Na_2 dimers.
Ionisation laser intensity ● $I_1 = 2 \times 10^6$ W cm^{-2}
□ $I_1 = 5.1 \times 10^4$ Wcm^{-2}

This effect may be attributed to a small proportion (less than 10 %) of other dissociating levels of Na_2^+ with different cross sections.

PHOTODISSOCIATION OF Na_2^+ PRODUCED BY RESONANT IONISATION OF THE Na VAPOUR

With the ionisation laser tuned to one of the $3S_{1/2}$-$3P_{1/2,3/2}$ resonances of Na, the photodissociation of the Na_2^+ ions created in the vapour has been studied in the density range 10^{12}-10^{13} cm^{-3} where the fractional ionisation of the vapour remains lower than 10^{-3} [12].

In figure 3, the survival Na_2^+ fraction F is plotted against I_{2p} for three different intensities I_1 of the ionisation laser which is tuned to the $3S_{1/2}$-$3P_{3/2}$ resonance. The maximum intensity I_1 corresponds to a pulse energy of 10 μJ. In these measurements the vapour atomic density is 7×10^{12} cm^{-3}. For comparison the survival

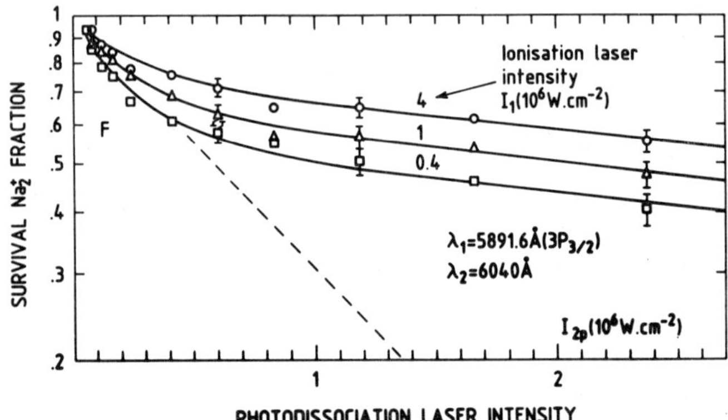

FIGURE 3. Survival fraction in photodissociation at 6040Å of Na_2^+ ions produced by resonant ionisation of the Na vapour at 5891.6Å ($3S_{1/2}$-$3P_{3/2}$). For comparison the straight line of fig.2 has been represented (---). ——, best fit obtained from the sum of two exponential terms.

fraction for Na_2^+ ions in the lowest vibrational levels from figure 2 has been reported in figure 3. A general view of the data suggests several remarks to be noted.

(i) For a given photodissociation intensity the survival fraction of the Na_2^+ ions produced at resonance is significantly different from that of the Na_2^+ ions in the lowest vibrational levels. The difference is more pronounced for high photodissociation intensities. For instance at $I_{2p} = 2 \times 10^6$ W cm^{-2} only about 50 % of the Na_2^+ ions produced at resonance are photodissociated while about 85 % of those in the lowest vibrational levels were photodissociated in the preliminary experiment. The difference is well beyond the uncertainty in the measurements of F which is estimated to be ± 0.03 in figure 3.

(ii) The photodissociation of the Na_2^+ ions produced at resonance is dependent on the ionisation laser intensity I_1.

(iii) In the semilogarithmic diagram of figure 3 deviation of the survival Na_2^+ fraction from the straight line already observed in the preliminary experiment and attributed to a distribution of further photodissociated levels, is much more pronounced for the Na_2^+ ions produced at resonance.

In figure 4 the survival Na_2^+ fraction has been compared at the two resonances for the same ionisation laser intensity. The photodissociation is less efficient for the Na_2^+ ions created at the $3P_{1/2}$ resonance.

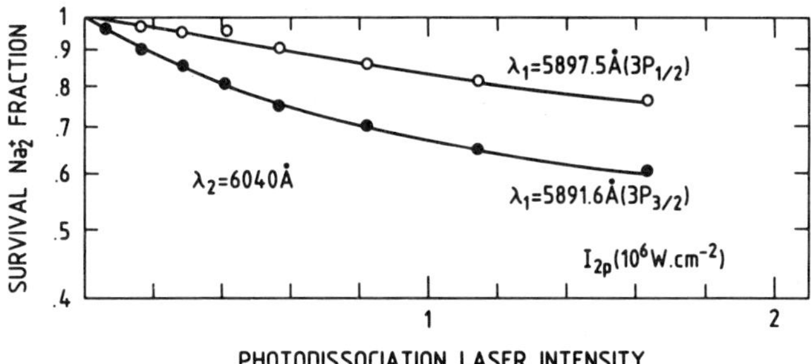

FIGURE 4. Survival fraction in photodissociation at 6040 Å of Na_2^+ ions produced by ionisation of the Na vapour at the two resonances 5897.5 Å ($3S_{1/2}$ - $2P_{1/2}$) and 5891.6 Å ($3S_{1/2}$-$3P_{3/2}$) for the same ionisation laser intensity.

In view of the previous remarks we have attempted to fit the three sets of experimental points in figure 3 with a weighted sum of exponential terms corresponding to the assumed distribution of dissociating levels and to write, instead of expression (2) the survival fraction in the form :

$$F = \sum_i c_i \exp(-\sigma_i I_{2p} t_E) \qquad (3)$$

where the relative abundances c_i of the levels satisfy

$$\sum_i c_i = 1 \qquad (4)$$

A numerical analysis has shown that two weighted exponential terms may be determined unambiguously from the data. This result does not mean that the assumed distribution of dissociating levels is reduced to two components, but rather it means that two groups of dissociating levels with distinct averaged cross sections and relative abundances may be distinguished from the data. A numerical analysis using a larger number of weighted terms in expression (3) is not possible with reasonable confidence.

An interesting result of the numerical analysis lies in the fact that the three sets of experimental points in figure 3 may be fitted by using the same couple of cross sections σ_a, σ_b and by varying only the relative abundances c_a, c_b. Thus one is led to the conclusion that the ionisation laser intensity I1 controls the relative proportion of the two groups of photodissociating levels.

The full curves in figure 3 represent the best fit obtained with the couple of cross sections :

$$\sigma_a = (3.1 \pm 0.6) \times 10^{-17} \text{ cm}^2$$
$$\sigma_b = (9.2 \pm 2) \times 10^{-19} \text{ cm}^2.$$

and with the relative abundances given in Table 1.

TABLE 1 - The relative abundances

I_1 ($W.cm^{-2}$)	C_a	C_b
4×10^6	0.25 ± 0.03	0.75 ± 0.03
1×10^6	0.35 ± 0.02	0.65 ± 0.02
4×10^5	0.44 ± 0.05	0.56 ± 0.05

It is noted that the cross section σ_b is much smaller than σ_a.

In order to investigate the origin of the observed variations of the survival Na_2^+ fraction with the ionisation laser intensity I_1 (figure 3), photodissociation effects due to I1 have to be examined. As two groups of levels with very different photodissociation cross sections are formed, the ionisation laser will preferentially photodissociate the levels with the larger cross section, thus changing the apparent relative abundances of the two groups.

In view of these considerations it is found that the relative abundances are expected to exhibit a change of 0.1 when the intensity I1 varies from 4.1×10^6 to 4.1×10^5 W cm^{-2}. The observed variations of c_a and c_b (table 1) are somewhat larger (0.19 ± 0.06) but the difference is not sufficiently significant to conclude that there is a direct I_1 dependence in the distribution of dissociating levels.

The variation of the survival Na_2^+ fraction with I_1 may be interpreted, for a large part, in terms of the photodissociation of a fraction of the Na_2^+ ions in the ionisation laser. Under these conditions the c_a and c_b values measured at the lowest intensity $I_1 = 4.1 \times 10^5$ W cm^{-2} ($c_a = 0.44$, $c_b = 0.56$) should be retained as the best estimates of the true relative abundances, since for this intensity the I_1 photodissociation does not exceed a few per cent in the first laser beam.

Discussion

Carré et al [6] have discussed in detail the possible ionising processes producing molecular ions at resonance.

Associative ionisation in collisions between two Na(3p) atoms has been proposed by von Hellfeld at al [4] as the most probable process :

$$Na(3p) + Na(3p) \rightarrow Na_2^+ + e \qquad (a)$$

According to Valance and Nguyen[13] a $^1\Sigma g$ potential energy curve correlated to Na(3p) + Na(3p) exhibits a crossing point with the $^2\Sigma g(Na_2^+)$ potential energy curve of the molecular ion at the internuclear distance $R = 8.5\ a_o$ (point A in figure 5). Using the dissociation energy of Na_2^+ ($^2\Sigma g$, v = 0) from the work of Leutwyler et al [14], and the spectroscopic constants of the $^2\Sigma g$ state, from Kirby-Docken et al [10], it is found that the crossing point is situated at the energy of the vibrational level v = 13 of Na_2^+. In the presence of an intense radiation field a laser-induced associative ionisation process is also possible :

$$Na(3p) + Na(3p) \xrightarrow{\hbar\omega_1} Na_2^+ + e \qquad (b)$$

According to the classical approximation of the Franck-Condon principle the molecular ions formed in this process are in high vibrational levels $v > 14$.

The data obtained in this experiment seem consistent with these theoretical predictions since a large number of the Na_2^+ ions produced at resonance do not photodissociate in the same way as Na_2^+ ions in the lowest vibrational levels.

A theoretical analysis of the Na_2^+ photodissociation process shows that the transition occurs via a resonant coupling, at internuclear distance R_c with the $^2\Sigma u$ dissociative state of Na_2^+ (Beswick and Durup[15c]) (figure 5). It is found that the transition, for the wavelength 6040 Å and the level v = 13, J = 0 occurs at $R_c = 6.9\ a_o$ which is about half way between the turning points of the vibrational motion of the nuclei (5.45 and 8.65 a_o). Kirby-Docken et al[10]

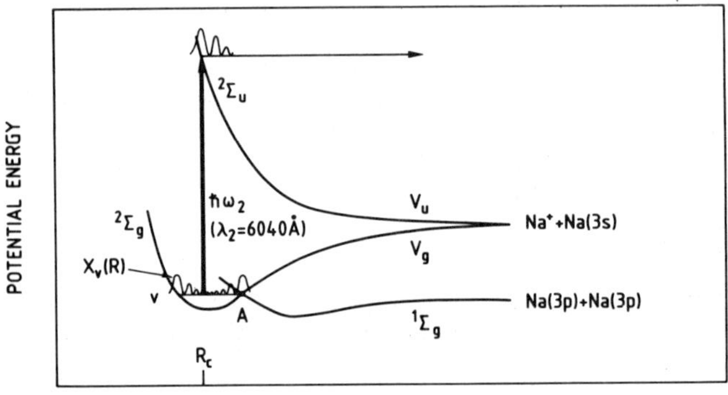

FIGURE 5. Potential energy curves (schematic) for the states $^2\Sigma g$ and $^2\Sigma u$ of Na_2^+ and for the state $^1\Sigma g$ of Na_2 correlated to Na(3p) + Na(3p). The crossing point A is situated at the vibrational level v = 13 of Na_2^+. The resonant transition for photodissociation of the level v = 13 at 6040 Å occurs at R_c = 6.9 a_o.

have calculated the photodissociation cross section of Na_2^+ only for the rovibrational level v = 0, J = 0. For an excited vibrational level the photodissociation cross section is rapidly oscillating with the wavelength of the light, due to nodes and maxima in the vibrational wavefunction $\chi_v(R)$. An estimate of the maxima of the cross section in the vicinity of Rc may be obtained by using the well known δ-function approximation for the wavefunction of the final repulsive state. Although the phase of the cross section oscillation is not reliable in this approximation, the general magnitude and, in particular, the maxima of the cross-sections are well reproduced[16]. The expression for the cross section in this approximation is

$$\sigma_v(cm^2) = 2.69 \times 10^{-18} \frac{\hbar\omega_2}{(dV_u/dR)_{R_c}} |D_e(R_c)|^2 |\chi_v(R_c)|^2 \tag{13}$$

In the present situation, the energy of the photon is $\hbar\omega_2$ = 0.151 Ryd, the derivative of the $^2\Sigma u$ repulsive potential at the transition if $(dV_u/dR)_{Rc}$ = 3.45 x 10^{-2} Ryd a_o^{-1}. The electron dipole moment $D_e(R)$ has been tabulated by Kirby-Docken et al [10] ; an interpolation gives $D_e(R_c)$ = 2.446 ea_o. Using the vibrational wavefunction of the harmonic oscillator as an approximation for $\chi v(R)$ in the vicinity of R_c, we found that the photodissociation cross section for the vibration level v = 13 oscillates in the vicinity of 6040 Å between 0 and a maximum value of 2.9 x 10^{-17} cm^2. This calculated maximum value is very close to the cross section $\sigma_a = (3.1 \pm 0.6) \times 10^{-17}$ cm^2 deduced from the data in figure 3.

At first sight this agreement appears to be rather surprising because the photodissociation cross section of a given rovibrational level of Na_2^+ has no reason to be maximum precisely at the wavelength λ_2 = 6040 Å chosen in the present experiment. However the Na_2^+ ions formed by associative ionisation in collisions of two Na(3p) atoms present a distribution of rotational levels J as well as vibrational levels. The distribution of J levels is related to the impact parameters of the collisions and may be quite large. Due to the J dependence of the Na_2^+ energy level, the internuclear distance R_c for the dissociative transition is slightly shifted with J. Consequently the oscillations of the photodissociation cross section of a high vibrational level are also shifted on the photon energy scale when J increases. This effect has been theoretically studied in detail [17] in the case of H_2^+ ions. A similar effect may be envisaged in the present situation for the Na_2^+ ions. If we assume that the rotational distribution of the created Na_2^+ ions is sufficiently spread, the photodissociation cross section of the rotational levels at 6040 Å may vary in large proportion and is expected to pass through maximum and minimum values according to the J level. In that case statistical considerations show that the most probable photodissociation cross sections to be observed will be precisely the stationary values corresponding to the maximum and to the minimum, in agreement with the data in figure 3 where two groups of states with different cross sections σ_a and σ_b have been determined. The conclusion of this discussion is that the data in figure 3 are consistent with the photodissociation of Na_2^+ ions in high vibrational levels in the vicinity of v = 13 and that they present a large rotational distribution. Hence field-free and field-assisted associative ionisation processes in collisions between two Na(3p) atoms are strongly suggested to be the main formation channels for Na_2^+ in the Na vapour irradiated at resonance.

INFLUENCE OF PHOTODISSOCIATION IN THE PRODUCTION OF Na^+ IONS AT RESONANCE

The direct measurements of Na_2^+ photodissociation in the present experiment associated with the previous results of

Carré et al [6] suggest the existence of a direct channel independent of photodissociation to explain quantitatively the formation of Na^+ ions in the vapour at resonance for atomic densities of the order of 10^{12} cm^{-3}. At this density electron-atom superelastic collisions and electron impact ionisation processes are negligible.

A comparison between the Na^+ and Na_2^+ yields plotted against the laser intensity (figure 2(b) of 6a) shows that the two ion yields were found to be equal at the laser intensity 8 x 10^4 $W.cm^{-2}$. In view of the effective time of the Na_2^+ ions in the laser beam waist and of the slightly different experimental conditions the corresponding Na_2^+ fraction surviving is estimated to be 0.73 for this laser intensity in comparison with the present experiment. Hence the Na^+ ion yield resulting from photodissociation did not exceed 37 % of the Na_2^+ yield showing that about 60 % of the Na^+ signal should be attributed to a direct process under the experimental conditions of atomic density (2.4 x 10^{12} cm^{-3}) and of laser intensity (8 x 10^4 W cm^{-2}).

Direct formation of Na^+ ions via energy-pooling collisions, proposed by Allegrini et al [18]

$$Na(3p) + Na(3p) \rightarrow Na(nl) + Na(3s) \qquad (c)$$

followed by photoionisation of the highly excited state Na(nl) may be suggested. Recent estimates of the cross sections for the states Na(4d), Na(5s) and Na(3d) by Le Gouet et al (1982)[19] , and Huennekens and Gallagher [20] are of the order of 10^{-15}-10^{-14} cm^2 at thermal energy and could explain the direct formation of atomic ions in the Na vapour irradiated at resonance.

Another process suggested by Geltman [21] is the laser-assisted Penning ionisation :

$$Na(3p) + Na(3p) \xrightarrow{\hbar\omega} Na^+ + Na(3s) + e \qquad (d)$$

The calculated cross section is only 10^{-18} cm^2 for a laser intensity of 10^6 W cm^{-2}.

REFERENCES

1. T.B. Lucatorto and T.J. Mc Ilrath, Phys. Rev. Lett. 37, 428 (1976)
2. C.H. Skinner, J. Phys. B : At. Mol. Phys. 13, 55 (1980)
3. G.H. Bearman and J.J. Leventhal, Phys. Rev. Lett. 41, 1227 (1978)
4. A.V. Von Hellfeld, J. Caddick and J. Weiner, Phys. Rev. Lett. 40, 1369(1978)

5. A.De Jong and F. van der Valk, J. Phys. B : At. Mol. Phys. 12, L561 (1979)
6. B. Carré, F. Roussel, P. Bréger and G. Spiess
a) J. Phys. B : At. Mol. Phys. 14, 4271 (1891)
b) 14, 4289 (1981)
7. T. Stacewicz, Opt. Comm., 35, 239 (1980)
8. R.M. Measures, J. Appl. Phys. 48, 2673 (1977)
9. F. Roussel, P. Bréger, G. Spiess and C. Manus Proc. 11th Int. Conf. on Physics of Electronic and Atomic Collision ed. K. Takayanagi and N. Oda (Kyoto 1979, Society for Atomic Collision Research) abstracts p. 290
10. K. Kirby-Docken, C.J. Cerjan and A. Dalgarno, Chem. Phys. Lett. 40, 205 (1976)
11. D.E. Nitz, P.B. Hogan, L.D. Shearer and S.J. Smith, J. Phys. B : At. Mol. Phys. 12, L 103 (1979)
12. F. Roussel, P. Bréger, G. Spiess, C. Manus and S. Geltman J. Phys. B : At. Mol. Phys. 13, L631 (1980)
13. A. Valance and Q. Nguyen Tuan, J. Phys. B: At. Mol. Phys. 15, 17 (1981).
14. S. Leutwyler, M. Hofman, H.P. Harri and E. Schumaker, Chem. Phys. Lett. 77, 257 (1981)
15. J.A. Beswick and J. Durup, Proc. Summer School on Chemical Photophysics, Les Ouches, ed. P. Glorieux et al. (Paris, 1979, CNRS) p. F1
16. G.H. Dunn, Phys. Rev. 172, 1 (1968)
17. J.D. Argyros, J. Phys. B : At. mol. phys. 7, 2025(1974)
18. M. Allegrini, G. Alzetta, A. Kopystynska and L. Mol Opt. Comm. 19, 96 (1976)
19. J.L. le Gouët, J.L. Picque, F. Wuilleumier, J.M. Bizau, P. Dhez, P. Koch and D.L. Ederer, Phys. Rev. Lett. 48, 600 (1982)
20. J. Huennekens and A. Gallagher, Phys. Rev. (1982)
21. S. Geltman, J. Phys. B : At. Mol. Phys. 10, 3057 (1977)
22. F. Roussel, B. Carré, P. Breger and G. Spiess, J. Phys.B At. Mol. Phys. 16, 1749 (1983).
23. T. Uzer and A. Dalgarno, Chem. Phys. Let. 61, 213 (1979)

ATOMIC AND MOLECULAR COLLISIONS IN LASER IRRADIATED SODIUM VAPOR: PULSED AND CW EXPERIMENTS

M. ALLEGRINI,* W. P. GARVER, V. S. KUSHAWAHA, and J. J. LEVENTHAL,
Department of Physics, University of Missouri-St. Louis, Missouri, U.S.A.
* Permanent address: Istituto di Fisica Atomica e Molecolare del C.N.R., Pisa, Italy

Abstract We have obtained strikingly different results in experiments in which Sodium vapor at low density was irradiated with either cw or pulsed laser light in the wavelength range 570-610 nm. To facilitate comparison of the results for the two kinds of excitation the data were taken under identical experimental conditions (except of course for the laser). The apparatus used in this work permits simultaneous mass analysis of the ions formed in the vapor and spectral analysis of the fluorescence emitted by the excited atoms.

INTRODUCTION

Since the discovery of complete ionization[1] and excitation transfer in collisions between excited atoms[2] in laser-irradiated Sodium vapor a wide variety of multiphoton, energy pooling and laser induced effects have been reported.[3] This work has shown that both the laser power density and the atom and molecule concentrations are crucial factors in determining the dominant effect. In the work that will be described here we show that the temporal characteristics of the laser are also important. In the case of cw irradiation, excited species can be continuously produced, so there is a steady-state concentration of potential reactants. However when pulsed lasers are employed, either the pulse duration or the excited state lifetime become key parameters. If the laser pulse is shorter than the lifetime of the excited state then it is the lifetime that determines the time dependence of the potential reactants. If the laser pulse is longer than the lifetime of the excited state then it is the laser pulse that is important. Thus, in order to observe the effects of interactions involving excited reactants either the excited state lifetime or the pulse duration should be comparable to, or longer than, the average time between collisions.

In order to investigate the effects of the temporal characteristics of the reactant pool we have performed experiments in Sodium vapor using both cw and pulsed laser irradiation over the wavelength range 570-610 nm. Of course such an experiment necessarily introduces extremes of laser power density, so the results also provide information on the intensity dependence. Here we report the preliminary results of these experiments emphasizing comparison of the data obtained with the two different methods of excitation.

EXPERIMENT AND RESULTS

The basic apparatus used for these experiments has been described elsewhere;[4] modifications and improvements have been made through the course of the work, however the general features of the techinque are unchanged. For one method of excitation a Nitrogen laser-pumped dye laser (operated with Rh-6G) with pulse duration ~10 nsec and peak power ~5mJ was used; the laser power density in the vapor was $\sim 5 \times 10^3$ W/cm^2. For the other method of excitation a cw laser was employed;[4] it was also operated with Rh-6G and provided a power density ~1 W/cm^2. Because we wish to compare data taken under precisely the same experimental conditions, the laser path through the reaction cell was exactly the same for both cw and pulsed excitation.

The Sodium atom density was measured using an accurate new method based on trapping of the D-line.[5] As in previous experiments from this laboratory,[4] the atom density was intentionally kept in the range 10^{12} - 10^{13} cm^{-3} to minimize the possibility of multiple collision effects. Under these conditions the neutral dimer concentration in the vapor was about one percent of the atom density. These Na_2 molecules should not be confused with molecular ions formed by associative ionization in collisions involving two atoms;

processes such as these are of course the very interactions we wish to study.

The apparatus has the capability to permit simultaneous detection of ions and fluorescence. Ions were identified by mass analysis and excited states by spectral analysis of the fluorescence. The 330 nm fluorescence line, corresponding to the 4p→3s transition is especially convenient because both the 5s and 4d levels radiatively decay to the 4p state; as a consequence, the presence in the spectrum of the 330 nm line constitutes proof that energy pooling excitation transfer

$$Na(3p) + Na(3p) \rightarrow Na(5s,4d) + Na(3s) \quad (1)$$

is occurring. Therefore, most of our data were acquired with the monochromator setting fixed at 330 nm while either the laser wavelength was fixed and ions mass analyzed, or the mass spectrometer setting was fixed (at Na^+ or Na_2^+) and the laser wavelength scanned.

With the cw laser the results are well known:[4] Na_2^+, formed by associative ionization,

$$Na(3p) + Na(3p) \rightarrow Na_2^+ + e \quad (2)$$

is the dominant ionic product and then only when the laser is tuned to a D-line. Small yields of Na^+ ions may be detected, especially under multimode laser irradiation, but these atomic ions result from photodissociation of nascent Na_2^+ that are produced by energy pooling associative ionization.

In contrast to cw excitation, the pulsed laser produces copious yields of both Na^+ and Na_2^+ over the entire range of wavelengths. In fact, as shown in Figure 1, if the

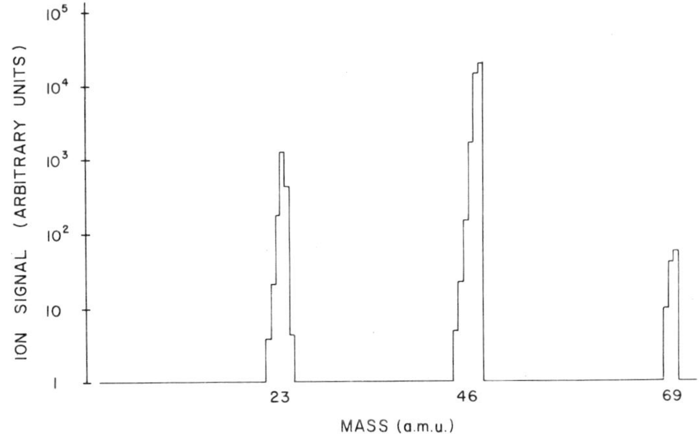

FIGURE 1. Mass spectrum taken with the cw laser tuned to the D_2 line. The atom density is higher than is typically used in these experiments, $\sim 10^{14}$ cm^{-3}. The signal has not been corrected for the efficiency of the mass filter as a function of mass.

atom/molecule density is increased above our usual operating values of $\sim 10^{12} - 10^{13}$ cm^{-3}, Na_3^+ can also be observed. The formation of Na_3^+ in the laser-irradiated vapor is however a subject for future study. The observation of Na^+ and Na_2^+ at all laser wavelengths has been observed by other authors,[6,7] although under slightly different experimental conditions, and with laser pulses of longer duration (~1 μsec). Figure 2 shows mass spectra taken for three different pulsed laser wavelengths, the D_2 - line, 580 nm and 600 nm. Inspection of this figure shows that no spectacular increase in either the Na^+ or Na_2^+ signals

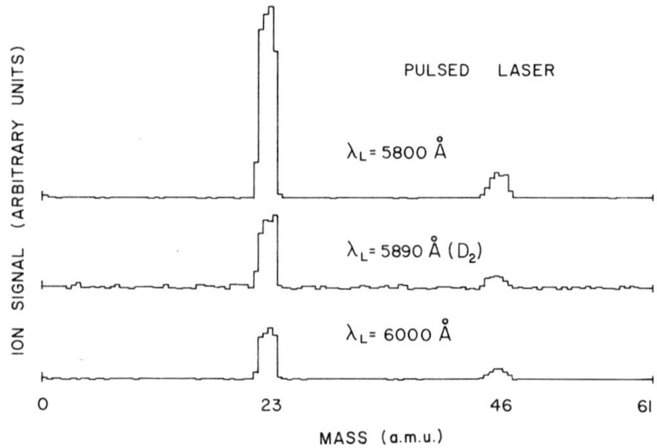

FIGURE 2. Mass spectra taken with the pulsed laser set at the three indicated wavelengths. The atom density is 5×10^{12} cm^{-3}. The spectra are not corrected for the efficiency of the mass filter as a function of mass.

occurs as the laser is scanned through the D-lines. This is better illustrated in Figure 3 which shows the Na^+ and Na_2^+ signals as functions of pulsed laser wavelength for a range that includes both D-lines; also shown is the 330 nm fluorescence. These fluorescence data demonstrate that the laser step size is sufficiently small to ensure data acquisition at the D-lines, and that excitation transfer in

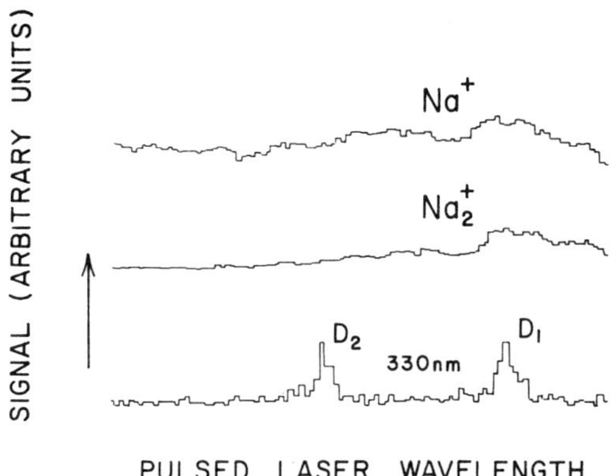

FIGURE 3. Na^+, $Na_2{}^+$ and 330 nm signals as functions of pulsed laser wavelength; the atom density is 5×10^{12} cm^{-3}.

Na(3p/Na(3p) energy pooling collisions, Reaction (1), is occurring. However, it is obvious that the associative ionization channel, Reaction (2), is not prominent.

Because the absence of enhancements of either the Na^+ or $Na_2{}^+$ signals at the D-lines was unexpected, a condition under which ion enhancement did occur was sought in order to authenticate proper apparatus operation. Verification was achieved when substantial enhancements of both the Na^+ and $Na_2{}^+$ signals were observed at wavelengths corresponding to the 3s→4d and 3s→5s two photon resonances, 579 and 602 nm respectively. At these wavelengths the Na^+ yield increases dramatically, a result of resonance enhanced three photon ionization of Na(3s). The origin of the Na_2^+ enhancements will be discussed below. Figure 4 shows typical data for the 3s→4d resonance.

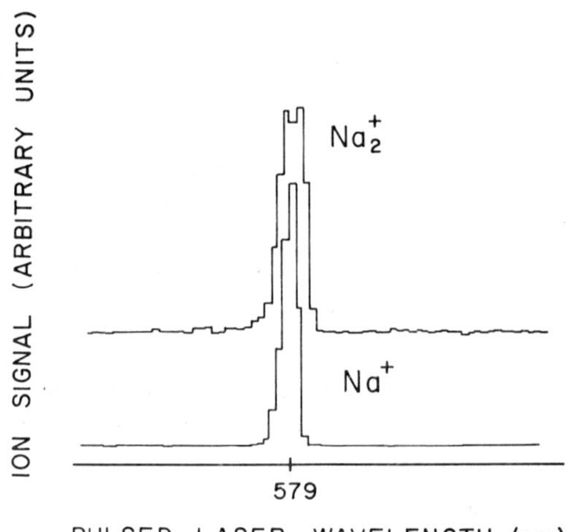

FIGURE 4. Na^+ and Na_2^+ signals as a function of pulsed laser wavelength. The peaks occur at 579 nm, the 3s→4d two photon resonance. The atom density is 5×10^{12} cm^{-3}.

DISCUSSION

Since the atomic component of the vapor is transparent to all light in this wavelength range except the D-lines and the two photon resonances, there is little doubt that the copious production of Na_2^+ over the entire range of laser wavelength is due to three photon ionization of the neutral dimer component of the vapor. This multiphoton ionization is enhanced by the fact that there are numerous two photon resonant intermediate excited states of the molecule. No such production was observed with the cw excitation because the low power density of this laser virtually precludes multiphoton processes. The atomic ions produced by pulsed laser excitation are a consequence of molecular ion dissociation, that is, Na_2^+ is formed in a repulsive state.

The enhancement of the Na^+ signal at the 3s→5s and 3s→4d two photon atomic resonances is, as noted previously, the result of resonance enhanced three photon ionization of Na(3s).[8] Although these resonance enhanced processes are clearly distinguishable at 579 and 602 nm, resonance enhanced three photon ionization at the D-lines is not distinguishable from the "background" Na^+ that is produced by (dissociative) multiphoton ionization of the dimers. However, these are different types of resonance enhanced three photon ionization, one requiring single photon ionization from either the 5s or 4d states, the other two photon ionization of the 3p state. The absence of an enhanced Na^+ signal from three photon ionization via the 3p level is consistent with theoretical pictures of multiphoton ionization.[8,9]

Because associative ionization in collisions between two Na(3p) atoms, Reaction (2), is known to occur with a cross section that is roughly gas kinetic,[10,11] it is, initially at least, surprising that the Na_2^+ signal is not appreciably enhanced as the laser is scanned through the D-lines. However, examination of the appropriate rate equations shows that, if the spontaneous decay rate, A, is greater than the product of the energy pooling rate coefficient and the Na(3p) density at the time of the laser pulse, then, for the parameters of this experiment, there should be roughly 10^3 ions/laser pulse from associative ionization. The <u>observed</u> signal however is $\sim 10^7$ ions/pulse so that the absence of an enhancement of the Na_2^+ signal at the D-lines is simply a signal-to-noise problem. In this case the "noise" is the Na_2^+ that results from multiphoton ionization of the dimer component of the vapor, which is obscuring the Na_2^+ produced by Reaction (2). On the other hand, there is no alternate way of producing 330 nm radiation at these laser wavelengths, so the 330 nm signal from Reaction (1) is not obscured.

As noted above we have observed enhancements of the Na_2^+ signal as the laser is scanned through the $3s \rightarrow 5s$ and $3s \rightarrow 4d$ two photon resonances. It is possible that 3s/5s and 3s/4d associative ionization

$$Na(4d,5s) + Na(3s) \rightarrow Na_2^+ + e \quad (3)$$

are responsible for these enhanced signals. However, since the rate coefficient for 3s/5s associative ionization is considerably smaller than that for 3p/3p associative ionization[4] we consider this possibility to be unlikely. Rather, it is probable that electron transfer in Na^+/Na_2 encounters produces these Na_2^+ ions. These two photon resonances provide increased concentrations of atomic ions which serve as reactants for this charge transfer process. Since the $Na_2(X)$ ionization potential is ~0.2 eV lower than that of Na(3s) the proposed charge transfer process is exothermic, and can therefore occur at thermal energies.

ACKNOWLEDGEMENTS

This work was supported by NATO under Research Grant No. 101.82 and by the U.S. Department of Energy, Division of Chemical Sciences, under Contract No. DE-AS02-76-ER02718.

REFERENCES

1. T. B. Lucatorto and J. J. McIlrath, Phys. Rev. Lett. 37, 428 (1976).

2. M. Allegrini, G. Alzetta, A. Kopystynska, L. Moi and G. Orriols, Opt. Commun. 19, 96 (1976).

3. N. K. Rahman and C. Guidotti, Photon-Assisted Collisions and Related Topics (Harwood Academic Publishers, Chur, London, New York, 1982).

4. J. J. Leventhal "Atomic Collision Processes in Laser-Excited Sodium Vapor" p. 217 in Reference 3.

5. W. P. Garver, M. R. Pierce and J. J. Leventhal, J. Chem. Phys. 77, 1201 (1982).

6. P. Polak-Dingels, J. F. Delpech and Weiner, Phys. Rev. Lett. 44, 1663 (1980).

7. F. Roussel, B. Carre, P. Breger and G. Spiess, J. Phys. B: Atom. Mol. Phys. 14 L 313 (1981).

8. C. Laughlin, J. Phys. B: Atom. Mol. Phys. 11 1399 (1978).

9. M. H. Nayfeh, private communication.

10. V. S. Kushawaha and J. J. Leventhal, Phys. Rev. A 25, 346 (1982).

11. A. de Jong and F. Van der Valk, J. Phys. B 12, L 561 (1979).

STEADY-STATE ELECTRON-ATOM COLLISION IN A LASER FIELD*

Leonidas Dimou and Farhad H.M. Faisal
Fakultät für Physik, Universität Bielefeld,
D-4800 Bielefeld 1, Federal Republic of Germany

Abstract A theory of steady-state electron-atom scattering in a laser field is presented. The ensemble distribution of steady state target atoms in a laser field and the fundamental system of radiative scattering amplitudes are derived. The theory is applied to an exactly solvable multi-channel scattering model. Results for steady-state radiative electron-atom scattering cross sections are presented graphically. Radiative elastic and inelastic cross sections as functions of electron energy and field parameters are examined with special reference to the influence of a scattering resonance, a compound-state (a bound state of atom + electron) as well as a threshold-cusp, which occur in the absence of the field. The elastic differential cross section is compared with the full elastic differential signal and is found to differ significantly only in the resonance region.

INTRODUCTION

For the steady-state electron-atom scattering in a laser field, the scattering signal may be obtained in two steps; first, by obtaining the field-dependent ensemble distribution of the steady-state beam-atoms and second by calculating the fundamental system of amplitudes for radiative electron scattering against the ensemble atoms. The observed scattering signal dS, detected in the solid angle dΩ, may then be obtained from

$$\frac{dS}{d\Omega} = \mathrm{Tr}(\hat{\rho}\, \hat{f}^{+}\, \hat{f}) \tag{1}$$

where $\hat{\rho}$ is the normalized distribution matrix (of the states of the ensemble atoms in the field) and $\hat{f}$ is the (radiative electron scattering) amplitude-matrix of interest. Steady-

*Work partially supported by Deutsche Forschungsgemeinschaft Fa 160/1-1 OZ: 2825

-state experiments can be arranged, for example, by crossing a CW laser beam with the atomic and the electron beams, or by employing pulsed lasers provided, of course, the pulse-duration is much longer than a typical relaxation period involved. Here we give the theory of steady-state radiative scattering cross sections (or signals) and discuss explicit results obtained from an exactly solvable multi-channel scattering model.

STEADY-STATE ENSEMBLE DISTRIBUTION

The density matrix equation in presence of relaxation, decay, and feeding mechanisms may be written as[1]

$$\frac{\partial\rho(t)}{\partial t} = -i\ [H,\rho(t)] - \frac{1}{2}\,[\Gamma,\rho(t)]_{+} + \Lambda \qquad (2)$$

where H is the Hamiltonian of the interacting atom-laser sub-system, Γ is the decay matrix and Λ is the 'feeding' (or 'pump') matrix. The steady-state solutions of (2) satisfy:

$$\frac{\partial\rho(t)}{\partial t} = 0\ . \qquad (2a)$$

To avoid lengthy algebra and to be concrete we consider a 2-state atomic model (level-separation ω_o) in a circularly polarized laser field. Then we may write

$$H = \begin{vmatrix} \Delta/2 & \beta/2 \\ \beta/2 & -\Delta/2 \end{vmatrix},\ \Gamma = \begin{vmatrix} \gamma_1 = \gamma_{tr} & 0 \\ 0 & \gamma_2 = \gamma_{tr} + \gamma_{ion} + \gamma_s \end{vmatrix}$$

and

$$\Lambda = \begin{vmatrix} \lambda_1 + \gamma_s\rho_{22}(t) & 0 \\ 0 & 0 \end{vmatrix}$$

where $\Delta = \omega_o - \omega$ is the detuning, β is the dipole coupling strength. $\gamma_{tr} = \frac{1}{T}$ with T, the transit time of the beam atoms across the interaction volume. γ_{ion} is the rate of photo-ionization (when significant) and γ_s is the spontaneous relaxation rate of the upper state. λ_1 is the rate at which the beam atoms enter (are fed into) the interaction volume. From (2) and (2a) the normalized steady state ensemble distribution is easily obtained:

$$P_{ij} = \frac{\rho_{ij}}{\rho_{11} + \rho_{12}} \qquad i, j = 1, 2$$

$$P_{11} = \frac{\beta'^2+\Delta^2+\gamma_{12}^2}{2\beta'^2+\Delta^2+\gamma_{12}^2}, \quad P_{12} = \frac{-\beta(\Delta+i\gamma_{12})}{2\beta'^2+\Delta^2+\gamma_{12}^2} = P_{21}^* \tag{3}$$

$$P_{22} = \frac{\beta'^2}{2\beta'^2+\Delta^2+\gamma_{12}^2}$$

with $$\beta'^2 = (\beta/2)^2\left(1 + \frac{\gamma_1}{\gamma_2}\right), \quad \gamma_{12} = \frac{\gamma_1+\gamma_2}{2} .$$

This distribution contains, as special cases, the distribution derived by Mollow[3] when γ_{ion} and γ_{tr} are neglected and by Mittleman[4] when, further, the off-diagonal correlation terms $P_{12} = P_{21}^*$ are neglected and when the spontaneous width γ_s turns out to be small compared to the power broadening $(\beta/2)^2$. The rate-equation distribution used by Hertel and Stoll[5] is another special case of (3).

STEADY-STATE RADIATIVE SCATTERING AMPLITUDES

Note that the field-dependent ensemble distribution (3) is given in the unperturbed representation of the atomic reference Hamiltonian. The corresponding steady-state radiative electron scattering wave functions are therefore governed by the Schrödinger equation

$$i \frac{\partial}{\partial t} |\psi(t)\rangle = [H_a - \frac{1}{2}\nabla_r^2 + H'(t) + V(\underline{r},\underline{x})]\,|\psi(t)\rangle \tag{4}$$

where $H_a = \sum_j |\phi_j(\underline{x})\rangle\, \varepsilon_j\, \langle\phi_j(\underline{x})|$ is the atomic reference Hamiltonian without the field, $V(\underline{r},\underline{x})$ is the collisional interaction and $H'(t) = \frac{i}{c}\,\underline{\nabla}_r\cdot\underline{A}(t) + \frac{1}{2c^2}A^2(t)$ is the electron-field interaction, with $\underline{A}(t) = \frac{A_0}{2}\,[(\underline{\varepsilon}_x - i\underline{\varepsilon}_y)e^{-i(\omega t+\delta)} + c.c.]$ with $F_0 \equiv \frac{\omega}{c}A_0$ (the peak field strength).

The Schrödinger-Floquet Scattering Equations

It is most convenient to change to a set of stationary collisional equations which is obtained by a generalization of the Floquet expansion[6,7]. Thus we write

$$|\psi(t)\rangle = \sum_{n=-\infty}^{\infty} |\psi_n(\underline{r},\underline{x})\rangle\, e^{-iE't+in(\omega t+\delta)} \tag{5}$$

substitution of (5) in (4) and collection of coefficients of equal powers of $e^{i(\omega t+\delta)}$ gives at once

$$[E - H_n^{(o)}\,]\,\psi_n(\underline{r},\underline{x})\rangle = V(\underline{r},\underline{x})\,|\psi_n(\underline{r},\underline{x})\rangle \tag{6}$$

where

$$H_n^{(o)} = -\frac{1}{2}\nabla_r^2 + H_a + n\omega + \frac{1}{2}\,i\omega\alpha_0(L^+S_n^- + L^-S_n^+) \tag{7}$$

with $L^\pm = -i(\nabla_x + i\nabla_y)$, $\alpha_0 = \frac{F_0}{\omega^2}$, $E = E' - \frac{1}{2}F_0\alpha_0^2$ and the 'index shifters' $S_n^\pm$ are defined by $S_n^\pm|\psi_n\rangle = |\psi_{n\pm1}\rangle$. It is useful to note that the stationary (radiative scattering) eqs. (6) are completely analogous to the usual scattering equations without the field but for an extra channel index n. Hence (6) may be tackled essentially by any known method of treating the scattering equations in the absence of the field.

Radiative Close-Coupling Equations

Expanding

$$|\psi_n(\underline{r},\underline{x})\rangle = \sum_j |\phi_j(\underline{x})\rangle\, F_{jn}(\underline{r}) \tag{8}$$

substituting (8) in (6), and projecting on $<\phi_j(\underline{x})$ one easily finds the radiative c-c equations for the channel wave functions $F_{jn}(\underline{r})$:

$$[E - (H_n^{(o)} + \varepsilon_j)]\, F_{jn}(\underline{r}) = \sum_{j'} V_{jj'}(\underline{r})\, F_{j'n}(\underline{r}) \ . \qquad (9)$$

(We do not write down the exchange term explicitly, which is easily done.) The Green's function of the operator on the left of (9) satisfies

$$[E - (H_n^{(o)} + \varepsilon_j)]\, G_{nn'}^{(j)}(\underline{r},\underline{r}') = \delta(\underline{r}-\underline{r}')\, \delta_{nn'} \qquad (10)$$

which has the exact solution

$$G_{nn'}^{(j)}(\underline{r},\underline{r}') = \sum_{N=-\infty}^{\infty} \sum_{\underline{K}} J_{n-N}(K\alpha_0 \sin\Theta_K)\, e^{-in\phi_K} \ .$$

$$\cdot \frac{e^{i\underline{K}\cdot(\underline{r}-\underline{r}')}}{(E-\varepsilon_j-N\omega-K^2/2)} \cdot e^{in'\phi_K}\, J_{n'-N}(K\alpha_o \sin\Theta_K) . \qquad (11)$$

And a similar result holds in the case of linear polarisation with $J_{n-N}(K\alpha_0 \sin\Theta_K)\, e^{-in\phi_K}$ replaced by $J_{n-N}(K\alpha_0 \cos\Theta_K)$ (and similarly for the factor with n') in (11).
Using (11) in (9) one gets the solution

$$F_{jn}^{(i)}(\underline{r}) = e^{i\underline{K}_0\cdot\underline{r}}\, J_n(K_0\alpha_0 \sin\Theta_0)\, e^{-in\phi_0}$$

$$+ \sum_{j'n'} \int d\underline{r}'\, G_{nn'}^{(j')}(\underline{r},\underline{r}')\, V_{jj'}(\underline{r}')\, F_{j'n'}^{(i)}(\underline{r}) \qquad (12)$$

which satisfies the initial condition with the atom in state $|\phi_i(\underline{x})>$ and the incident electron momentum $\underline{K}_0$. Taking the limit $r \to \infty$ in (12), the scattered wave on the right hand side becomes

$$\sum_{jN} J_{n-N}(K_{jN}\alpha_0 \sin\Theta)\, e^{-in\phi}\, \frac{e^{iK_{jN}r}}{r}\, f_{i,0\to j,N}(\Omega_0,\Omega)$$

where the fundamental system of scattering amplitudes are (see also ref. (2))

$$f_{i,0\to j,N}(\Omega_0,\Omega) = -\frac{1}{2\pi}\sum_{j'n'} J_{n'-N}(K_{jN}\alpha_0\sin\Theta_K)\, e^{-in'\phi_K} \cdot \langle e^{i\underline{K}_{jN}\cdot\underline{r}}|V_{jj'}(\underline{r})|F^{(i)}_{j'n'}(\underline{r})\rangle \quad (13)$$

with $\Omega_0 = (\Theta_0,\phi_0)$, $\Omega = (\Theta,\phi)$, the incident and the scattered directions, respectively. The elementary differential scattering cross sections are thus

$$\frac{d\sigma_{i,0\to j,N}}{d\,\Omega} = \frac{K_{j,N}}{K_{i,0}}\,|\,f_{i,0\to j,N}(\Omega_0,\Omega)|^2. \quad (14)$$

The actual signal would be given by suitable combinations of such elementary cross sections (and also of interference terms when coherent scattering is relevant) according to (1).

AN EXACTLY SOLVABLE MULTICHANNEL MODEL

The Model Potential

We have applied the theory developed above to an exactly solvable multichannel radiative scattering model[2] in 3-dimension, defined by the collisional interaction

$$V(\underline{r},\underline{x}) = \sum_{jj'} |\phi_j(\underline{x})\rangle\, U_{jj'}(\underline{r})\, \langle\phi_{j'}(\underline{x})| \quad (15)$$

with $U_{jj'}(\underline{r}) = W_{jj'}\,\delta(\underline{r})\,\frac{\partial}{\partial r}\,r$ where $W_{jj'}$ are constants which may be obtained, if desired, phenomenologically with respect to non-radiative scattering or bound-state data. In this sense the pseudo-potentials (15) would obviate the need to include exchange explicitly.

The Model Amplitudes

Substituting (15) in (12) operating with $\phi_j^*(\underline{x})\delta(\underline{r})\frac{\partial}{\partial r}r$ and integrating one gets a set of J algebraic equations where J equals the total number of target states used. For a finite number of states this set is easily solved and the substitution of the result back into (13) yields exact expressions of the radiative amplitudes. For a two-state system for example, we easily find (see also ref. (2))

$$f_{i,0\to j,N}(\Omega_0,\Omega) = \frac{-1}{2\pi}\sum_n J_n(K_0\alpha_0\sin\Theta_0)\, e^{in(\phi_0-\phi)} \cdot g^{(n)}_{i\to j}(E_i)\, J_{n-N}(K_{jN}\alpha_0\sin\Theta) \quad (16)$$

where

$$g_{1\to j}(E\ ) = W_{j1}(1 + i(1-a)\, R_{2n}\delta_{j1})/d_n(E_1),\ j = 1, 2 \quad (17)$$

and

$$g_{2\to j}(E\ ) = W_{j2}(1 + i(1-a)\, R_{1n}\delta_{j2})/d_n(E_2),\ j = 1, 2 \quad (18)$$

$$d_n(E_i) = (1 + R_{1n}(E_i))(1 + iR_{2n}(E_i) + aR_{in}(E_i)\cdot R_{2n}(E_i)$$

$$s_{jn}(E_i) = \sum_m K_{j,n+m}\left[\frac{1}{2}\int_0^\pi d\Theta\sin\Theta\ J_m(K_{j,n+m}\alpha_0\sin\Theta)\right] \quad (19)$$

and

$$R_{jn}(E_i) = \frac{1}{2\pi}\, W_{jj}\, s_{jn}(E_i)\ ,\quad K_{j,N} \equiv [2(E_i-\epsilon_j-N\omega]^{1/2}$$

$$a = \frac{W_{12}\, W_{21}}{W_{11}\, W_{22}} \qquad = K_{j,N}(E_i).$$

RESULTS AND DISCUSSIONS

Previously[2] we have compared analytically the exact model solution (16) with the low-frequency approximation[8-10] and shown that the low-frequency approximation is properly speaking also a low-field approximation[2,11] and that a formula similar to that for the low-frequency approximation can be reobtained at the opposite high-frequency limit when the coupling in the intermediate states becomes actually weak. Analytical discussions of the scattering singularities have also been made. Here we shall concern ourselves with explicit results for radiative elastic and inelastic scattering cross-sections or signals, at different field strengths, frequencies and incident electron energies.

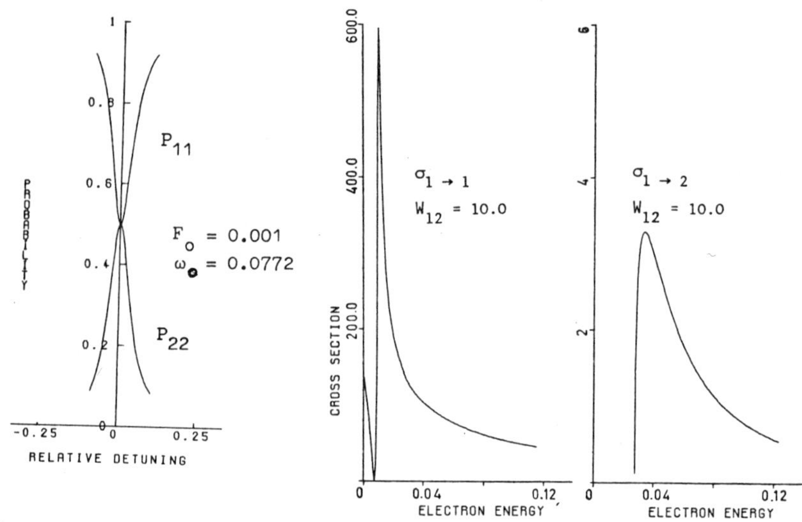

Figure 1 Figure 2

First we consider an example of the steady-state distribution for the present problem. Fig. 1 shows the diagonal elements of the ensemble distribution (13) as a function of the scaled detuning $\frac{\Delta}{\omega_0}$ at a field strength F_0 = 0.001 a.u. (All numerical results are reported here in a.u.). One notices that the occupation probability of the upper state, P_{22}, sharply decreases with increasing positive or negative detunings. It is only near the resonance frequency that the two-states are comparably occupied and both of them can take part significantly in the radiative scattering.

Fig. 2 exhibits the elastic $\sigma_{1\to1}$ and the inelastic $\sigma_{1\to2}$ scattering cross sections in the usual non-radiative case. The potential parameters in this case are $W_{11} = W_{22} = 34.5$ and $W_{12} = W_{21} = 10.0$. In the elastic channel a prominent scattering resonance occurs at $E_R = 0.0096$ which should be compared with the excited state threshold at $E_{th} = 0.0272$. Besides, there exists a negative energy compound-state (or a negative-ion-state) at $E_B = -0.025$ which does not appear in the diagram. The inelastic $\sigma_{1\to2}$ excitation cross section is shown on the right hand side as a function of incident electron energy. These cross sections are to be compared with the corresponding ones in presence of the field (for a field strength $F_0 = 0.0025$ and frequency $\omega = 0.0272 = \omega_0$) shown in Fig. 3. Note that the cross sections in Fig. 3 are weighted by the occupation probability, P_{11}, of the initial state of the target atoms. The top three curves show the radiative cross sections for which the target state is not changed (c.f. $\sigma_{1\to1}$ in Fig. 2) but the electron is scattered with a final energy E_f either equal to the incident energy E_i ($\sigma_{1,0\to1,0}$) or with an energy which is different from E_i by one photon energy; $E_f = E_i + \omega$ for $\sigma_{1,0\to1,-1}$ and $E_f = E_i - \omega$ for $\sigma_{1,0\to1,1}$. Thus, $\sigma_{1,0\ 1,0}$ shows the modific-action in purely elastic scattering, due to the 'dressing effect' of the field; $\sigma_{1,0\to1,-1}$ corresponds to the absorpt-ion of a field photon by the electron while it scatters from the steady-state ensemble atoms. Similarly $\sigma_{1,0\to1,1}$ corresponds to the cross sections for emission of a photon by the scattered electron. Note that there are two prominent structures in each of these cross sections. The narrower structure is due to a photon resonance with a negative-ion state i.e. with the compound-state which occurs at

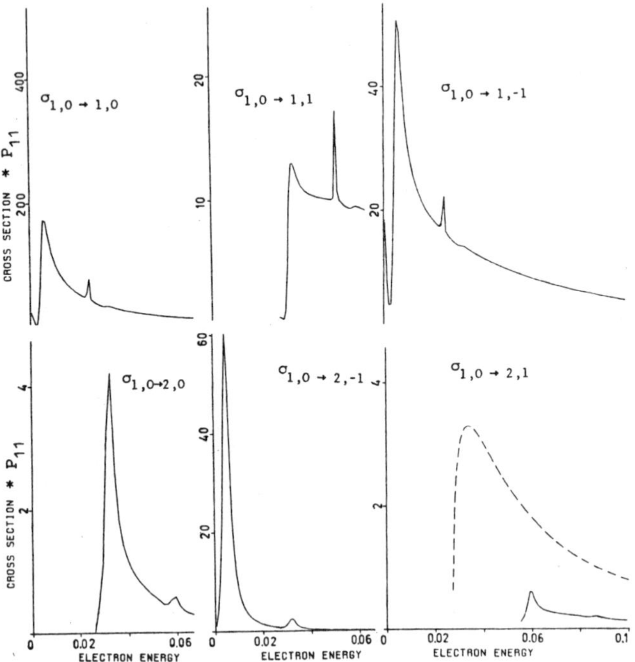

Figure 3

E_B = −0.025. The somewhat broader structure on the other hand is due to the (positive energy) scattering-resonance at E_R = .0096 or due to a one-photon resonance with it. It should be noted that these structures are actually shifted by small amounts, by dynamic Stark effect, from their unperturbed positions. The width of the narrower resonance is actually a measure of photo-ionization cross section of the corresponding negative-ion in the radiation field. This is at the same time related to the (stimulated) capture or recombination probability of the scattered electron, through the usual reciprocity relation. The lower three figures (Fig. 3) illustrate radiative cross sections (c.f. the non-radiative $\sigma_{1\to2}$, the dashed curve on the extreme right hand side) for the transfer of the initial state ensemble-population to their excited states by electron collision simultaneously with or without the emission or absorption of a photon by a scattered electron. $\sigma_{1,0\to2,0}$ corresponds to the modification of the cross section for ordinary electron impact excitation due to the effect of 'dressing' by the field. $\sigma_{1,0\to2,-1}$ corresponds to the collisional population excitation along with absorption of a photon, while $\sigma_{1,0\to2,1}$ corresponds to that with emission of a photon. Note that the inelastic energy threshold for the absorption assisted cross section is lowered by one photon energy requiring a lesser electron energy to initiate the population excitation. For the emission cross section the threshold is equally shifted upwards, making the ensemble population-excitation more 'endothermic'. A second higher order resonance peak is discernable in all the three cross sections.

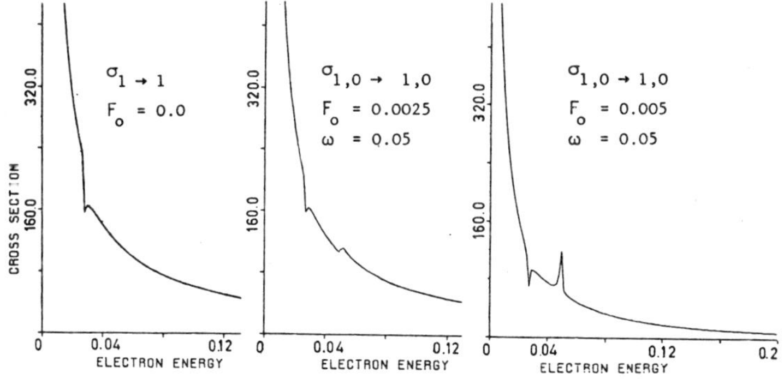

Figure 4

The case of a Wigner-Baz threshold-cusp[10] and its modification due to radiative scattering is shown in Fig. 4. The left hand curve shown the occurence of a cusp in the ordinary elastic scattering channel at the first inelastic threshold $E_{th} = .0272$. (In this case the potential parameters are chosen to be $W_{11} = W_{22} = 34.5$ and $W_{12} = W_{21} = 138.0$; and $\omega_0 = 0.0272$ and $\omega = 0.05$.) The middle and the right hand curves exhibit the modification of the cusp signlarity at the field strengths $F_0 = 0.0025$ and $F_0 = 0.005$, respectively. Besides the cusp there appears a resonance-peak at both the field strengths indicated. This is due to an one-photon resonance with a shallow compound-state at $E_B = -.00025$. One observes that the photon-resonance is more pronounced at the higher field strength due to a greater suppression of the 'potential-scattering' background (in presence of the stronger field) than to any absolute enhancement of the resonance at this field strength. We note that the enhancement of the cross-section below the threshold is associated with the occurence of an elastic scattering-resonance immediately under the threshold.

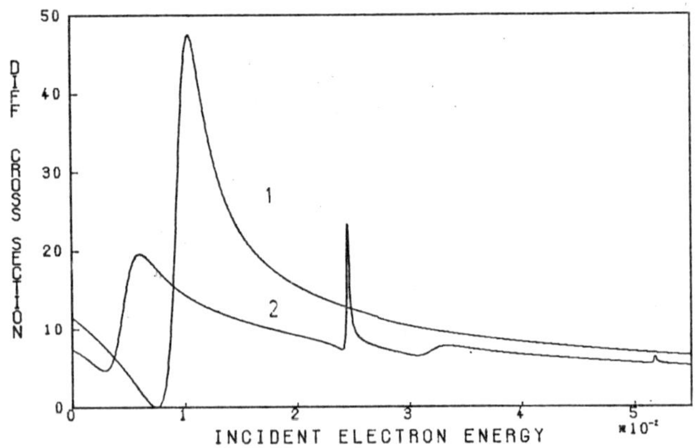

Figure 5 (a)

Im Fig. 5(a) we compare the differential scattering cross-sections at $\Theta = 45^{o}$, $\phi = \phi_{o}$ ($\Theta_{o} = 90^{o}$ to the direction of the propagation of the field). Curve 1 corresponds to the field-off elastic cross section $\frac{d\sigma_{1\to 1}}{d\Omega}$, while curve 2 shows the simply weighted field-on elastic cross section

$$\frac{d\sigma_{el}}{d\Omega} = P_{11}\frac{d\sigma_{1,0\to 1,0}}{d\Omega} + P_{22}\frac{d\sigma_{2,0\to 2,0}}{d\Omega} \tag{20}$$

(Here the potential parameters are chosen to be the same as in the case of Fig. 2, and $\omega_{o} = \omega = 0.0272$, $F_{o} = 0.0025$.) Curve 2 shows a significant modification of the resonance scattering in presence of the field. The ordinary resonance peak (in curve 1) is reduced and shifted to the left while another sharper peak arises on the right; the latter corresponds to a field-shifted photon resonance with the negative-ion or the compound state at $E_{B} = -0.025$. Further visible are the two smaller (higher order) peaks (one broad and one narrow) to the right, which are separated by $\simeq \omega$ from the respective dominant resonances on the left.

In Figs. 5(b) and (c) we compare the full elastic scattering signal

$$\begin{aligned}\frac{dS}{d\Omega} = {} & P_{11}(|f_{1,0\to 1,0}|^{2} + |f_{1,0\to 2,-1}|^{2}) \\ & + P_{22}(|f_{2,0\to 2,0}|^{2} + |f_{2,0\to 1,1}|^{2}) \\ & + 2\,\mathrm{Re}\,\{[f^{*}_{1,0\to 1,0}\, f_{2,0\to 1,1} \\ & + f^{*}_{1,0\to 2,-1}\, f_{2,0\to 2,0}]P_{12}\}\end{aligned} \tag{21}$$

calculated from (1), (3) and (16), with $\frac{d\sigma_{el}}{d\Omega}$ given by eq. (20).

We notice from Fig. 5(b) (the parameter values are the same as in Fig. 5 (a)) that $\frac{dS}{d\Omega}$ (curve 3) possesses the same overall structure and are essentially of the same magnitude $\frac{d\sigma_{el}}{d\Omega}$ except near the resonance regions. The quantitative difference in the resonance regions is significantly higher at the lower energies. We note that the full signal (eq.(21)) consists of a diagonal part

$$\frac{dS'}{d\Omega} = P_{11}\left(|f_{1,0\to 1,0}|^2 + |f_{1,0\to 2,-1}|^2\right) + P_{22}\left(|f_{2,0\to 2,0}|^2 + |f_{2,0\to 1,1}|^2\right) \tag{22}$$

plus a correlation part. In order to estimate the influence, if any, of the correlation part we have compared the elastic scattering signal $\frac{dS'}{d\Omega}$ without correlation with the full $\frac{dS}{d\Omega}$. The two curves are found to completely overlap at all energies in the scale of Figs. (5), showing the negligible contribution from the correlation term in (21). This is due to the smallness of P_{12} compared to P_{11} and P_{22} at zero detuning, $\Delta = 0$ (see eqs. (16)) at not too high field intensities. We expect therefore that for most intensities of practical interest the elastic scattering signal in presence of the field may be computed from $\frac{dS'}{d\Omega}$ (eq. (22)) (instead of the full $\frac{dS}{d\Omega}$, eq. (21)) but not simply from $\frac{d\sigma_{el}}{d\Omega}$ (eq.(20)) for all energies.

The modification of the elastic scattering signal at a higher field strength is shown in Fig. 5(c). At this field strength, $F_o = 0.005$, the resonances are further broadened and shifted to the left than in Fig. 5(b). Once again the difference between $\frac{dS}{d\Omega}$ (curve 3) and $\frac{d\sigma_{el}}{d\Omega}$ (curve 2) is significant near the resonance region only and $\frac{dS'}{d\Omega}$ completely overlaps $\frac{dS}{d\Omega}$ (curve 3) in the scale of the figure. We may also note that the Wigner-Baz cusp at the inelastic threshold, $E_{th} = 0.0272$, appears more prominently in the full elastic signal (curve 3) than in the approximate $\frac{d\sigma_{el}}{d\Omega}$ (curve 2).

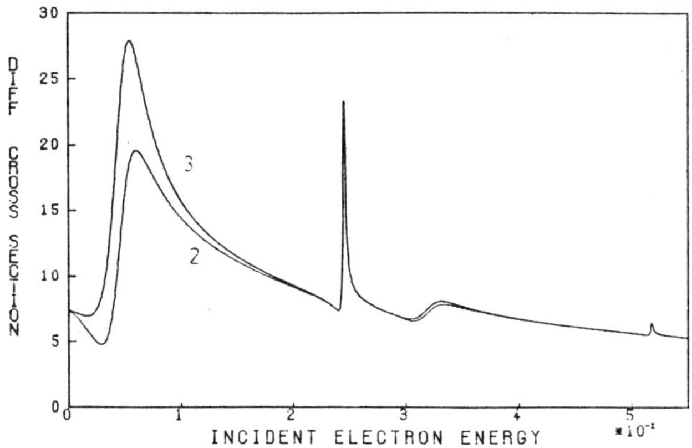

Figure 5(b)

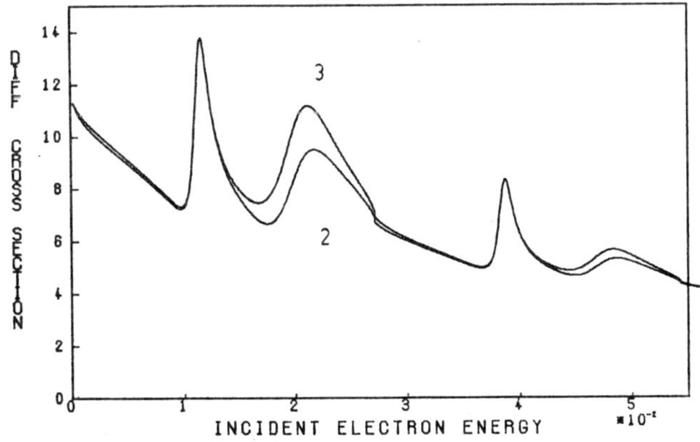

Figure 5(c)

REFERENCES

1. M. Sergent, M.O. Scully and W.E. Lamb, Laser Physics (Addison-Wesley, London 1974)
2. F.H.M. Faisal, in Photon-assisted Collisions and Related Topics eds. N.K. Rahman and C. Guidotti (Harwood Academic Publishers, London 1982), pp. 287 - 301
3. B.R. Mollow, Phys.Rev. 188, 1969 (1969)
4. M.H. Mittleman, Phys.Rev. A14, 1338 (1976) and Erratum Phys.Rev. A15, 1355 (1977)
5. I.V.Hertel and W. Stoll, J.Phys. B7, 570 (1974)
6. J.H. Shirley, Phys.Rev. 138, B979 (1965)
7. F.H.M. Faisal, Nuovo Cim. 33B, 775 (1976)
8. H. Kruger and C. Jung, Phys.Rev. A17, 1706 (1978)
9. M.H. Mittleman, Phys.Rev. A20, 1965 (1979)
10. L.D. Landau and E.M. Lifshitz, Quantum Mechanics (Pergamon Press, Oxford 1975) pp. 556 - 70
11. C. Jung and H.S. Taylor, Phys.Rev. A23, 1115 (1981)

* The work partly consists of the continuing doctoral research of one of us (L.D.).

THE INVERTED MULTIPHOTON PROBLEM

CHRISTOF JUNG
Fachbereich Physik, Universität, 6750 Kaiserslautern
West Germany

Abstract It is shown how much information about the spatial-temporal energy flux distribution in a laser beam is necessary in order to calculate multiphoton transition rates induced by this beam. In addition it is considered how much information about the flux distribution can be traced back from measurements of multiphoton transition rates. These two sets of information are the same.

INTRODUCTION

Calculations of multiphoton transitions are usually done for fixed values of the power density of the electromagnetic field. This corresponds to a plane wave field. However, it does not correspond to the experimental situation. A real high power laser produces localized light pulses and therefore the experimental power flux $F(\vec{x}, t)$ is space and time dependent. The detector collects signals from events occuring at different space-time points and therefore from events induced by different power densities. Accordingly, the observed count rates are some average of the transition probabilities over the spatial-temporal intensity distribution of the laser field. In some multiphoton experiments the power flux function $F(\vec{x}, t)$ is not known completely and in addition it is not known exactly how the atomic systems are located inside the beam. This lack of knowledge seems to prevent a quantitative comparison between experiment and theory. (These problems have occured in Refs. 1,2 for example.) Therefore, it would be a great help for the evaluation of multiphoton experiments, if we could answer the following questions:

I: What is the minimal amount of information about the power flux function F and about the position of the observed events inside the beam, which we need in order to calculate the observed count rates?

II: What is the maximal information about the power flux, which we can trace back from the knowledge of the count rates of a multiphoton experiment?

III. How can we check experiment against theory, even if we do not know the power flux function F used in the experiment?

This article gives some answers to these questions.

AVERAGING OF TRANSITION PROBABILITIES

For the moment let us assume that the process under consideration is well understood theoretically and that we know the formula for the transition probability TP(N,F,y) in a field of constant power flux F. N is the difference of the photon numbers in the final and in the initial state, i.e. the net number of photons absorbed/emitted. y is a collective set of additional variables, which determine a particular transition uniquely. As a particular example for multiphoton transitions we shall use free-free transitions described in low frequency approximation. In this special case the transition probability is given by

$$\frac{d\sigma}{d\Omega} = \frac{p_f}{p_i} J_N^2 (\sqrt{F} \cdot \alpha) \frac{d\sigma}{d\Omega}^{el} (\vec{p}_i', \vec{p}_f') \qquad (1)$$

This formula gives the cross section for a scattering event in which an electron comes in with momentum $\vec{p}_i$ and energy $E_i = p_i^2/2m$, scatters from a target, absorbs/emits N photons during the collision and goes out with momentum $\vec{p}_f$ and energy $E_f = p_f^2/2m = E_i + N\hbar\omega$. α is an abbreviation for $e\vec{\varepsilon}\ \sqrt{8\ c\pi}\cdot(\vec{p}_f - \vec{p}_i)/(mc\hbar\omega^2)$, e is the electron charge, m is the electron mass, $\vec{\varepsilon}$ is the polarization vector of the laser, c is the speed of light, ω is the laser frequency, J_N is the Bessel function of order N.

$\frac{d\sigma}{d\Omega}^{el}$ is the corresponding elastic scattering cross section for radiationless electron-target scattering taken at the shifted momenta $\vec{p}\,'_{i/f} = \vec{p}_{i/f} - mN\hbar\omega\vec{\varepsilon}/(\vec{\varepsilon}\cdot(\vec{p}_f - \vec{p}_i))$. For various derivations and discussions of eq. (1) see Refs. 3,4,5. In the case of free-free transitions the quantities $\vec{\varepsilon}$, ω, $\vec{p}_i$, $\vec{p}_f$ play the role of the variables y in the general formula TP(N, F, y).

In the following we assume that the variations of Γ are slow compared to atomic distance and time scales, so that the outcome of each individual event can be described by the formula for TP(N, F, y) where F is the flux value at the particular space-time point at which the event occurs. Let us denote the space-time density of observed events by $w(\vec{x}, t)$. For convenience we normalize w according to

$$\int_{R^4} d^3x\, dt\ \ w(\vec{x}, t) = 1 \qquad (2)$$

w may contain a factor taking care of the different efficiency of the detector for events occuring at different space-time points. Then, the relative count rate for a process in which N photons have been absorbed/emitted in the end is

$$Q_N = \int_{R^4} d^3x\, dt\, w(\vec{x}, t)\, TP(N, F(\vec{x}, t), y) \tag{3}$$

CONNECTION BETWEEN COUNT RATES AND POWER MOMENTS OF THE FLUX

For any fixed value of y we may consider Q_N to be a map which assigns a sequence of numbers to the functions F and w. To make the action of this map more transparent we would like to split it into two steps. In the first step the relevant information about F and w is condensed into a sequence of numbers μ_N and this first step should be independent of y. $\{\mu_N\}$ should be chosen in such a way, that in the second step this sequence $\{\mu_N\}$ can be mapped onto $\{Q_N\}$ by an invertible map, such that each particular sequence $\{Q_N\}$ corresponds to an unique sequence $\{\mu_N\}$ and vice versa. The second step depends parametrically on y. If all these conditions are fulfilled, then $\{\mu_N\}$ is that information about F and w which is relevant for the calculation of multiphoton transitions.

The easiest way to find the construction of μ_N is to look at the particular example of free-free transitions in low frequency approximation. We insert (1) into (3) and write Q_N in the form

$$Q_N = \frac{p_f}{p_i}\, R_N\, \frac{d\sigma}{d\Omega}^{el} \tag{4}$$

where

$$R_N = \int_{R^4} d^3x\, dt\, w(\vec{x}, t)\, J_N^2(\sqrt{F(\vec{x}, t)} \cdot \alpha) \tag{5}$$

The problem of finding μ_N is identical with the problem of finding such combinations of Q_N or of R_N, which are independent of α. Noting that (see eq. 8.536 in Ref. 6)

$$\sum_{k=0}^{\infty} \frac{(N+k)2(2N+k-1)!}{k!} J_{N+k}^2(z) = \frac{(2N)!}{(N!)^2} (z/2)^{2N} \tag{6}$$

for $N > 0$ and

$$J_0^2(z) + 2 \sum_{k=1}^{\infty} J_k^2(z) = 1 \tag{7}$$

we see that the solution is given by

$$\frac{(N!)^2\, 2^{2N}}{(2N)!\, \alpha^{2N}} \sum_{k=0}^{\infty} \frac{2(N+k)\,(2N+k-1)!}{k!} R_{N+k} =$$

$$= \int_{R^4} d^3x\, dt\, w(\vec{x}, t)\, (F(\vec{x}, t))^N = \mu_N \tag{8}$$

for $N > 0$ and

$$\sum_{N=-\infty}^{\infty} R_N = \int_{R^4} d^3x\, dt\, w(\vec{x}, t) = 1 = \mu_0 \tag{9}$$

Eq. (9) leads to the sum rule for free-free transitions, which is in complete agreement with the experiment (see Refs. 1,2).

To prove that μ_N are the desired quantities, we have to show that eq. (8) can be inverted, so that the mapping between the sequences $\{Q_N\}$ and $\{\mu_N\}$ is one to one. A power series expansion of the Bessel functions (see eq. (9.1.14) in Ref. 7) in eq. (5) gives

$$R_N = \sum_{k=0}^{\infty} \frac{(-1)^k (2N+2k)!\ \alpha^{2N+2k}}{2^{2N+2k}(2N)!(N+k)!(2N+k)!k!} \mu_{N+k} \tag{10}$$

Therefore, μ_N, the weighted averages of the powers of the flux function, are the minimal information about F and w which we need in order to calculate the values of R_N for a given value of α. Any more detailed knowledge of F or w would be of no use in the calculation of R_N.

So far we have considered the case of free-free transitions in low frequency approximation. In the same way as in eq. (10) we can expand the transition probability for any multiphoton process in powers of the flux and obtain the following generalization of eq. (10)

$$Q_N = \sum_{k=0}^{\infty} a_{N,k}(y)\ \mu_k\ , \quad N=0,1,\ldots \tag{11}$$

Therefore it is obvious, that the knowledge of $\{\mu_N\}$ is sufficient to calculate count rates for any process in an inhomogeneous field, if the transition probabilities for the individual event in a homogeneous field are known. The reverse step, to calculate μ_N in explicite form from a knowledge of $\{Q_N\}$ is more complicated in the general case, where we do not have such a simple formula like eq. (6) for Bessel functions. In any case eq. (11) can be inverted approximately by numerical methods.

If there is a function g which fulfils the equations

$$\int_0^\infty dF\ F^N\ g(F) = \mu_N \tag{12}$$

for all $N \geq 0$, then g(F) gives the relative weight of those space-time points, at which the flux has a value F. Then we can rewrite the quantities R_N in the form

$$R_N = \int_0^\infty dF\ g(F)\ J_N^2\ (\alpha \sqrt{F}\ ) \tag{13}$$

and for the general case we obtain

$$Q_N = \int_0^\infty dF\ g(F)\ TP(N,F,y) \tag{14}$$

In many cases it is more convenient to work with g(F) than with μ_N.

AN EXAMPLE

Next let us look at a realistic example for the connection between $F(\vec{x},t)$, $w(\vec{x},t)$, μ_N and g(F). (Additional examples can be found in Ref.8). We take a continuous wave laser with the beam running in z-direction and with a Gaussian profile perpendicular to the beam direction, i.e.

$$F(\vec{x},t) = F_0\ \exp\ [-\ \lambda(x^2+y^2)] \tag{15}$$

The spatial distribution of the atomic systems is assumed to be a 3-dimensional Gauss function. We open the detector for a time interval of length T. Then, the normalized weight function w is given by

$$w(\vec{x},t)=(\frac{\xi}{\pi})^{3/2} \exp [- \xi(x^2+y^2+z^2)] \frac{1}{T} \Theta(\frac{T}{2} - |t|) \qquad (16)$$

where Θ is the unit step function. Inserting (15) and (16) into (8) gives

$$\mu_N = F_0^N \; \xi/(\xi+ \lambda N). \qquad (17)$$

The function g(F) fulfilling eq.(12) with μ_N from eq.(17) is given by

$$g(F) = \Theta(F_0-F) \; (F/F_0)^{\xi/\lambda} \; \xi/(\lambda \cdot F) \qquad (18)$$

APPLICATION TO THE EVALUATION OF EXPERIMENTS

Now let us consider the application of these ideas to the evaluation of experimental data. In an experiment only a finite number of Q_N values are essentially different from zero. Therefore it is sufficient to observe the first K values of Q_N only. Accordingly, we can determine the values of μ_N for $0 \leq N \leq K$ by inverting eqs.(11) for $0 \leq N \leq K$. From the knowledge of the first K values of μ_N numerical approximations for g(F) can be constructed. This is the classical moment problem in mathematics. For the mathematical background see Refs. 9,10 and for a particular numerical procedure see Refs. 11,12.

Now we are ready to give the answer to question III raised in the introduction. Let us assume that a theoretical formula for the transition probability TP(N,F,y) for some multiphoton process is to be tested and that we can make measurements of Q_N for all N values for which Q_N is essentially different from zero. However, we do not know

the functions F and w in our experiment. What can we do to check the measured Q_N values against the formula for TP nevertheless? A possible procedure is the following:
We perform the experiment for several values $y_1, y_2, y_3, \ldots$ of the parameters y in such a way that it is guaranteed that F and w stay the same for all measurements. E.g. in the case of free-free transitions we could change the direction of the polarization of the laser and leave everything else the same. For each value y_i we obtain a sequence of count rates $Q_{N,i}$ from which we can construct a sequence of $\mu_{N,i}$ values or an approximate weight function $g_i(F)$ by inverting eq.(11) where we have inserted the particular theoretical formula for TP(N,F,y) to be tested. If the formula for TP is correct, then the various results for $\mu_{N,i}$ or $g_i(F)$ should be the same for all i within the experimental accuracy. If they are not the same within the experimental error limit, then the theory to be tested is not compatible with the experimental results.

The essence of these ideas is to perform the first set of measurements in order to probe the laser beam, assuming a particular theory to be correct, and then to use this new information about the laser as input for the evaluation of further experiments. Thereby, in the first step we reverse the usual procedure of multiphoton experiments and deal with an "inverted multiphoton problem".

So far we have used the special example of free-free transitions for the explanation of the ideas. Two other multiphoton processes to which these ideas can be applied immediately are the spectroscopy of dressed states of polar molecules (see Ref.13 for the theory, Ref.14 for experimental results and Ref.15 for the general background) and the absorption of additional photons after multiphoton ionization of atoms (see Refs.16,17,18). For the last-named processes there is also another interesting method to avoid the averaging problem (see Ref.19).

REFERENCES

1. A. Weingartshofer, J. Holmes, G. Caudle, E. Clarke, and H. Krüger, Phys.Rev.Lett.39, 269 (1977).
2. A. Weingartshofer, E. Clarke, J. Holmes, and C. Jung, Phys.Rev.A19, 2371 (1979).
3. N. M. Kroll and K. M. Watson, Phys.Rev.A8, 804 (1973).
4. M. H. Mittleman, Phys.Rev.A19, 99 (1979).
5. L. Rosenberg, Phys.Rev.A20, 275 (1979).
6. I. S. Gradshteyn and I. M. Ryzhik, Table of Integrals Series and Products (Academic,New York,1965).
7. M. Abramowitz and I. A. Stegun, Handbook of mathematical functions (Dover,New York,1965).
8. C. Jung, Phys.Rev.A24,360 (1981).
9. J. A. Shohat and J. D. Tamarkin, The Problem of Moments Mathematical Surveys 1(American Mathematical Society, Providence,1950).
10. N. Akhiezer, The Classical Moment Problem and Some Related Questions in Analysis (Oliver and Boyd, Edinburgh,1965).
11. R. G. Gordon, J.Math.Phys.9,655 (1968).
12. J. C. Wheeler and R. G. Gordon, in The Pade Approximant in Theoretical Physics, edited by G. A. Baker and J. L. Gammel (Academic,New York,1970),Chap.3.
13. B. Macke and J. Legrand, J.Phys.B7,865 (1974).
14. E. Arimondo and P. Glorieux, Phys.Rev.A19,1067 (1979).
15. J. E. Bayfield, Phys.Rep.51,317 (1979).
16. P. Agostini, F. Fabre, G. Mainfray, and N. K. Rahman, Phys.Rev.Lett.42,1127 (1979).
17. Y. Gontier and M. Trahin, J.Phys.B13,4383 (1980).
18. P. Kruit, J. Kimman, and M. J. Van der Wiel, J.Phys.B14,L597 (1981).
19. P. Kruit, J. Kimman, H. G. Muller, and M. J. Van der Wiel, J.Phys.B16 ,937 (1983).

CONDITIONS FOR THE OBSERVATION OF THE PREDICTED SPLITTING OF THE LASER ASSISTED ELECTRON IMPACT TRIPLE DIFFERENTIAL CROSS SECTION OF ATOMS

M. ZARCONE* and M. R. C. McDOWELL
Department of Mathematics, Royal Holloway College (University of London), Egham Hill, Egham, Surrey TW20 OEX, England.

Abstract The observational conditions under which the predicted splitting of the laser assisted electron impact ionisation of atoms triple differential cross section occurs are investigated for the usual simple laser model and when the model is modified to include temporal or spatially inhomogeneity or when the laser has a particular multi-mode structure. Particular attention is paid to the variation with laser polar isation, and it is found that the effect is always present when the polarisation is perpendicular to the momentum transfer.

INTRODUCTION

The presence of a strong electromagnetic field ("laser") changes significantly the conditions of the electron-atom scattering processes. In fact the photons of the laser, exchanging energy and momentum can play the role of a third body, opening new channels and allowing the observation of electron-atom scattering parameters which would not otherwise be observable. Many of the processes studied, such as elastic and inelastic potential scattering,[1] electron

*On leave of absence from Istituto di Fisica Dell'Università, Via Archirafi 36, 90123 Palermo, Italy.

impact ionisation,[2,3] x-ray ionisation,[4,5] Compton effect[5,6] etc. lead to, in the first order treatment and when the laser is treated in dipole approximation, a cross section given by a product of a squared Bessel function (carrying all the information about the laser field) times the field free cross section calculated at a shifted final energy that takes into account the number of photons absorbed (see Eq.(1) below). The modulating effect of the oscillating squared Bessel function as a function of the scattering angles gives a significant modification of the angular distribution of the free particle in the final state. This effect is particularly prominent in electron impact ionisation of atoms. Recently Cavaliere et al[2] for hydrogen atoms and Zarcone et al[3] for helium atoms have evaluated the triple differential cross section (TDC) in the first Born approximation (FBA) for different laser polarisations. In both cases the way that the shape of the angular distribution of the ejected electron is changed is shown to depend strongly both on the parameters characterising the laser (polarisation, intensity, frequency) and those of the electron-atom scattering process (incident and final electron energy, scattering angle). Moreover, the contribution to the cross section given by the process where no photons are absorbed exceeds almost everywhere the contribution given by process with $n \neq 0$. More recently Zangara et al[7] have reinvestigated the ionisation problem using a more realistic laser model. For spatial or temporal inhomogeneity[8] and for a multimode[9] laser the cross section preserves the factorised structure of the homogeneous case. However, the squared Bessel function is replaced by a more complicated expression. For the cases considered they show that neither the inhomogeneities nor the multimode structure cancel or drastically alter the predictions made on the basis of the homogeneous, single mode laser. In this paper we give a complete analysis of the dependence of the shape and the intensity of the TDC on the different parameters involved in the collision process for the case of a single mode homogeneous laser and also for the inhomogeneous and the multimode case. In particular, we will emphasize situations where the TDC for $n \neq 0$ photons dominate the zero photon one. This work is intended to give a useful guide to the choice of the collision and laser parameters involved in setting up a possible experiment.

HOMOGENEOUS, SINGLE MODE LASER

The theory of laser assisted electron impact ionisation with a single mode homogeneous laser has been discussed by different authors.[10,11] Here we give only the main assumptions and the limitations assumed in the derivation of their result.

The laser field is treated classically as single mode and homogeneous. The field is taken in the dipole approximation. Further the effect of the laser on the atomic states is neglected, so that the laser electric field ε_L is assumed considerably smaller than the characteristic interatomic field $\varepsilon_{at} = Z^3 e/a_o{}^2 = 5 \times 10^9$ V/cm, and in addition it is supposed off-resonance with any atomic transition frequency ω_{ij}. For the initial and final fast electron wavefunctions they assume laser modulated plane waves and for the slow ejected electron laser modulated continuum atomic states. The latter states are approximate solutions with positive energy of the Schroedinger equation of an atom in a laser field. This approximation is valid when the momentum provided to the free particle by the laser

$p_L = \frac{e}{c} A(t)$ is smaller than the particle's own momentum $p = \hbar\kappa$, i.e. when $\delta = \varepsilon_L/\omega\kappa << 1$.

Within these approximations, the FBA to the TDC may be written as

$$\frac{d^3\sigma(L)}{d\Omega_1\, d\Omega_2\, d\varepsilon} = \sum_{n=-\infty}^{\infty} J_n^{\,2}(\lambda_n) \frac{d^3\sigma(0)}{d\Omega_1\, d\Omega_2\, d\varepsilon} = \sum_{n=-\infty}^{\infty} \sigma_n(L) \qquad (1)$$

where the left-hand side is the TDC with the laser on and the right-hand side is a sum over the number of photons absorbed (n < 0) or emitted (n > 0) of the field-free cross section (TDCO) at energy

$$k_f^{\,2} = k_i^{\,2} - \kappa^2 + E + n\hbar\omega\ , \qquad (2)$$

times a squared Bessel function of argument

$$\lambda_n = \frac{1}{\omega^2}(\underline{k}_{if} - \underline{\kappa})\cdot\underline{\varepsilon}_L\ . \qquad (3)$$

In Eq. (2)-(3) $\underline{k}_{if} = \underline{k}_i - \underline{k}_f$ is the momentum transfer, $\underline{\kappa}$ is the slow ejected electron's momentum, E is the atomic ionisation potential, and ε_L is the laser electric field intensity. As shown in our previous paper[3] the TDCO is a slowly varying function of the final energy. For processes involving a small number of photons, it can be regarded as a constant over a wide range of n. All the features of the cross section due to the laser field are buried in the Bessel functions. Moreover, for H and He, the TDCO presents the well known two-peak structure[3] showing that the maximum ionisation probability is given when the ejected electron leaves the atom along the directions $\pm\,\underline{k}_{if}$ and is zero in the direction perpendicular to it. Due to the factorised structure of the cross section, the TDC will also be zero in the direction perpendicular to $\underline{k}_{if}$; consequently any change in the shape of the TDC can occur only in a small angular range around the directions $\pm \underline{k}_{if}$. For this reason we can restrict our analysis by fixing the ejected electron angle at $\theta_k = \hat{\underline{k}} \cdot \hat{\underline{k}}_{if} = 0$.. The Bessel function is an oscillating function with the different low-order zeros ordered in the following way[12]

$$\lambda^{I}_{n\neq 0} = 0\,, \qquad \lambda^{I}_{0} = 2.405\,, \qquad \lambda^{II}_{1} = 3.832$$
$$\lambda^{II}_{2} = 5.136 \qquad \lambda^{II}_{0} = 5.520 \qquad \lambda^{II}_{3} = 6.382 \tag{4}$$

etc., where the upper index is the order of the zero and the lower is n .

The change in the structure of the TDC and the relative importance of contributions from different numbers of photons n depends strongly on the particular range of variation of the Bessel function argument λ_n Eq.(3). For very small values of λ $(\lambda \simeq 0)$ the main contribution to the TDC comes, of course, from the $n = 0$ process, since $\lambda^{I}_{n\neq 0} = 0$. Increasing λ the $n = 0$ process loses importance, so that when $\lambda \simeq \lambda^{I}_{0}$, the $n = 1$ process is more important for by $\lambda \simeq 1.45$ we have

$$J_1(1.45) \simeq J_0(1.45)\,. \tag{5}$$

For values of $\lambda >> 1$ the situation becomes more complicated, and since the zeros of the Bessel function become much closer we find a more complex and modulated structure for the TDC.

The variation of λ as a function of the electron ejected angle depends strongly on the collision parameters and on the laser parameters. For example, in our previous calculations for Helium[3] with the following parameters:

$E_i = 256.5$ eV, $E_{e_J} = 3$ eV, $\theta_s = 4°$

$\omega = 1.17$ eV and $\varepsilon_L = 5\ 10^7$ V/cm we have

we have

$$|\lambda| \simeq (0.4,\ 4.4) \quad \text{for} \quad \underline{\varepsilon}_L \,||\, \underline{k}_{if}$$

and (5)

$$|\lambda| \simeq (0.0,\ 2.4) \quad \text{for} \quad \underline{\varepsilon}_L \perp \underline{k}_{if}$$

while, in the case considered by Cavaliere et al[2] for hydrogen with $E_i = 150$ eV, $E_{e_J} = 5$ eV, $\theta_s = 5°$, $\omega = 1.17$ and $\varepsilon_L = 10^6$ V/cm we have

$$|\lambda| \simeq (0.008,\ 0.1) \quad \text{for} \quad \underline{\varepsilon}_L \,||\, \underline{k}_{if}$$

and (6a')

$$|\lambda| \simeq (0.0,\ 0.064) \quad \text{for} \quad \underline{\varepsilon}_L \perp \underline{k}_{if}\ .$$

The difference in the range of λ in Eq.(6) and (6a) gives a complete explanation of the different shape of the TDC for the two polarisations and/or the two examples considered. In fact in Cavaliere et al's[2] example $\lambda << 1$ and the only Bessel function zero contributing to the change in the structure of the TDC is the $\lambda = 0$ zero, the $n = 1$ photon process being much smaller than the $n = 0$ one.

One interesting experimental situation is when processes with $n \neq 0$ dominate the $n = 0$ process. In this context it is useful to investigate for which values of the laser parameters (intensity, frequency and polarisation) we can have $J_0(\lambda) = 0$ and some $J_{n\neq 0}(\lambda) \neq 0$. For reasons discussed above, we confine our analysis to the region around $\theta_K = 0$, and we use the collision parameters given above (Eq.5) for electron-Helium ionisation.

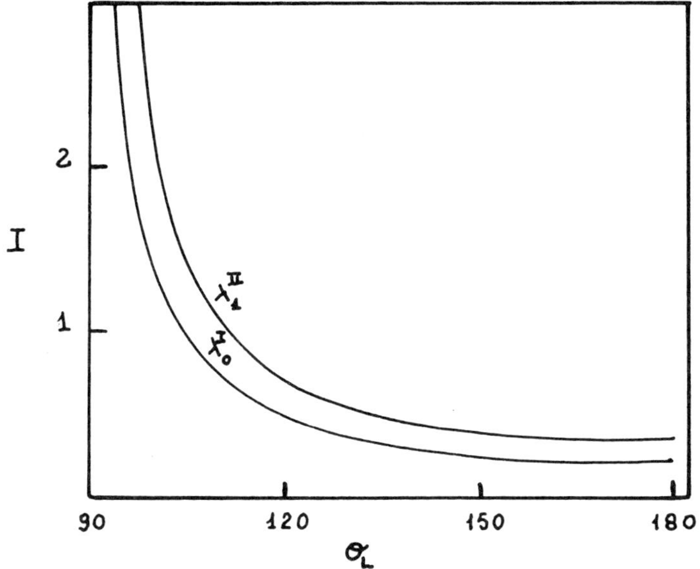

FIGURE 1. The curves relate the values of the laser electric field intensity I [10^9 V cm^{-1}] and laser polarisation θ_L (measured from the momentum transfer $\underline{k}_{if}$ for which $\lambda = \lambda_0^I$ the first zero of J_0 or $\lambda = \lambda_1^{II}$ the second zero of J_1, in the simple laser model, with the geometry specified in the text).

In Figure 1 we plot the value of the electric field laser intensity and the laser polarisation angle $\theta_L = \hat{\underline{\varepsilon}}_L \cdot \hat{\underline{k}}_{if}$ which give the zeros $\lambda_0^I = 2.405$ and $\lambda_1^{II} = 3.832$ respectively. We see that for not too large a given laser intensity it is possible to select a laser polarisation corresponding to the $\lambda = \lambda_0^I$ so that the probability of a zero photon process with the slow electron ejected along $\underline{k}_{if}$ is zero. Accordingly, the probability for the $n = 1$ process will be $J_1^2(\lambda_0^I) \simeq 0.27$, which implies a large cross section.

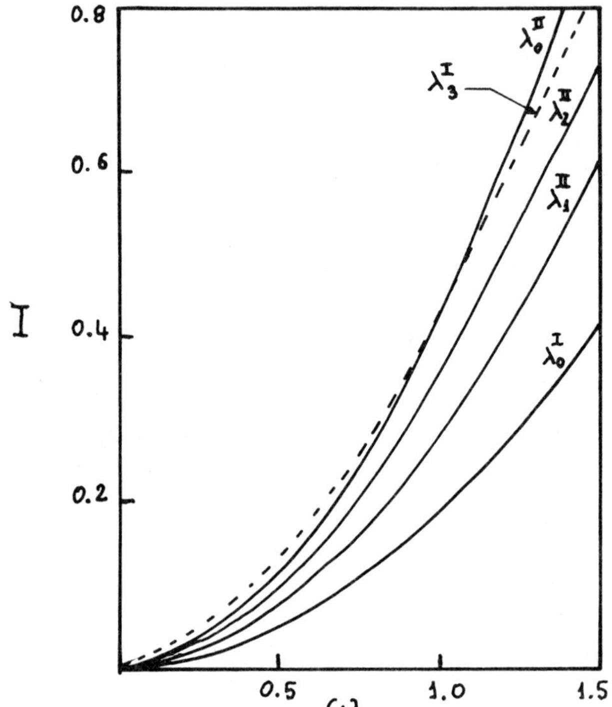

FIGURE 2. As Figure 1, but now relating the intensity to the frequency ω (eV) for λ_0^{I}, λ_1^{II}, λ_2^{II}, λ_3^{II}, λ_0^{II} for $\theta_L = 0$.

It is important to note that, due to the small difference between the two curves, this effect can be observed only with a laser with a field intensity determined to better than 10%. We see also that the most favourable situation for this geometry is at low intensity for $\theta_L = 0$, while for $\theta_L = 90°$ the intensity must be very large, and the model is inapplicable.

In Figure 2 we instead show for the more convenient polarisation $\theta_L = 0$, the laser intensity and frequency for several values of λ. Again, for a given value of frequency we can choose a value of the intensity corresponding to the curve with λ_0^I so that $J_0^I = 0$. We note that for low frequency, or high frequency and high intensity we have a situation where it is not possible to select a particular zero of $J_0(\lambda)$, because this would require a laser with an unrealistic homogeneity (for the low frequency case) or an unrealistic resolution for the high frequency and intensity case.

In the next sections we will examine in more detail the influences of the laser properties (inhomogeneity and multimode structure) on the intensity and shape of the TDC.

INHOMOGENEOUS SINGLE MODE LASER

A real laser normally has a strong macroscopic inhomogeneity associated with the decision to focus the radiation in order to increase the field in the collision chamber. Consequently different target atoms will be found in regions of different field strength and the measured differential cross sections may be affected by such an inhomogeneity.

The theory of particle-atom collisions in the presence of an inhomogeneous laser has been developed in recent years.[7,8] The inhomogeneity may be both temporal and spatial, and in both cases the factorised structure of the cross section of Eq.(1) is still valid. However, the squared Bessel function is now replaced by a more complicated function $F^2(n,\lambda)$.[7,8]

The basic assumptions of the theory are (i) the laser is supposed single mode, with an amplitude depending on the spatial or temporal coordinates; (ii) the scale of variation of the field amplitude is assumed very small compared with the typical lengths defined by the collision event. In this scheme the cross section is calculated following the same procedures used for the single mode homogeneous laser with constant amplitude, and then an average is performed over all the relevant space-time points where or when the particular collision occurs.

The resulting averaged cross section depends on a parameter γ given by the ratio of the electron beam dimension s to the laser beam dimension α, for a spatially inhomogeneous laser, and to the ratio of the detector response time τ_0 to the duration of the laser pulse τ_L for a temporarily inhomogeneous laser. For the spatially inhomogeneous case with $\gamma >> 1$ we are in a situation where few of the scattering events are in the laser field. With temporal inhomogeneity where the detector is not able to respond separately to two collisions within τ_0, in which time $(\tau_0 >> \tau_i)$ the laser intensity may have changed appreciably. In the usual Gaussian model of inhomogeneity, the average electric field involved in the collision is smaller than that for a homogeneous laser, so many-photon events should be less important.

When $\gamma << 1$ the laser field intensity for any collision in the scattering chamber may be considered constant during the collision, and the inhomogeneity of the laser does not affect the process.

The most difficult case to discuss (as always) is when the scales are comparable, i.e. $\gamma \sim 1$, and we examine this below, first quoting the results of the earlier papers.[7,8]

For a spatially inhomogeneous laser with a laser field intensity distribution characterised by a Gaussian

$$\phi(x,y) = \exp\{-1/2\ \alpha(x^2 + y^2)\} \tag{7}$$

the squared Bessel function J^2 is replaced in the factorised TDC, Eq.(1) by[7]

$$F_s^2(n,\lambda) = \sum_{k=0}^{\infty} G_\nu(\lambda_n,k) \frac{1}{(\nu + k)} \{1 - \exp\{-(\nu + k)\}\} \quad (8)$$

where $\nu = |n|$ and

$$G_\nu(\lambda_n,k) = \frac{(-1)^k (2\nu + 2k)!(1/2\lambda_n)^{2(\nu+k)}}{k![(\nu + k)!]^2 (2\nu + k)!} . \quad (9)$$

For a temporal inhomogeneity with a field intensity given by a Gaussian function

$$f(t) = \exp\{- 1/2 \frac{\pi}{\tau_L} t^2) \quad (10)$$

We have similarly[7]

$$F_t^2(n,\lambda) = \sum_{k=0}^{\infty} G_\nu(\lambda_n, k)\left\{\frac{1}{(\nu + k)^{1/2}} \mathrm{erf}\{0.5[\pi(\nu + k)]^{1/2}\}\right\} \quad (11)$$

in terms of the error function.

The behaviour of these complex expressions is far from obvious, but the behaviour is readily computable for λ_n not too large ($\lambda_n \leqq 10$).

In Figures 3 and 4 we compare the behaviour of $F_s^2(n,\lambda)$ and $F_t^2(n,\lambda)$ with J_n^2 as a function of λ for the cases $n = 0$ and $n = 1$ respectively. As expected the inhomogeneity decreases the contribution from the $n = 1$ process in favour of the $n = 0$ process, especially for small values of λ. For both the temporal and spatial inhomogeneity the curves keep the oscillatory behaviour of the Bessel function, even though they are smoother and the zeros are dumped out. Moreover, the distance between successive minima is larger than the corresponding distance between two successive zeros of the homogeneous case. From this fact we

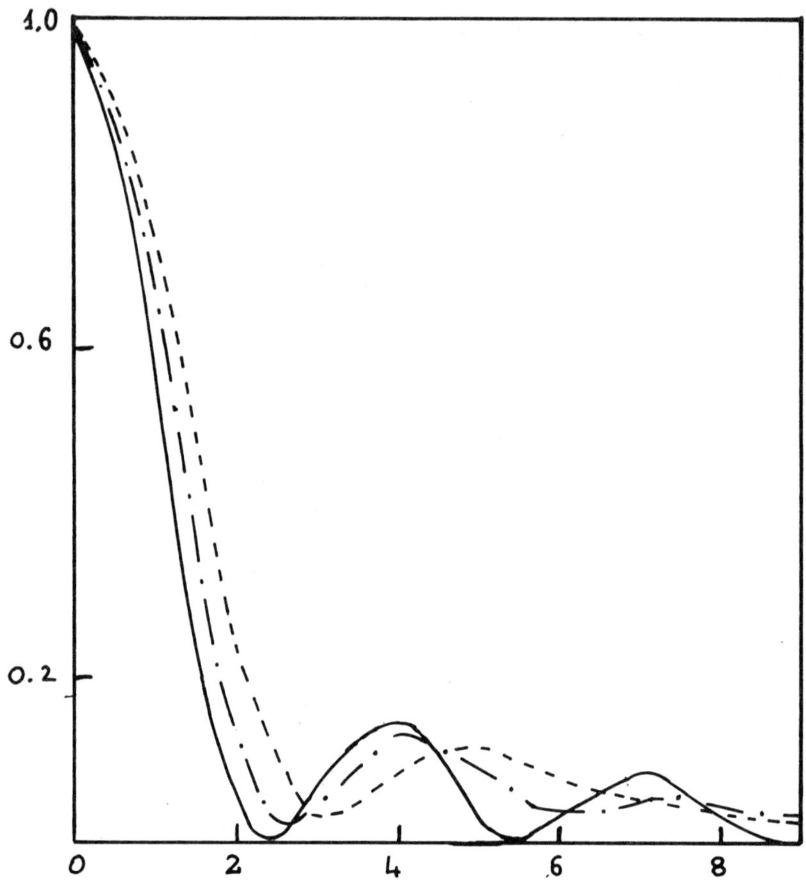

FIGURE 3. —— $J_n^{\ 2}(\lambda)$; --- $F_s^{\ 2}(n,\lambda)$; -·- $F_t^{\ 2}(n,\lambda)$ compared for $0 \leq \lambda \leq 8$ and $n = 0$.

can deduce that the TDC will present a laser modulated behaviour, but with a less rich and pronounced structure. Under conditions when a splitting of the forward and backward peak is predicted (n = 1 process) the resulting lobes will be rather further apart than in the homogeneous case.

In the homogeneous case it is always possible to choose parameters of the scattering process and/or of the laser so

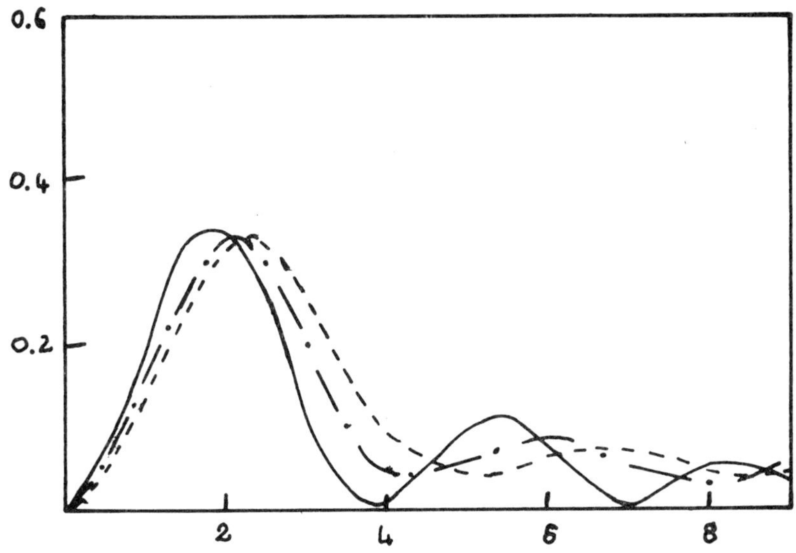

FIGURE 4. As Figure 3 but $n = 1$.

that for a particular ejected electron angle the $n \neq 0$ photon processes are more important than the $n = 0$ process. In the present case this will occur for slightly larger values of λ than in the homogeneous case. In conclusion we can say that for small values of λ the behaviour of the inhomogeneous functions F_s^2 and $F_t^2(n)$ are very similar to that of the squared Bessel function. For increasing λ F_s^2 and F_t^2 rapidly get out of phase with J_n^2, showing a phase advance and strong damping. Thus the behaviour of the system with inhomogeneities is increasingly different from that of the homogeneous system. However, these inhomogeneities will only be important at low frequencies and/or high intensities as we have seen in section 2.

MULTIMODE LASER

In this section we will examine the modifications expected

in the laser assisted electron impact ionisation TDC if the laser is multimode. Following Zoller,[9] the radiation of a multimode laser with a large number of uncorrelated modes may be represented by a chaotic field. Using the Gaussian properties and the first-order correlation function, and allowing the laser frequency band-width $\Delta\omega$ to become vanishingly small, it is possible again to get a factorised TDC of the same form as the homogeneous single mode case, with

$$F^2_{MM}(n,\lambda) = \exp\{-1/2\,\lambda_n^{\,2}\} I_n(1/2\,\lambda_n^{\,2}) \tag{12}$$

where I_n is the modified Bessel function (Abramowitz and Stegun, ref.(12) Eq.9.6.3). A comparison of the function $F^2_{MM}(n,\lambda)$ with the square Bessel function is made in Figures 5 and 6 for $n = 0$ and $n = 1$ respectively. The function $F^2_{MM}(n,\lambda)$ is nonotonic, decreasing with increasing λ and shows no oscillations or zeros. Consequently, we expect the same behaviour as in the single mode homogeneous case for very small values of λ, $\lambda \ll 1$. Moreover, in this case it is not possible to have a situation where the $n = 1$ contribution to the cross section is larger than the $n = 0$ one. The only zero present is for $\lambda = 0$ and $n \neq 0$, so again for a laser polarisation angle $\theta_L = 90$ we have a splitting of the forward and backward peak, but it is not possible to have such a splitting for any other polarisation.

CONCLUSION

More realistic laser models than used in the derivation of the Kroll-Watson result, which allow for spatial or temporal inhomogeneity or multimode effects, do not change the shape

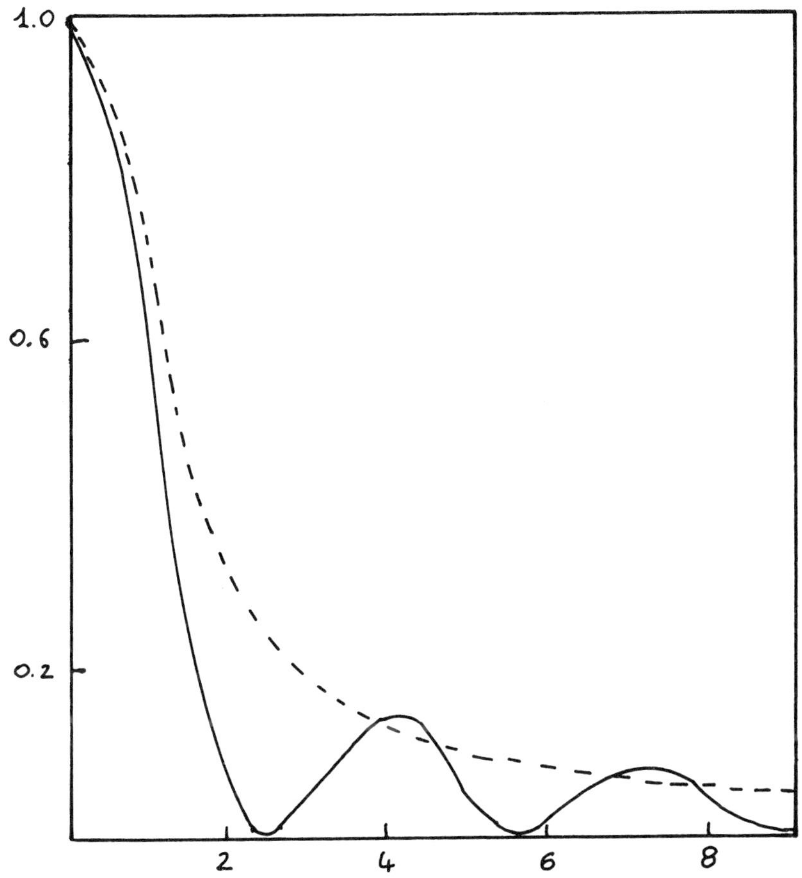

FIGURE 5. —— $J_n^2(\lambda)$; --- $F_{MM}^2(n,\lambda)$ for n = 0.

nor strongly affect the magnitude of the predicted TDC obtained using a simple single mode homogeneous laser model, provided the laser parameters (frequency, intensity and polarisation) and the collision parameters (incident energy scattering angle, final energy and target atom used) are chosen in order that the value of the argument of the relevant Bessel function is restricted to $\lambda \ll 1$.

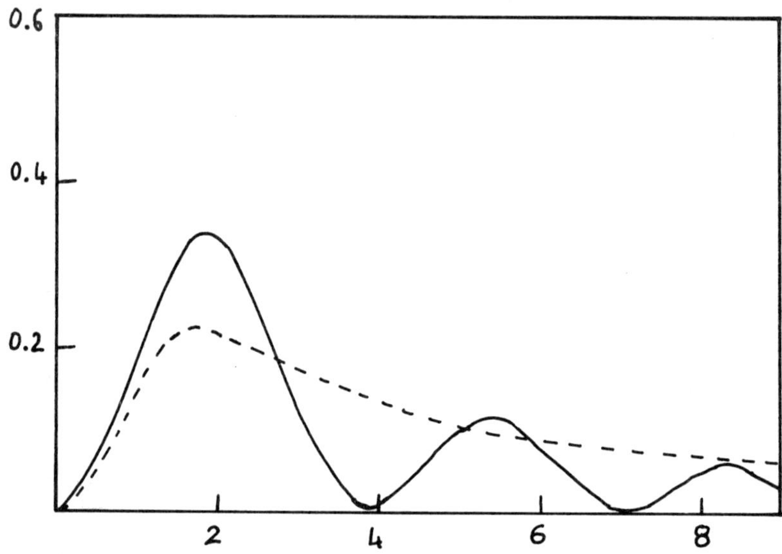

FIGURE 6. As Figure 5 but n = 1.

For the more interesting experimental situation when for $\lambda >> 1$ processes with $n = 1$ are predicted to be more likely than that with $n = 0$ in the simple model, the more realistic laser models predict an attenuation of the effect, and in some models the predicted splitting is wiped out, except in the case of when the laser polarisation is perpendicular to the momentum transfer.

REFERENCES

1. N. M. Kroll and K. M. Watson. Phys. Rev. A 17 1706 (1973).
2. P. Cavaliere, C. Leone, R. Zangara and G. Ferrante. Phys. Rev. A 24 910 (1981).
3. M. Zarcone, D. L. Moores and M. R. C. McDowell. J. Phys. B: At. Mol. Phys. 16 L11 (1983).
4. G. Ferrante, E. Fiordilino and M. Rapisarda. J. Phys. B: At. Mol. Phys. 14 L497 (1981).

5. M. Jain and N. Tzoar. Phys. Rev. A 15 147 (1977); A 18 538 (1978).
6. G. Ferrante and E. Fiordilino. J. Phys. B: At. Mol. Phys. 15 2157 (1982).
7. R. Zangara, P. Cavaliere, C. Leone and G. Ferrante. J. Phys. B: At. Mol. Phys. 15 3881 (1982).
8. R. Daniele, G. Ferrante and S. Bivona. J. Phys. B: At. Mol. Phys. 14 L213 (1981).
9. P. Zoller. J. Phys. B: At. Mol. Phys. 13 L249 (1980).
10. P. Cavaliere, G. Ferrante and C. Leone. J. Phys. B: At. Mol. Phys. 13 4495 (1980).
11. J. Banerji and M. H. Mittleman. J. Phys. B: At. Mol. Phys. 14 3717 (1981).
12. M. Abramowitz and I. A. Stegun. Handbook of Mathematical Functions, Dover Publications Inc. N.Y. (1965).

PHOTON CORRELATIONS AND BANDWIDTH EFFECTS IN LASER-ASSISTED ELECTRON-ATOM COLLISIONS

F. TROMBETTA[x], C.J. JOACHAIN[x] and G. FERRANTE[+]
[x]Physique Théorique, Université Libre de Bruxelles, Belgium
[+]Istituto di Fisica dell'Università, Palermo, Italy

Abstract We discuss photon correlation effects in electron-atom collisions occurring in the presence of strong chaotic laser fields. In particular, we analyze the influence of the laser bandwidth on the scattering cross section and study the enhancement factor between the chaotic field cross section and that corresponding to a purely coherent field.

INTRODUCTION

Most of the theoretical treatments of atomic collisions in strong electromagnetic (laser) fields proposed until now have used the assumptions of homogeneous monochromatic and purely coherent radiation. However, lasers do not always satisfy these assumptions as their bandwidth and departure from coherence is not necessarily negligible. In particular, high-power pulsed lasers are generally operated in multimode configurations, with bandwidths larger than typical atomic linewidths and intensities undergoing substantial fluctuations. Real lasers undergo intensity, phase and frequency fluctuations which, depending on experimental circumstances, should be properly treated as stochastic processes. These problems have been thoroughly investigated in the related area of the interaction of radiation with bound states of atoms and molecules[1]. Instead, in the case of atomic collisions in strong electromagnetic fields, investigations of this kind have only been initiated very recently[2-8].

In both spectroscopic problems and field-assisted atomic collisions, consideration of the field properties is aimed at establishing closer contact with experiment. However, in the case of collisions, additional interest arises from the fact that it is possible to include in a rather simple way the properties of the field to an accuracy which at present does not seem attainable in the problems of interaction of radiation with atoms and molecules. This remarkable result is achieved through the use of scattering states embedded in the assisting field. Thus, the inclusion of radiation properties in collision processes is also interesting in order to understand in general the meaning of these properties in strongly nonlinear atomic processes.

In this article, we consider the case of electron-atom collisions in the presence of a chaotic field, namely a complex Gaussian stochastic process with both amplitude and phase fluctuations. After an outline of the formal development and of the derivation of the differential cross section, we discuss in some detail two results, whose interest is believed to go beyond this particular context. The first one arises from the comparison of the laser bandwidth $\Delta\omega$ with the width Γ_i of the incoming particle energy distribution. We find that even when Γ_i is much larger than $\Delta\omega$, the latter has a strong broadening effect on the scattering cross section, due to the highly nonlinear interaction between the field and the particles. The second result concerns an enhancement factor, defined as the ratio, for a given multiphoton process, of the chaotic field cross section to that of a purely coherent field. We find, contrary to expectations generally based on perturbative treatments of the field, that the enhancement factor can be much smaller than unity for some values of the parameters and for particular multiphoton processes. For simplicity, hydrogenic targets will be considered, and exchange effects neglected (see, however, refs. 4 and 9).

THEORY

The wave function of the free particle (an electron, of charge -e, mass m and wavevector $\underset{\sim}{k}$) embedded in the field is the nonrelativistic Volkov wave

$$\chi(\underset{\sim}{r},t) = \exp(i\underset{\sim}{k}.\underset{\sim}{r} - \frac{i}{\hbar}\varepsilon t)\exp(-i\lambda\int^{t} A(t')\,dt') \qquad (1)$$

where $\varepsilon = \hbar^2k^2/2m$, $\lambda = (e/mc)\underset{\sim}{k}.\underset{\sim}{n}$, $\underset{\sim}{n} = \underset{\sim}{A}(t)/A(t)$, and the $A^2(t)$ term has been removed by a contact transformation. For atomic states with principal quantum number $n > 1$, we use a wave function obtained as a first-order AC Stark mixing of the degenerate states :

$$\phi(\underset{\sim}{r},t) = \psi(\underset{\sim}{r})\exp\{-\frac{i}{\hbar}E^{(o)}t\}\exp\{-i\alpha F(t)\}\exp\{-\frac{i}{\hbar}e\underset{\sim}{n}.\underset{\sim}{r}F(t)\} \qquad (2)$$

where $\psi(\underset{\sim}{r})$ is a linear combination of unperturbed states $\psi_j^{(o)}$ in the degenerate multiplet of energy $E^{(o)}$, α is a root of the determinant

$$\| \hbar\alpha\,\delta_{ij} - \langle \psi_i^{(o)} | e\,\underset{\sim}{n}.\underset{\sim}{r} | \psi_j^{(o)} \rangle \| = 0,$$

and $F(t) = \int^{t} E(t')dt'$, E(t) being the magnitude of the electric field at time t.

Treating the collision to first order of perturbation theory and following refs 2 and 7, we find the average transition probability for the transition $i \rightarrow f$:

$$\langle P_{fi}\rangle = (|V_{fi}|/\hbar)^2\int dt_1 dt_2 \exp\{i(E_f^{(o)}+\varepsilon_f-E_i^{(o)}-\varepsilon_i)$$
$$\times (t_1-t_2)/\hbar\} I(t_1,t_2)$$

where

$$V_{fi} = \int d^3r_1 d^3r_2\, e^{-i\underset{\sim}{k}_f\cdot\underset{\sim}{r}_2}\, \phi_f(\underset{\sim}{r}_1)[r_{12}^{-1}-r_2^{-1}]\, e^{i\underset{\sim}{k}_i\cdot\underset{\sim}{r}_2}\, \phi_i(\underset{\sim}{r}_1)$$

and

$$I(t_1,t_2) = \exp\{-\tfrac{1}{2}[\alpha_{fi}^2 \iint_{t_2}^{t_1} d\tau_1 d\tau_2 \,\langle E(\tau_1)E(\tau_2)\rangle$$
$$+ \lambda_{fi}^2 \iint_{t_2}^{t_1} d\tau_1 d\tau_2\, \langle A(\tau_1)A(\tau_2)\rangle]\} .$$

We will assume that the Gaussian process is Markovian and that the first-order correlation function for the electric field is given by

$$\langle E(\tau_1)E(\tau_2)\rangle = \frac{E_o^2}{2} \cos \omega|\tau_1-\tau_2| \exp\{-\Delta\omega|\tau_1-\tau_2|\},$$

where E_o is the variance of the field-amplitude distribution, ω is the central angular frequency and $\Delta\omega$ the bandwidth. Accordingly, for the vector potential we have :

$$\langle A(\tau_1)A(\tau_2)\rangle = \frac{E_o^2 c^2}{2(\omega^2+\Delta\omega^2)} \cos[\omega|\tau_1-\tau_2| -\phi]\, e^{-\Delta\omega|\tau_1-\tau_2|}$$

where $\cos\phi = (\omega^2-\Delta\omega^2)/(\omega^2+\Delta\omega^2)$, $\sin\phi = 2\omega\Delta\omega/(\omega^2+\Delta\omega^2)$. Now, substituting in $I(t_1,t_2)$ and integrating over τ_1 and τ_2 we have

$$I(t) = \exp\{-\rho\} \exp\{-\gamma t + L\cos(\omega t-\theta)\}$$

where $t = t_1-t_2$ and

$$\rho = \alpha^2 \cos\phi + \lambda^2 \cos2\phi,\ \gamma = \alpha^2\Delta\omega+\lambda^2(\Delta\omega \cos\phi+\omega \sin\phi)$$

$$\alpha^2 = \alpha_{fi}^2 E_o^2/2(\omega^2+\Delta\omega^2) \qquad \lambda^2 = \lambda_{fi}^2 E_o^2 c^2/2(\omega^2+\Delta\omega^2)^2$$

$$\alpha_{fi} = \alpha_f - \alpha_i \qquad \lambda_{fi} = \lambda_f - \lambda_i = (e/mc)(\underset{\sim}{k}_f - \underset{\sim}{k}_i)\cdot\underset{\sim}{n}$$

$$L = (\lambda^4+\alpha^4+2\lambda^2\alpha^2\cos\phi)^{1/2} \qquad \theta = \text{arctg}[(\alpha^2\sin\phi+\lambda^2\sin2\phi)/\rho]$$

The quantities λ and α account, respectively, for the field coupling with the free and the bound initial and final states. When $\Delta\omega \neq 0$, λ and α are seen to interfere, while when $\Delta\omega \to 0$, $L \to (\lambda^2+\alpha^2)$. The transition probability for unit time is

$$W = (|V_{fi}|^2/\hbar^2)\; e^{-\rho}\; 2\text{Re}\int_0^\infty dt\; \exp\{i(E_f^{(o)}+\varepsilon_f-E_i^{(o)}-\varepsilon_i)t/\hbar\}$$

$$e^{-\gamma t}\; \exp\{L\; e^{-\Delta\omega t}\cos(\omega t-\theta)\} \qquad (3)$$

This result corresponds to sharp-line excited atomic level and delta-like energy distribution of the incoming beam.

To obtain a more realistic expression and for purposes of later comparisons, we average the expression given by eq. (3) over the energy distribution of the incident beam and the spread of the excited atomic level. For simplicity, we take the two distributions to be Lorentzians, having respectively bandwidths Γ_i and Γ_f. Thus, we must calculate[10]

$$\overline{W} = \int d\varepsilon \int dE\; L_i(\varepsilon_i+\varepsilon)\; L_f(E_f^{(o)}+E)\; W(\varepsilon_i+\varepsilon, E_f^{(o)}+E)$$

where

$$L_i(\varepsilon) = \pi^{-1}\,[\hbar\Gamma_i/((\hbar\Gamma_i)^2 + (\varepsilon-\varepsilon_i)^2)]$$

and similarly for $L_f(E)$. Assuming that V_{fi} and λ do not change significantly within the energy width Γ_i, we can write

$$\overline{W} = (|V_{fi}|^2/\hbar^2)\; e^{-\rho}\; 2\text{Re}\int_0^\infty dt\; \exp[i(E_f^{(o)}+\varepsilon_f-E_i^{(o)}-\varepsilon_i)t/\hbar]$$

$$e^{-\Gamma t}\; \exp(L\; e^{-\Delta\omega t}\cos(\omega t-\theta)) \qquad (4)$$

where now $\Gamma = \gamma + \Gamma_i + \Gamma_f \equiv \gamma + \Gamma_o$. Using the expansion

$$\exp\{L\, e^{-\Delta\omega t} \cos(\omega t-\theta)\} = \sum_{N=-\infty}^{+\infty} \sum_{k=0}^{\infty} f_{\nu k}(L)\, e^{(\nu+2k)\Delta\omega t}\, e^{iN(\omega t-\theta)}$$

where

$$\nu = |N|, \qquad f_{\nu k}(L) = (1/k!(\nu+k)!)(L/2)^{\nu+2k}$$

and performing the integral in eq. (4), we find that

$$\overline{W} = (2\pi/\hbar) \sum_N |V_{fi}|^2\, G_N(E_o^2, \omega, \Delta\omega)$$

where the field-dependent weight of the N-channel is

$$G_N(E_o^2, \omega, \Delta\omega) = e^{-\rho} \sum_k f_{\nu k}(L)\, \pi^{-1} [(\overline{\Gamma}_{\nu k} \cos N\theta + \overline{E}_N \sin N\theta)/(\overline{\Gamma}^2_{\nu k} + \overline{E}^2_N)] \tag{5}$$

with

$$\overline{\Gamma}_{\nu k} = \hbar[(\nu+2k)\Delta\omega + \gamma + \Gamma_o], \qquad \overline{E}_N = E_f^{(o)} + \varepsilon_f - E_i^{(o)} - \varepsilon_i + N\hbar\omega \tag{6}$$

The corresponding double differential cross section is given by

$$\frac{d^2\sigma}{d\Omega d\varepsilon_f} = \frac{k_f}{(2\pi\hbar)^3} \frac{W}{\hbar k_i \left(\frac{i}{m}\right)} \equiv \sum_N \left(\frac{d^2\sigma}{d\Omega d\varepsilon_f}\right)_N \tag{7}$$

where

$$(d^2\sigma/d\Omega d\varepsilon_f)_N = G_N(E_o^2, \omega, \Delta\omega)(d\sigma/d\Omega)^{FBA} \tag{8}$$

and the symbol $(d\sigma/d\Omega)^{FBA}$ refers to the first Born approximation to the differential cross section, namely :

$$(d\sigma/d\Omega)^{FBA} = (k_f/k_i)(m/2\pi\hbar^2)^2\, |V_{fi}|^2 \tag{9}$$

As the field enters only through the quantity G_N, we shall analyze it for obtaining informations on the photon correlation effects. The quantity G_N given by eq. (5) has the structure of "generalized Lorentzians", each centered on the energy $\overline{E}_N$, with a broadening $\overline{\Gamma}_{\nu k}$ (eq. (6)) which depends on all the parameters of the process : the number ν of exchanged photons, the laser bandwidth $\Delta\omega$, the coupling of the laser with the particles (through γ) and the field-independent source of broadening Γ_o. The k-index is a direct consequence of our non-perturbative treatment of the field; in fact, it enters through $f_{\nu k}$ and gives a dependence on the intensity of the field as $I^{\nu+2k}$ which, for k = 0, yields the characteristic perturbative I^{ν} dependence of the ν-photons processes.

We remark that in the adiabatic limit ($\Delta\omega \ll \omega$) we obtain

$$G_N^{AD}(E_o^2,\omega,\Delta\omega) = \exp\{-\beta^2/2\} \sum_{k=0}^{\infty} f_{\nu k}(\beta^2/2)\pi^{-1} \Gamma_{\nu k}/(\Gamma_{\nu k}^2 + \overline{E}_N^2) \tag{10}$$

where $\beta^2 = \alpha_o^2 + \lambda_o^2$, $\alpha_o^2 = \alpha_{fi}^2 E_o^2/\omega^2$, $\lambda_o^2 = \lambda_{fi}^2 E_o^2 c^2/\omega^4$, and

$$\Gamma_{\nu k} = \hbar[(\nu+2k+\alpha_o^2/2+3\lambda_o^2/2)\Delta\omega + \Gamma_o].$$

We see that also in this case, which is probably the closest to the experimental conditions, all the previous comments remain true. Besides, we now obtain a superposition of Lorentzians, whose bandwidths have a strong nonlinear dependence on the field. This is illustrated in Fig. 1, where a scattering line-width much larger than Γ_o is obtained.

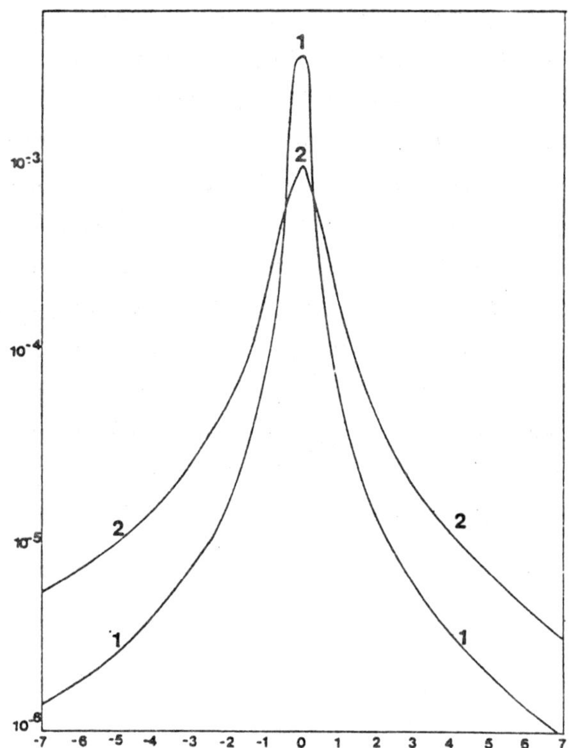

FIGURE 1. $(d^2\sigma/d\Omega d\varepsilon_f)$ vs $(\varepsilon_f-\varepsilon_i)/\hbar\omega$ for an incident energy $\varepsilon_i = 100$ eV and an atomic transition between two Stark mixed states with n = 2. $\underset{\sim}{K}_i \parallel \underset{\sim}{E}_o$, $E_o = 5.10^7$ V/cm, $\hbar\omega = 1.17$ eV, scattering angle 45°. Curve 1 - $\Delta\omega = 10^{-2}\ \Gamma_i$, curve 2 - $\Delta\omega = 10^{-1}\ \Gamma_i$, for for $\Gamma_i = 100$ meV.

In the limit $\Delta\omega \to 0$, neglecting field-independent broadening factors and using the δ-function representation

$$\lim_{x\to o} x/(x^2 + x_o^2) = \pi\ \delta(x_o)$$

the N-channel scattering line-shape becomes :

$$G_N^{(o)}(E_o^2,\omega) = \exp(-\beta^2/2)\ I_N(\beta^2/2)\ \delta(\overline{E}_N)$$

where I_N is the modified Bessel function. We recall that for the same process, with the same assumptions, taking the field as homogeneous and single-mode, one obtains

$$G_N^H(E_o^2,\omega) = J_N^2(\beta)\ \delta(\overline{E}_N)$$

J_N being the ordinary Bessel function. Thus, the scattering differential cross section accompanied by an exchange of N photons in the presence of a chaotic field is

$$(d\sigma/d\Omega)_N^C = \exp(-\beta^2/2)\ I_N(\beta^2/2)\ (d\sigma/d\Omega)^{FBA} \tag{11}$$

while in the presence of a homogeneous and single-mode field

$$(d\sigma/d\Omega)_N^H = J_N^2(\beta)\ (d\sigma/d\Omega)^{FBA} \tag{12}$$

where $(d\sigma/d\Omega)^{FBA}$ is given by eq. (9).

THE ENHANCEMENT FACTOR

Using eqs (11) and (12) let us define an enhancement factor as

$$\chi(N,\beta) = (d\sigma/d\Omega)_N^C/(d\sigma/d\Omega)_N^H = \exp(-\beta^2/2)I_N(\beta^2/2)/J_N^2(\beta) \tag{13}$$

accounting for the photon correlation effects on the multiphoton exchanges. We remark that $\chi(N,\beta)$ has the form (13) both in the case of first-order potential and atomic scattering if, as in our case, the atomic degenerate states are modified according to a first-order AC Stark treatment and the field modifications of the ground state are neglected. We also note that the argument of J_N^2 is actually a function of $\mathcal{E}_o$, the constant amplitude of the homogeneous field, while the numerator is a function of $E_o = (\langle |E|^2\rangle)^{1/2}$, the variance of the chaotic field amplitude. As usual, for the sake of comparison, we will take $E_o = \mathcal{E}_o$.

From the known behaviour of the Bessel functions I_N and J_N some properties of $\chi(N,\beta)$ are readily established : 1) it is an oscillating function of β; 2) it is an even function of β ; 3) conservation of energy implies that $\beta = \beta(N)$ through k_f. Hence, although apparently $\chi(N,\beta) = \chi(-N,\beta)$, actually we will have $\chi(N,\beta(N)) \neq \bar{\chi}(-N,\beta(-N))$, the channels N and -N being enhanced in a different way.

The behaviour of χ in various regions of interest is also readily obtained, namely :

a) $$\chi(N,\beta) \approx N! \tag{14 a}$$

for $\beta \ll 1$ (weak fields) and arbitrary N;

b) $$\chi(N,\beta) \approx \sqrt{\pi}/2 \cos^2(\beta + N\pi/2 - \pi/4) \tag{14 b}$$

for $\beta \gg 1$ (strong fields) and $N^2 \ll \beta^2$. We remark that eq. (14 b) yields $\sqrt{\pi}$ if we average over the fast oscillations.

The result (14 a) is well known from the related area of multiphoton photoexcitation and photoionization process[1,11] while the result $\chi \approx \sqrt{\pi}$, arising from eq. (14 b), has been obtained in refs 2 and 11. These results suggest an N! behaviour for weak fields of the enhancement factor, and a strong reduction of it for strong fields, although it remains larger than one. However, the following limiting expressions of $\bar{\chi}$, and its numerical evaluation, show that the above conclusions are not generally true and that in particular χ may be smaller than unity.

c) $$\chi(N,\beta) \approx N! \tag{14 c}$$

for $\beta \gg 1$ (strong fields) and $N^2 \gg \beta^2$. Eq. (14 c) shows the N! behaviour also in the strong field limit, but in multiphoton processes;

d) $\chi(N,\beta) \approx \sqrt{\pi}\, e^{-N^2/\beta^2} (1 - N^2/\beta^2)^{1/2}$ (14 d)

for $\beta \gtrsim N$ and $\beta, N \gg 1$. Eq. (14 d) is a fastly decreasing function of the ratio N/β, and at $\beta \approx N$ may be much smaller than unity, as shown by the next limiting expression

e) $\chi(N,\beta) \approx 2(\Gamma(1/3))^{-2}\, e^{-N^2/\beta^2} (6/\beta)^{1/3}$ (14 e)

for $\beta \approx N$ and $\beta, N \gg 1$. We remark that the minimum of χ, i.e. $N \approx \beta$, is actually an implicit condition on N because $\beta = \beta(N)$. Thus, fixing all the parameters of the process ($\underset{\sim}{E}_o, \omega, \underset{\sim}{k}_f, \underset{\sim}{k}_i$ and the atomic transition), we find a particular N for which $\beta \approx N$ and whose channel (and the closest ones) suffers the strongest reduction in the chaotic field. This property is illustrated in Table 1, in which we present calculations for the case of electron scattering by hydrogen atoms and in Figure 2, where we show the dependence of χ on N and β. It is worth noting that the relevant parameter for the behaviour of χ is β and not only E_o. The minimum of χ

TABLE 1. Values of the enhancement factor χ. Caption as in Fig. 1, but left table for $E_o = 5.10^7$ V/cm, right table for $E_o = 10^6$ V/cm.

N	β (N)	χ	N	β (N)	χ
0	4.39	1.10	0	0.88(-1)	1.00
1	4.45	2.60	1	0.89(-1)	1.00
2	4.50	2.20	2	0.90(-1)	1.99
3	4.56	0.45	3	0.91(-1)	5.98
4	4.62	0.43	4	0.92(-1)	23.92
5	4.68	0.79	5	0.93(-1)	119.58

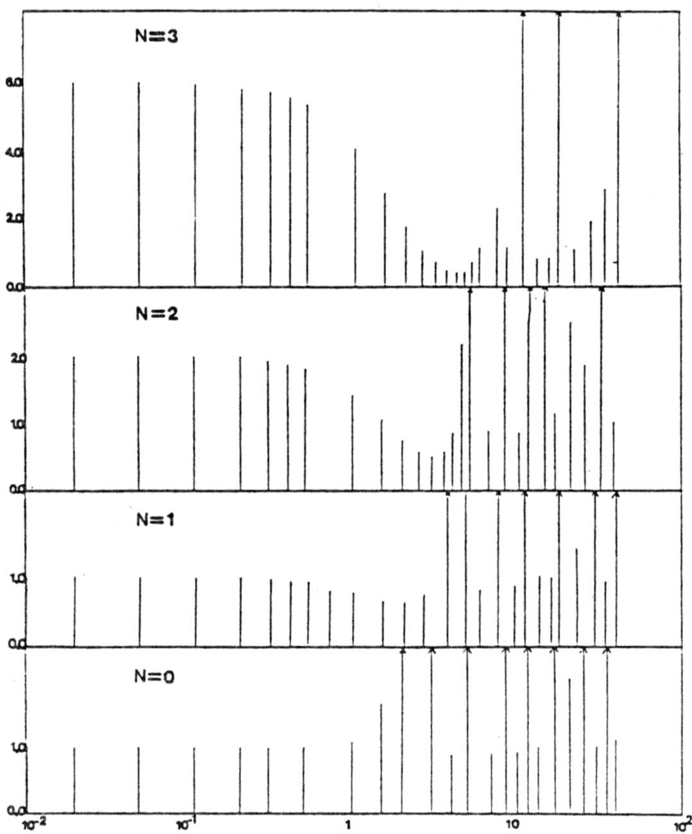

FIGURE 2. The enhancement factor χ vs β for N = 0,1,2 and 3.

in the region $N \approx \beta$ is explained by the behaviour of the Bessel functions; in fact, $J_N(\beta)$ has its maximum value when $N \approx \beta$, so that, fixing the amplitude of the field in the homogeneous case, some particular multiphoton processes will be the most probable. On the other hand, $\exp(-\beta^2/2) I_N(\beta^2/2)$ is a smooth function, because in the chaotic field all the amplitudes are simultaneously present. Thus, it is expected that the multiphoton processes for which $N \approx \beta$ will be the most reduced by the chaotic field. We observe that the informations reported here about the enhancement factor are also supported by results obtained in ref 12 in a different context.

ACKNOWLEDGEMENTS

This work was supported in part by the Italian Ministry of Education, the National Group of Structure of Matter and the Sicilian Regional Committee for Nuclear and Structure of Matter Researches.

REFERENCES

1. See for example : P. Lambropoulos and P. Zoller, in Invited Papers of the 2d Int. Conf. on Multiphoton Processes, edited by M. Janossy and S. Varro (Central Research Institute for Physics, Budapest, 1980), p. 193.
2. P. Zoller, J. Phys. B : At. Mol. Phys. 13, L 249 (1980).
3. R. Daniele and G. Ferrante, J. Phys. B : At. Mol. Phys. 14, L 635 (1981).
4. G. Ferrante, C. Leone and F. Trombetta, J. Phys. B : At. Mol. Phys. 15, L 475 (1982).
5. G. Ferrante and F. Trombetta, in Proceedings of XI Nat. Conf. On Nonlinear and Coherent Optics, vol. I (USSR Academy of Sciences : Yerevan 1982), p. 359.
6. E.L. Beilin and B.A. Zon, Kvant. Elektronika 9, 1692 (1982), in Russian.
7. R. Daniele, F.H.M. Faisal and G. Ferrante, J. Phys. B : At. Mol. Phys. (in press).
8. H.W. Lee, P.L. De Vries and T.F. George, J. Chem. Phys. 69, 2596 (1978).
9. F. Trombetta, C.J. Joachain and G. Ferrante, in Proceedings of the XIII ICPEAC (Berlin, 1983).
10. B.R. Mollow, Phys. Rev. 175, 1555 (1968).
11. V.A. Kowarskii, N.F. Perel'man, I. Sh. Averbukh, S.A. Baranov and S.S. Todirashku, Non-adiabatic Transitions in Strong Electromagnetic Fields, (in Russian), Kishinev, Shiintsa, 1980, p. 129.
12. A.T. Georges and P. Lambropoulos, Phys. Rev. A 20, 991 (1979).

TRANSITIONS BETWEEN RESONANCES: SOME PROPOSALS FOR EXPERIMENTS

R. STEPHEN BERRY
Dept. of Chemistry and the James Franck Institute, The University of Chicago, Chicago, Illinois 60637, U.S.A.

Abstract Several classes of scattering resonances are especially good candidates for experimental study of optical transition between resonances. The three sorts considered in some detail are: vibrational transitions in electron-diatomic, particularly homonuclear diatomic, molecules; electronic transitions in electron-diatomic molecule resonances, particularly with regard to the transient, vibrationally-bound HCl^- that might be useful as a short-wavelength laser; and electronic transitions between the low-lying electron-rare gas (Ne, Ar, Kr and Xe) resonances, as a means to probe the possible validity of the collective, molecule-like picture for these resonances.

INTRODUCTION

Radiation-induced processes in collisions can be looked upon as events that catch a system for almost an instant, at some nearly specific interaction distance. This concept is useful whatever the duration of the collision, and has been used successfully for predicting and interpreting phenomena simultaneously involving two collision partners and electronic radiation. If the collision partners remain close for a time long enough to give an identity to the complex - as, for example, when a collision complex has time to rotate or vibrate a few hundred periods - then a different way of viewing the radiation-induced process may sometimes be more useful. In this latter interpretation, the radiation-induced process is conceived as an optical transition between resonances or quasi-bound states. Let us classify some of the relevant kinds of resonances, and then select three that would be particularly interesting for experimental study by electromagnetic interactions. Each sort of potential experiment has its own special motivation. In the case of transitions between vibrational states of the H_2^- or N_2^- transient ion, it is the possibility of using an electric-dipole-allowed transition in a homonuclear diatomic ion to study the properties of those very important transient species. The electronic transition between a vibrationally bound state of HCl^- and the dissociative lower state of $H + Cl^-$ would be important because of the possibility of obtaining light amplification at about 155nm. In the case of transition among resonances of Ne^- and Ar^-, it is the determination of oscillator strengths and selection rules, and the inference of the degree of molecule-like electron correlation that is the challenge. Other processes appearing in the classification, particularly between resonances in heavy-particle collisions, may be equally interesting candidates but will not be discussed here.

CLASSIFICATIONS AND EXAMPLES

In order of complexity, the first sort of compound-state resonance in our list is the electron-atom scattering resonance. Examples are the e-H transient H^- negative-ion states, the e-He states such as the "$1s2s^2$" and "1s2s2p" transient states, and the corresponding transient states of Ne^- and the higher rare gases. These, which will get the greatest attention here, will be the last to be discussed. The cases of interest here are all Feshbach resonances, in which a specific kind of excitation of the target is recognizable.

The second kind of resonance is also an electron scattering resonance, in this case associated with transient deposition of energy from the projectile electron into vibrational excitation of the target. The classic examples of this phenomenon are the transient H_2^- and N_2^- species in which the long lifetime of the resonant compound system comes from the sharing of the electron's energy with the H-H or N-N vibrations.

It is useful to distinguish here between resonances associated with vibrationally excited states of the target and electronically excited states of the target molecules. The latter are Feshbach resonances that owe their long life to sharing of the projectile electron's energy with electronic degrees of freedom. The example that will be discussed here is the transient HCl^- formed by collisions of HCl with electrons having about 8.8eV of energy.

Some resonances are not easily identified with single excited states of the target molecules, but rather with a broad superposition of target states, to form a well-defined compound state. Electrons trapped temporarily by polarization fields or electric dipole moments, as with the transient, dissociating HF^- or $NaCl^-$ molecules, may be representative of this class called shape resonances.

The final class to be listed here are resonances in heavy-particle scattering. Examples that may be observable and possible candidates for radiation-induced processes are $F{\cdot}H_2$, $Na{\cdot}O_2$ and $Na{\cdot}SO_2$, either by induction of enough anisotropy to make the vibrational transition $\Delta V=1$ observable in H_2 and O_2, or by measurable shifts in the electronic spectrum of either of the partners.

MOTIVATIONS FOR EXPERIMENTS

Vibrational Transitions in N_2^-, CO^- and H_2^-

The first possible experiment we shall consider here is the vibrational excitation of H_2^-, CO^- or N_2^- by infrared radiation. The scattering of slow electrons from these molecules[4,5] shows clear evidence of resonances in both elastic and inelastic scattering. The inelastic scattering shows that the parent molecule can be left in any of several excited vibrational states, e.g.v' up to 4 for e-H_2 scattering. Moreover the inelastic e-N_2 scattering cross sections for any one vibrational channel shows resolved peaks, as a function of collision energy, that can be attributed to individual vibrational states of the transient N_2^-. Hence the vibrational frequency of the transient N_2^- is known, and is approximately $1600cm^{-1}$. The peaks for CO^- are distinguishable but are not well resolved and those of H^- are completely blended into a single broad resonance, about 3eV wide.

Part of the interest in trying to observe vibrational transitions in these species is simply the possibility of observing an electric dipole transition in a <u>homonuclear</u> diatomic molecule whose vibrational transitions are electric-dipole allowed. This is because of the presence of the electric charge of the extra electron. The symmetry condition that requires the electric dipole moment of N_2 or any other homonuclear diatomic molecule to vanish rests not only on the spatial symmetry of the molecule but also on the vanishing of all moments lower than the dipole moment, in this case, only the monopole moment or net charge. The condition that the dipole moment vanishes on grounds of symmetry is inapplicable if the molecule has a monopole moment. (Recall that if the 2^n-pole moment is nonvanishing, then the value of any higher moment depends on the choice of origin of the coordinate system.) More relevant here is that the derivative of the electric dipole moment of a homonuclear diatomic ion, with respect to internuclear distance, is nonvanishing, so that such ions may have infrared transitions. Hence the transitions $\Delta v = \pm 1$, especially, of N_2^- are observable in the infrared at a bit above 1600 cm^{-1}, and with widths of about 1200cm^{-1}. Clearly it is natural to first study vibrational transitions in stable ions of homonuclear diatomics such as O_2^-, the diatomic halide ions such as Cl_2^-, and positive ions such as N_2^+ or O_2^+. These would suffice to demonstrate the allowed nature of the transitions. One would expect the intensities of the infrared transitions in positive ions to be proportional simply to the square of the mean change in internuclear distance from one vibrational state to the next, i.e. to the quantum number itself. One would expect a negative molecular ion, with its weakly bound extra electron, to be far more polarizable than a positive molecular ion, with a hole in its valence shell. If this supposition is correct, and if the extra electron is able to exert enough force on the internuclear vector to couple its motion to that of the electrons, then vibrational transitions of O_2^-, Cl_2^- and other stable homonuclear diatomic negative ions should be considerably more intense than the corresponding transitions of the positive ions, roughly in the ratio of the square of the polarizabilities of the negative and positive ions. This relation should hold approximately for those cases in which the hole of the positive ion occupies an orbital with the same bonding power as the extra electron of the negative ion. This is the case for O_2^+, and O_2^-, where the hole of O_2^+ and the extra electron of O_2^- are both in the $\pi_g 2p$ antibonding orbital. The O_2^+ ion is more tightly bound than is O_2, while the O_2^- ion is more weakly bound than O_2.

For N_2^+ and N_2^-, the situation is just a little different. The hole of N_2^+ is in the $\pi_u 2p$ bonding orbital, which is, in magnitude, roughly the same in bonding power as the $\pi_g 2p$ antibonding orbital of the lowest N_2^- resonance, but opposite in character. Both the hole in N_2^+ and the extra electron of N_2^- tend to increase the length of the N-N bond with respect to that of N_2, and with about the same force.

The responses of the homonuclear diatomic resonances to resonant radiation can be expected to be somewhat different from those of the stable homonuclear ions. If O_2^- or O_2^+ is subjected to radiation in the region of the vibrational frequency, we can expect to see a conventional vibration-rotation band system corresponding to the usual propensity rules of an unharmonic oscillator. The homonuclear diatomic resonances can be expected to show something quite different: first, they are capable of losing their electrons right out to infinity, so their static polarizabilities are infinite. Second, their rotation-vibration transitions are broad so that they will respond to any radiation in the frequency range of their vibrational transitions, probably even radiation corresponding to large values of Δv. Consequently we might expect the N_2^-, CO^- and H_2^- transient molecule-ions to exhibit elastic and inelastic scattering of radiation, in a frequency band about their fundamental vibrational frequencies about as wide as their electron resonance scattering envelopes, with total oscillator strengths of order unity, perhaps an order of magnitude less. That is, the quasibound electrons of N_2^-, CO^- and H_2^- should couple extremely effectively with a radiation field of any frequency in the range in which the electron's motion can also couple through its bonding force to the nuclei. This phenomenon differs from conventional photodetachment because in photodetachment, the electron's final state is a true continuum state so the electron escapes immediately upon reaching the continuum and can no longer couple with the radiation field.

The experiments suggested by this rather qualitative analysis are simply the vibrational absorption and fluorescence spectroscopy of stable positive and negative ions and the light scattering, elastic and inelastic, by the negative ion resonances of the homonuclear N_2^- and H_2^-, and the closely related CO^-. The natural frequency range for N_2^- would be 1600 to about 16,000 cm^{-1}, the same approximately for CO^- and about 2000 to 40,000 cm^{-1} for the much more short-lived and therefore much less accessible H_2^-.

ELECTRON-MOLECULE COMPOUND, ELECTRONICALLY-EXCITED RESONANCES: HCl^-

Electron scattering from HCl shows resonance-like peaks[6], one in elastic scattering with a maximum around an electron energy of 8.8eV, probably associated with one of the peaks in the H^- yield from dissociative attachment which occurs at 9.3eV. This resonance probably corresponds to the capture of an electron into a $d\pi$ or π_g orbital, giving a short lived $HCl^-(^2\Pi_g)$ state with several vibrationally bound levels. The relation between the potential curve of this state and those of the state correlating with H + Cl^- and of the ground state of HCl are shown in Figure 1. The relevant point here is that the vibrationally-bound

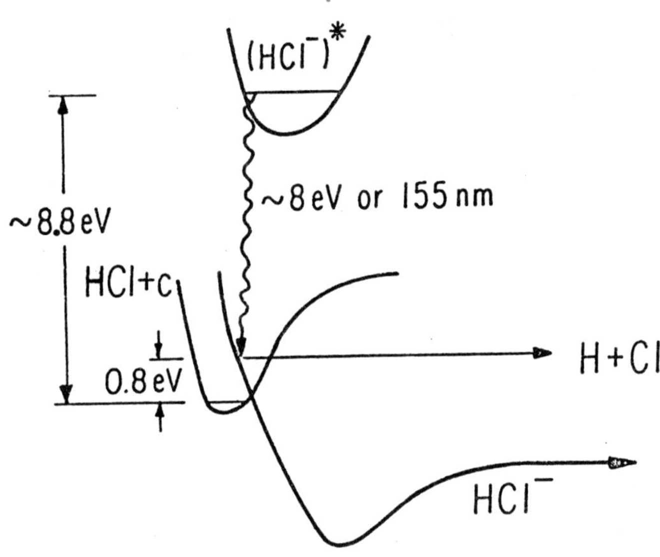

FIGURE 1. Potential curves and transitions corresponding to the possible stimulated emission from the upper HCl^- resonance.

electron scattering resonance, particularly if created with a rich source of electrons such as those from the photoeffect on n-doped Si,[7,8] can be populated and that dissociative resonance, optically connected to the upper state, lies below the vibrationally bound resonance. Because dissociative states can never maintain populations, the excitation of the upper resonance automatically constitutes the creation of an inverted population. Consequently we can imagine that light in the appropriate frequency range for favorable Franck-Condon factors might be amplified by passage through HCl vapor in which the 8.8ev resonance has been populated.

The fact that H^- has been detected from the upper resonance demonstrates that its lifetime toward electron loss is considerably longer than the time required for dissociation, ca. 10^{-13} sec. A current of 10μa for 10nsec into a target thick with respect to excitation of the resonance will produce about 6×10^{14} excited, transient HCl^- ions. The frequency emitted in the transition from the upper state of HCl^- to the lower, as shown in Fig. 1, is about 15.5 nm, corresponding to about 0.8 mJ per 10ns pulse or 80 watts of power during a flat 10ns pulse, at that short wavelength, if all the energy is drained by stimulated emission.

It appears that HCl is a better candidate for this process than is HF because of the strong dominance of the dissociative channel that gives $e + HF \rightarrow H + F^-$. The HI molecule may be a suitable alternative to HCl, perhaps through excitation of the channel that gives $H^- + I(\,^2P_{1/2})$ The maximum cross section for production of H^- in this channel is at about 6.5 eV; the best collision energy for production of the quasibound HI^- excited species lies a bit lower, below the dissociation limit of the appropriate state.[9]

TRANSITIONS BETWEEN NEGATIVE-ION RESONANCES OF Ne, Ar, Kr, AND Xe

Langendam and Van der Wiel[10] reported optical transitions between electron scattering resonances of neon, monitored by the emission at 73.6 and 74.3nm from the upper states of the transitions. Subsequently Van der Wiel and Read have raised some questions about whether artifacts might have been responsible for the emission. Without making any exploration into the interpretation of possible artifacts, we wish to point here to the considerable importance of the experiment that Langendam and Van der Wiel set out to do, and to the need for new experiments to study precisely the process they so wisely chose to investigate. Specifically, what is needed are independent investigations of the optical transitions between the electron scattering resonances of the rare gases, particularly those one might characterize by the configurational designations $(\ldots ns^2np^5(n+1)\ell(n+1)\ell')$ - that is, scattering resonances that derive their long lifetimes because the projectile electron shares its incoming energy by exciting a bound electron from the valence shell to the next empty shell and, in so doing, drops into that same shell of excited orbitals.

The background that explains the current significance of the Langendam-Van der Wiel experiment comes from two streams. One is that set of experiments done by Read and his collaborators[11-15] to identify and classify the low-lying electron scattering resonances of Ne, Ar, Kr and Xe, and the related theoretical work to interpret these resonances.[16,17] Extensive earlier work demonstrated the existance of the resonances, and these are cited in Ref. 12; here our interest is in the high-resolution results and the assignments of the best-resolved structure in the cross sections. The essential consequence of this work is that many of the resolved resonances can be classified according to a coupling scheme in which the two "outer" electrons couple strongly to one another as do the electrons in doubly-excited helium or other two-electron systems, while all the other electrons in normally filled shells couple in the manner usual for the rare gas singly-positive ions. Then these cores, in $^2P_{3/2}$ or excited $^2P_{1/2}$ states interact weakly with the Russell-Saunders-coupled outer electrons. The most apparent symptom of this coupling is the repeated appearance of the characteristic doublet splitting intervals of the cores, among the scattering resonances. This was probably the first indication, to Read, Brunt and King, that the outer two electrons are coupled much more strongly to one another than either is to the core. The rest of the case comes from comparisons of the patterns of energy levels of the allocative earth atoms, suitably scaled, to the patterns of resolved resonances of the "Ne^-", "Ar^-" and "Kr^-" transients. As we shall see, the very pattern of the energy levels and their assignments in all these systems becomes an important clue, suggestive of the possibilities that make the Langendam-Van der Wiel experiment especially important. The pattern of energy levels as laid out by Read, Brunt and King is shown in Fig. 2.

The second stream of ideas that makes the electron-rare gas resonances crucial is the interpretation of the resonances - the doubly-excited, autoionizing quasibound states - of the neutral helium atom. Many of these resonances were computed quite accurately and several were observed and assigned, for He itself and for H^- and other He-like ions. They could be related to single configurations or to rather simple and specific combinations of configurations. However, it was a major step toward understanding the physical nature of these resonances when

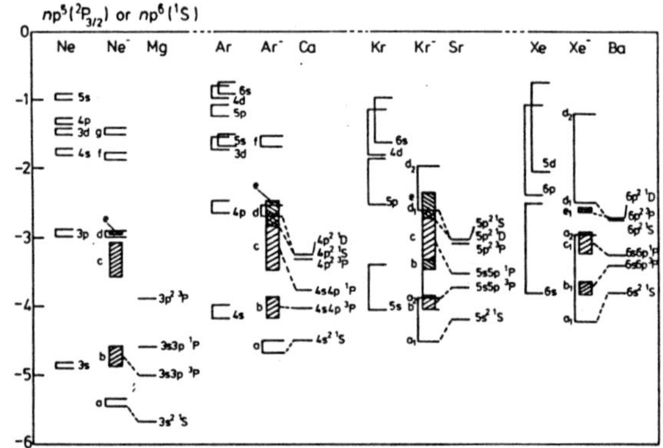

FIGURE 2. The pattern of energies of resonances in e-rare gas collisions, as adapted from Read, Brunt and King.[12]

Kellman and Herrick[18] first pointed out that the "intra-shell" resonances, in which the dominant configurations have the same principal quantum numbers n_1 and n_2 for the two electrons, contain finite multiplet series that conform remarkably well to rotor series. This was followed by three other papers that showed how the rotor series were part of supermultiplets,[19,20] and finally how the particular supermultiplets to which the He-like series belonged were precisely those of a linear three-body system with collective, near-rigid-body rotations and bending vibrations.[21] The Kellman-Herrick proposal was confirmed by examination of the simultaneous spatial distributions of the two electrons in a representation especially appropriate for probing the degree of collective vibration of independent-particle motion.[22,23] This representation uses a kind of internal axis system: one plots the conditional probability distribution for finding electron 2 at distance r_2 from the nucleus and at angle θ_{12} with respect to the axis defined by the position vector $\vec{r}_1$ of electron 1, when electron 1 is at distance r_1 from the nucleus. [The representation of probability density in terms of arctan (r_1/r_2) and θ_{12}, for fixed $(r_1^2+r_2^2)$, is a related form, particularly if the range of the arctangent is restricted to the nonredundant range of 0 to $\pi/2$. However the molecule-like characteristics are clearer if the real distance r_1 is fixed at a series of values, rather than if the dilatation variable $(r_1^2+r_2^2)$ is fixed.] Some examples of these conditional probability distributions are given in Figures 3 and 4.

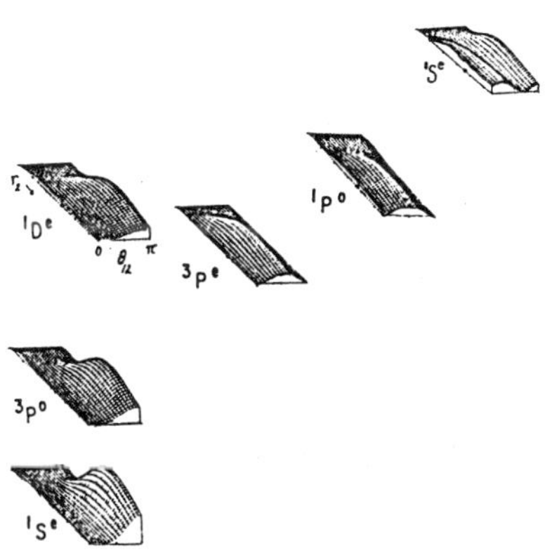

FIGURE 3. Some conditional probability distributions for doubly-excited states of He, with $n_1=n_2=2$. One axis is that of the distance r_2; the other is the relative angle θ_{12}. The distances r_1 are indicated by the heavy dot on the axis of $\theta_{12}=0$; these distances are chosen to be the most probable. The similarities

of the $^1S^e$, $^3P^0$ and $^1D^e$ rotor series and of the $^1P^0$ and $^3P^e$ "bending" states are evidence for the validity of the molecular model.

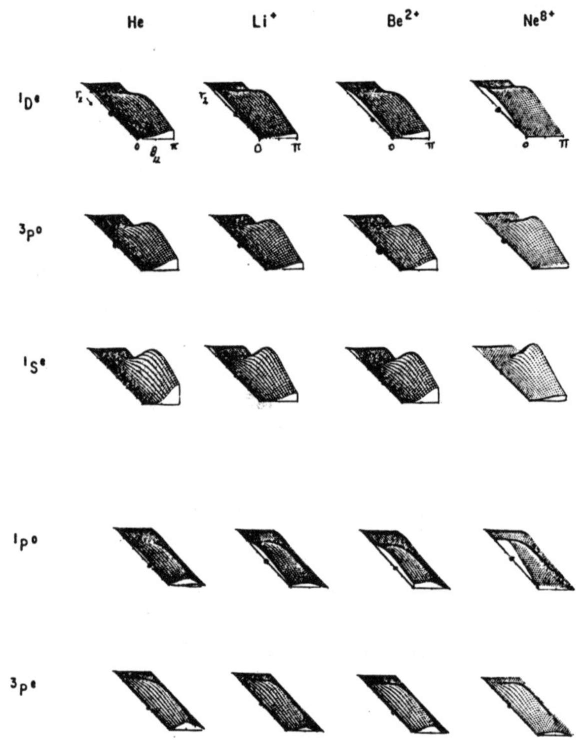

FIGURE 4. Conditional probability distributions for several doubly excited states of He and He-like ions, all built on states with $n_1=n_2=2$. With higher Z, the collective, molecular picture transforms into the independent electron picture. This is especially clear with the "$2p^2$" $^1D^e$ state.

TRANSITIONS BETWEEN RESONANCES

The great importance of the recognition of collective, molecule-like behavior in doubly-excited helium and other helium-like ions is the implications of such correlation regarding the relationships among excited states and the transferability of all our understanding of the spectra and structure of triatomic molecules to these atoms. Consider first the notion that the states with $n_1=n_2=2$, "$2s^2$" $^1S^e$, "2s2p"$^3P^0$ and "$2p^2$"$^1D^e$ form a truncated rotor series. If this is correct, the three functions should give very similar probability distributions in the r_1, r_2, θ_{12} space, differing only by small effects of centrifugal distortion; the principal differences among them should be only in their rotational wave functions which ought to be essentially separable rotor functions of the Euler angles that specify the orientation of the three-particle system in laboratory coordinates. Similarly, if the molecular picture is correct, the states corresponding to excited states of the bending vibration should occur in nearly degenerate pairs having very similar spatial distributions in the internal axis system, and the angular behavior of these states should correspond in form to successively higher excitations of a bending vibration. That is, the first excited partners, the "2s2p"$^1P^0$ and "$2p^2$"$^3P^e$ states should have similar spatial distributions, with a node at $\theta_{12}=\pi$ when r_1 and r_2 are equal, at or near their most probable values. By contrast, in the independent particle picture, the "$2s^2$"$^1S^e$, "2s2p"$^3P^0$and "$2p^2$"$^1D^e$ states have no reason to have similar spatial distributions, nor do the "2s2p"$^1P^0$ and "$2p^2$"$^3P^e$, which have different orbital configuration, spin and parity. In the one complete intrashell set of states that have been examined for these properties,[23] it is clear that well-converged wave functions do exhibit the similarities that the molecular picture implies. The same holds for resonances of H^- and Li^+, but for ions with high enough nuclear charge, the kinetic energy of the electrons and the electron-nuclear attraction overcomes the correlating influence of the electron-electron repulsion, and many states of Ne^{+8}, for example, are much more independent-particle-like than collective-molecular in their behavior. In Fig. 4, one can see how the conditional probability distribution for the "$2p^2$"$^1D^e$ is almost symmetric about $\theta_{12}=\pi/2$, with a form almost like the square of a single spherical harmonic. Even the antisymmetric stretching mode of doubly-excited He can be identified: its first excited state is the "2s3s"$^3S^e$ state, with a node at $r_2=r_1$ when r_1 is at its most probable value, a maximum at $\theta_{12}=\pi$ and small r_2 when r_1 is greater than its most probable value and a maximum at $\theta_{12}=\pi$ and large r_2 when r_1 is smaller than its most probable value.

Now we draw the two strands together. The patterns of the energies of the resonances of electrons scattering from Ne, Kr and Ar have been interpreted by Read, Brunt and King as due to two electrons outside the valence shell, somewhat analogous to the doubly-excited states of He. Many of these resonances were assigned by Read, Brunt and King; theoretical analyses of e + Ne by Clark and Taylor[16] and by Ojha, Burke and Taylor[17] have confirmed some of these assignments and have suggested some reassignments among the closely-spaced peaks. It is an intriguing conjecture, that these resonant states might correspond to collective, molecule-like behavior; this would be a natural extension of the analogy to doubly-excited helium. It would require some reassignments among peaks, and naturally some experimental tests, which might well include optical resonance-resonance transitions.

The observed resonances obtained by Read, Buckman and King[15] are shown in Fig. 5, in which the signal is the current of metastable neutral neon atoms arriving at the detector. The assignments of these and the lowest

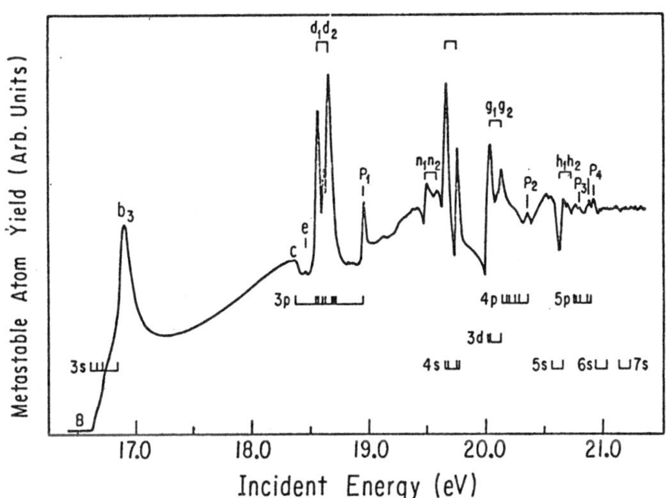

FIGURE 5. Electron scattering resonances of Ne, as found by Read, Buckman and King.[15]

resonance (not shown in the figure) as made by Read, Brunt and King (RBK) and by Ojha, Burke and Taylor (OBT) are as follows:

a: $3p^5(4s^2\ ^1S^e)$- both RBK and OBT;

b: $3p^5(4s\ 4p\ ^3P^0)$- both RBK and OBT;

c: $3p^5(4s4p^1P^0)$ - RBK; $3p^5(4s3d^1D$ and $4p^{21}D)$ - OBT;

d: $3p^5(4p^2\ ^1S^e)$ - both RBK and OBT;

ĕ: $3p^5(4p^2\ ^1D)$ - RBK; $3p^5(4s4p\ \ ^1P^0)$ and
$3p^54p\ ^3S^e4s$ - OBT

f: $3p^5(5p^2\ ^1S)$ - RBK

Here we propose a somewhat different possible set of assignments that would be likely if the collective, molecule-like picture is a good first approximation for the behavior of the rare gas resonances. That is, if it is collective, molecule-like dynamics that underlies the coupling characterized by Read, Brunt and King as jLS coupling, and gives rise to the "grandparent" model as the model just described is called, then the intervals and state labels should correspond to the pattern of a near-rigid rotor with a bending vibration and hence to a set of levels $E_L \sim L(L+1)$ superposed on a set of equally-spaced levels $E_v \sim v$, with the (2L+1)-degeneracy necessary for the rotation levels, a very appropriate degeneracy $g_v \sim v+1$ corresponding to the strictly harmonic, 2-dimensional oscillator, which breaks down to nearly degenerate pairs for odd v and to pairs plus a singlet for even v.

This assignment is shown in Fig. 6a, with the positions of the observed resonances marked on the left ordinate. Predicted but unobserved resonances are shown with dots. The leftmost column of assignments corresponds to the rotor series; the spacing of the a and b resonances is almost precisely half that of the b and d resonances, and very close to one third of that of the d and p_2,p_3 resonances. Hence we suggest the possibility that the d resonances correspond to $3p^5(4s4d + 4p^2, ^1D^e)$ and that the p_2,p_3 resonances correspond to $^3F^0$ coupling of the outer two electrons, by a mixture of 4s4f, 4p4d

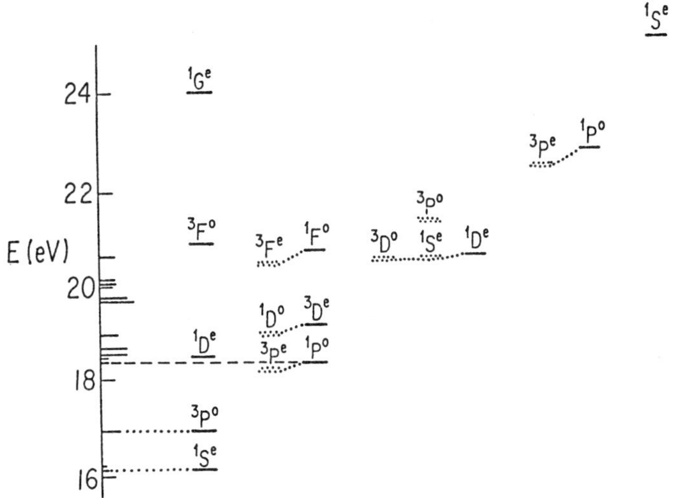

FIGURE 6a. Energy levels as observed (left axis) and the tentative assignments, according to the collective molecular model.

and perhaps higher configurations, and that a $^1G^e$ resonance, a $^3H^0$ and a $^1K^e$ resonance, corresponding to L=4,5 and 6, also might be found at the corresponding higher levels. Based on the spacing of the a and b resonances, the rotational constant for this series is slightly greater than 0.3 eV, and the spacings from level (L-1) to level L are 2×0.3 ×L. The broad c and small narrow e resonances would correspond to the $^1P^0$ and $^3P^e$ states of the first excitation of the bending vibration. The more likely assignment of these two would attribute the broad, strong c resonances to the $^1P^0$ state based in significant part on the $3p^5(4s4p\ ^1P^0)$ configuration and the e, to a state containing a good deal of $3p^5(4p^2\ {}^3p^e)$. This would put these states in the order found for molecules and for particles on a sphere,[24] but in the reverse of the order for the corresponding states of He. Then the p_1 resonance would correspond to one D-component of the pair with one quantum of bending vibration and one quantum of rigid-body rotation, in the separable limit. The more likely assignment would make this the ${}^3D^e$, but this would again be in reverse order to the corresponding helium states. The successively higher states are indicated in Fig. 6a.

The optically allowed intrashell transitions based on the collective-molecular scheme are indicated in Fig. 6b. Several of these, notably the lowest ${}^1S^e \rightarrow {}^1P^0$

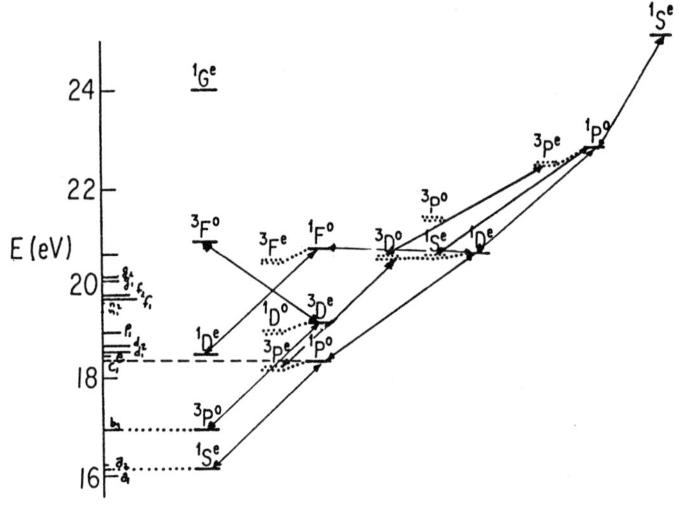

FIGURE 6b. The same levels, with optically allowed transitions that would be detectable as resonance-resonance transitions.

$^3P^0 \rightarrow {}^3D^e \rightarrow {}^2F^0$ and $^1P^0 \rightarrow {}^1D^e$, lie in the easily-accessible 500-600 nm range. The assignments of the c and d resonances, in particular, could be checked by optical excitation of the a resonance; if the assignment of Read, Brunt and King and that presented here is correct, the $^1S^e \rightarrow {}^1P^0$ transition should be broad and centered at about 551 nm; if the assignment of the c and d resonances is as made by Ojha, Burke and Taylor, the a resonance sould give no absorption in the 500-550 nm region. The assignment of the d resonances to the $^1D^e$ rotor level can be checked by looking for an optical transition between that resonance and the $^1F^0$ level at about 571.3 nm. The c resonance is probably too broad to give rise to optical resonance-resonance transitions; this is unfortunate because its assignment as a $^1P^0$ state could otherwise be confirmed by a $^1P^0 \rightarrow {}^1D^e$ transition at about 539 nm.

To conclude this section, we summarize by saying that there is enough suggestion of possible collective-molecular character to the low-lying resonances of e + Ne, Ar, Kr and Xe to make this possibility important to check experimentally. (It will no doubt be studied theoretically as well.) This picture, if strictly applied, implies some assignments of the resonances different from those made previously. Consequently it becomes important to check the assignments of the resonances by experimental means as powerful and unambiguous as possible. One such means is the investigation of optical transitions between the resonances, which are governed by electric dipole selection rules. Some of the specific predictions of the collective-molecular model and of the previous assignments are made explicitly.

Space limitations prevent the fuller exploration here of the predictions of the collective-molecular model.

ACKNOWLEDGMENT. This research was supported by a Grant from the National Science Foundation. The author would like to thank Drs. R. Abouaf and R. Azria for helpful discussions regarding electron-hydrogen halide resonances.

REFERENCES

1. R. Azria, Invited Papers of the XII International Conference on the Physics of Electronic and Atomic Collisions, Gatlinburg, Tenn., July, 1981, S. Datz, ed. (North-Holland, Amsterdam, 1982).
2. R. Abouaf, J.P. Gauyacq, D. Teillet-Billy and R. Azria, J. Phys. B(to be published, 1983).
3. J.P. Ziesel, R. Azria and D. Teillet-Billy, Phys. Rev. A (to be published, 1983).
4. H. Erhardt and K. Willmann, Z. Phys. 204, 462 (1967).
5. H. Erhardt, L. Langhans,F. Linder and H.S. Taylor, Phys. Rev. 173, 222 (1968).
6. R. Azria, Y. LeCoat, D. Simon and M. Tronc, J. Phys. B 13, 1909 (1980).
7. D.T. Pierce, R.J. Celotta, G.-C. Wang, W.N. Unertl, A. Galejs, C.E. Kuyatt and S.R. Mielczarek, Revs. Sci. Inst. 51, 478 (1980).
8. R. Azria, private communication.
9. R. Azria, "Angular Distribution of Negative Ions from Electron Attachment Processes," (to be published).
10. P.J.K. Langendam and M.J. Van der Wiel, J. Phys. B 11, 3603 (1978).
11. J.N. Brunt, G.C. King and F.H. Read, J. Phys. B 9, 2195 (1976).
12. F.H. Read, J.N. Brunt and G.C. King, J. Phys. B 9, 2209 (1976).
13. J.N. Brunt, G.C. King and F.H. Read, J. Phys. B 10, 1289 (1977).
14. J.N. Brunt, G.C. King and F.H. Read, J. Phys. B 10, 3781 (1977).
15. F.H. Read (private communication; to be published).

16. C.W. Clark and K.T. Taylor, J. Phys. B 15 L213 (1982).
17. P.C. Ojha, P.G. Burke and K.T. Taylor, J. Phys. B 15, L507 (1982).
18. M.E. Kellman and D.R. Herrick, J. Phys. B. 11, L755 (1978).
19. D.R. Herrick and M.E. Kellman, Phys. Rev. A 21, 418 (1980).
20. D.R. Herrick, M.E. Kellman and R.D. Poliak, Phys. Rev. A 22, 1517 (1980).
21. M.E. Kellman and D.R. Herrick, Phys. Rev. A 22, 1536 (1980).
22. H.-J. Yuh, G. Ezra, P. Rehmus and R.S. Berry, Phys. Rev. Lett. 47, 497 (1981).
23. G.S. Ezra and R.S. Berry, Phys. Rev. A 28, xxxx (1983).
24. G.S. Ezra and R.S. Berry, Phys. Rev. A 25, 1513 (1982).

ABOVE THRESHOLD IONIZATION
IN THE LOW INTENSITY LIMIT

G. PETITE, F. FABRE, P. AGOSTINI
Centre d'Etudes Nucléaires de Saclay, Service de Physique des Atomes et des Surfaces,
91191 Gif-sur-Yvette Cedex, France

Abstract Recent measurements of above-threshold ionization cross sections of Xenon and Cesium are reviewed and discussed within the framework of perturbation theory.

INTRODUCTION

Multiphoton ionization (MPI) of atoms and molecules has been investigated for almost two decades[1]. It is well known that an atom can be ionized by a strong monochromatic electromagnetic field even when the photon energy is much smaller than the ionization potential. The atom needs to absorb a minimum number of photons which is given by the integer N immediatly larger than the ratio $V_I/\hbar\omega$ of the ionization potential to the photon energy. The electron is therefore ejected with a kinetic energy given by the generalized photoeffect law

$$E_N = N\hbar\omega - V_I \qquad (1)$$

However, whenever the photoelectrons energy spectrum was recorded, kinetic energies much larger than allowed by (1) were observed [2,3]. There are two physical mechanisms which could possibly be invoked to explain this fact. The first one is the multiphoton inverse Bremsstrahlung[4], a special case of photon-assisted electron-atom collision[5,6]. Since this process implies a collision, it is important only when the atomic density is high or when the irradiation time is long enough. Under the typical MPI experimental conditions it was shown to be negligible[7]. The second one is a special case of MPI which had been pointed out in one of the very first theoretical MPI papers[8] but was experimentally demonstrated only recently[9,10]. It is now usually referred to as Above-Threshold Ionization (ATI). It is the process by which, an atom which could normally be ionized by absorbing N photons is ionized by absorbing (N+S) photons (i.e. S photons more than necessary). Quantum mechanically, it is a direct, (N+S)-photon transition between the ground state and the continuum. Clearly, it is a special case of MPI in which N "steps" are taken through the discrete part of the spectrum while S are taken through the continuum. Of course, if the (N+S)-order process is observable all the processes of lower order will, in general be observable too. In the "low" intensity limit this process will therefore give rise to electrons with kinetic energies given by:

$$E_{N+S} = (N+S)\hbar\omega - V_I \qquad (2)$$

The photoelectron spectrum will consist of the normal line of energy E_N plus a few extra-lines separated by the photon energy $\hbar\omega$. If one can further increase the intensity (without saturating the process), then many other interesting effects can appear : 1) the combined action of level shifts and ponderomotive force can induce drastic changes of the photoelectron spectrum and, at the same time the lowest-order perturbation theory, the usual theoretical tool for describing MPI, breaks down. These "high" intensity effects are demonstrated elsewhere in this book[11].2)Part of the energy of the (N+S) photons can be used to excite the ion and even to ionise it [12].

In this paper, ATI in the low intensity limit will be reviewed. After a brief account of the theoretical problems involved,we are presenting the most significant experimental results obtained in ATI of Xenon atoms at 0.53 μm : intensity dependence of the different lines in the spectra and the relative importance of the cross sections.

These results are found in qualitative agreement with the predictions of lowest order perturbation theory. The contribution of ATI to the total ion yield is also investigated in order to solve the following puzzle : in the lowest-order perturbation scheme the ATI rate $W_{N,S}$ is given by :

$$W_{N,S} = \sigma_{N,S}\, I^{N+S} \qquad (3)$$

where $\sigma_{N,S}$ is the generalized cross-section and I the intensity.

Therefore the total ionization rate is given by :

$$W = \sum_{S=0}^{\infty} W_{N,S} = \sum_{S=0}^{\infty} \sigma_{N,S}\, I^{N+S} \qquad (4)$$

On the other hand, older experiments, based on measurements of the ion yield, have shown that W depends on I according to :

$$W = \sigma_N \; I^N \tag{5}$$

with an excellent precision, leading to an apparent contradiction between (4) and (5).

Finally, for a complete understanding of the process, a quantitative comparison between theory and experiment is necessary. This comparison has been undertaken in the case of 5-photon ATI of Cesium and some results are reported here.

THEORY

Let us consider an atom coupled to the electromagnetic field through the dipole interaction. The transitions of interest here are those coupling the ground state to continuum states by absorption of N+S photons, where N is the minimum number of photons necessary to ionize the atom. The generalized cross-section of this process can be obtained via the standard perturbation theory and is given by :

$$\sigma_{N+S} = 2\pi \frac{\alpha c}{a_o} \left(\frac{1}{I_o} \right)^{N+S} \int_\Omega | M_{fg}^{(N+S)}(\omega)|^2 d\Omega \tag{6}$$

where $\alpha = 1/137$, $a_o = 0.5$ Å, $I_o = 1.4\ 10^{17}$ W.cm^{-2}, while $M_{fg}^{N+S}(\omega)$ is the (N+S)-order matrix element given by :

$$\sum\!\!\!\!\!\!\int \frac{\langle f|r^\lambda|i_{N-1}\rangle \; .... \; \langle i_1|r^\lambda|g\rangle}{(\omega_{i_{N+S-1}} - \omega_g - (N+S-1)\omega) ... (\omega_{i_1} - \omega_g - \omega)} \tag{7}$$

$|f\rangle$ is a continuum state, $|g\rangle$ is the ground state, $|i_1\rangle, ... |i_{N-1}\rangle$ intermediate state. ω_g, ω_i, ω are respectively the ground state, intermediate states and photon energies, with $r^\lambda = \vec{r}.\vec{\varepsilon}_\lambda$, the dipole interaction of the electron at position $\vec{r}$ and e.m. field of polarisation $\vec{\varepsilon}_\lambda$.

In (7) the sum goes over the discrete and continuum parts of the spectrum. The new difficulty, in calculating $M_{fg}^{N+S}(\omega)$, is to find a way to handle the (S-1) branchcuts of the function (7) lying in the continuum part of the spectrum This has been done by a number of authors[9, 14, 15, 16, 17] with a variety of techniques which will not described here.

The results of such a calculation for hydrogen are summarized in Table I, taken from ref.16. From Table I, it can be concluded that the probability depends roughly only on N+S, i.e. the probability of ATI is quite comparable to be probability of MPI of the same order. Here "comparable" means of the same order of magnitude. Of course, when comparing two processes involving two different wavelengths one must avoid the cases of intermediate resonances (or antiresonances[1]) which would give non-typical values.

TABLE I

S	N = 6	N = 8	N = 10	N = 12
0	1.39(-69)	1.49(-97)	4.51(-123)	3.46(-149)
1	2.84(-83)	9.85(-111)	7.78(-136)	9.81(-162)
2	2.92(-97)	2.53(-124)	5.35(-149)	1.10(-174)
3	2.80(-111)	5.84(-138)	2.61(-162)	1.08(-187)
4	2.66(-125)	1.35(-151)	1.89(-175)	9.87(-201)
5	2.32(-139)	2.75(-165)	1.04(-188)	8.91(-214)

Values of the quantity $W_{N+S}/I^{(N+S)}$, (the generalized cross-section with I in W/cm^2) for (N+S) ionization of the ground state of Hydrogen. (Ref.16). The numbers in parenthesis are the powers of 10.

An immediate consequence of the perturbative treatment is that the probability of the (N+S)-order ATI is proportional to I^{N+S}. This implies that, at some intensity, named threshold intensity[16], the probability of an (N+S)-order process would become equal to the probability of the (N+S-1)-order process if it were not for the depletion of the atom ground state (the so-called saturation of the transition). This problem has been handled in ref.18 where it is shown how the ratio of probabilities of consecutive orders, first increases linearly with intensity and then saturates at some intensity I_{sat} defined by the equation :

$$\int_{-\infty}^{t_{sat}} \sum_{S=0}^{\infty} W_{N+S}(I(t))\,dt = 1 \qquad (8)$$

which means that the saturation is reached at time t_{sat}:

the corresponding intensity $I(t_{sat})$ is I_{sat}. According to this calculation, if $I_{sat} < I_{th}$ (which is usually true) the maximum amplitudes of A.T.I components of the electron spectrum must be obtained at I_{sat}. In some cases, however, it seems that $I_{th} < I_{sat}$. (This should happen systematically for short enough pulse durations). In these cases, it is not expected that the perturbation calculation will remain reliable [11]. One has then to rely on more powerful techniques[18] like resolvent operator or non perturbative methods[19] which seem necessary to explain the results described in ref.11.

To conclude this part it can be said that, in the "low "-intensity limit, ATI is a "normal" MPI as far as the value of the probability and the intensity dependence are concerned. The maximum ATI will be observed at saturation.

ATI OF XENON AT 0.53 μm

Xenon can be ionized by absorbing 6 photons of the second harmonic of the YAG ($\hbar\omega$ = 2.34 eV). Due to the fine structure splitting of the ion ground state the electron spectrum is expected to show two series of lines with two characteristic intervals : namely the photon energy $\hbar\omega$ and the fine structure splitting 1.3 eV.

For this particular experiment[20] we have used a 20 ns, 10 pps, frequency doubled YAG laser with a peak intensity of about 10^{10}-10^{11} Wcm^{-2}. The electron analyzer (a standard hemispherical spectrometer) has been described elsewhere[8] together with details about the experimental set up. Fig.1 shows the energy spectrum obtained, displaying indeed the expected two series of lines labelled here (1/2, S) and (3/2, S) where S is the number of photons absorbed above the threshold and 1/2, 3/2 the possible angular momenta of the ion ground state.

Similar results have been obtained by P. Kruit and coworkers[11] in the case of 3-photon resonant, 5-photon ionization of Xenon. They have used a time of flight spectrometer which is described in this book[11].

The main problems related with the recording of such spectra are the ion space charge, the energy calibration and the transmission calibration. For details the reader is referred to original publication and to F.Fabre's thesis[21]. Here we will only state that :

i) the space charge problem can be solved either by an extracting electric field or by keeping the plasma Debye length, given by :

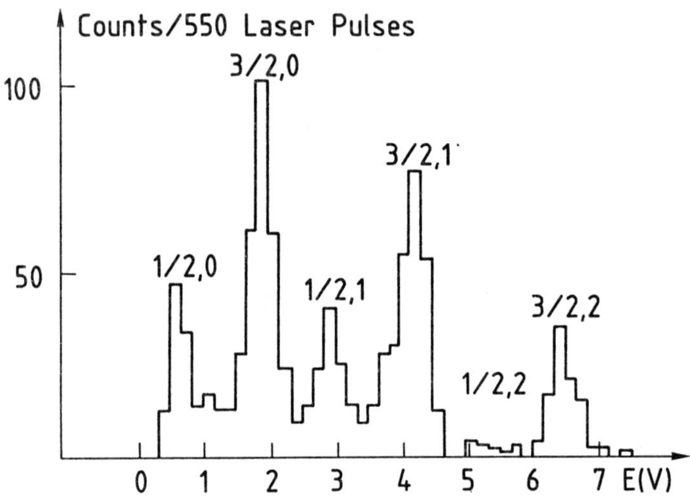

FIGURE 1. Photoelectron energy spectrum for Xenon at 0.53 μm.

$$\lambda_D(cm) = 600\sqrt{\frac{E\,(eV)}{n\,(cm^{-3})}} \qquad (9),$$

larger than the focal beam waist. (E is the electrons energy and n the density).

ii) the energy calibration is usually solved by Self-identification of the spectrum, whenever this is possible.

iii) since it is very difficult to keep the transmission constant for electron energies under one eV (due to stray fields) it is, at least, necessary to calibrate it.

For Xenon, it is, for the time being, impossible to calculate reliable MPI or ATI cross-sections. Therefore, in these experiments, only the general features of the theory can be tested. Among these are the intensity dependence of the different ATI orders and the relative cross-sections for the successive ATI orders.

We have measured[8] the intensity dependence for the lines of fig.1 and the results are reported in table II.

TABLE II

Line	(1/2,0)	(1/2,1)	(1/2,2)	(3/2,0)	(3/2,1)	(3/2,2)
N+S	6	7	8	6	7	8
Exp	X	7.3 0.1	7.8 0.3	5.5 0.7	6.5 1.6	5.7 1.8

the line "Exp" reports the experimental values of the slope

$$K = d(\text{Log Ne}) / d(\text{Log}I)$$

of the electron number Ne at a given energy as a function of the intensity I.
P. Kruit et al[11] have measured the dependence of the ratio of two consecutive ATI lines on the intensity and found it linear to a very good precision.

From Table II it can be seen that the theoretical values N+S are included in the experimental error bars, (excepted for the (3 /2, 2) line). Therefore, at intensities of about 10^{11} W.cm^{-2}, it does not seem that slope measurements can invalidate the predictions of lowest order perturbation theory. The case of the (3/2,2) line could be a hint of a possible breakdown, but the "average" result is that perturbation theory is still valid.

These results are drastically different from the ones reported in Ref.11 at higher intensities.

A second set of results in ATI of Xenon is about the relative ATI cross-sections. From this point of view, the qualitative aspect of fig.1 can be misleading because this spectrum has been taken at saturation intensity (i.e. it shows the maximum relative values for ATI signals). In order to obtain relative cross-sections, signals as a function of intensity for the lines (3/2, 0), (3/2,1), (3/2,2) have been fitted by computed values obtained from equations of the type :

$$\begin{cases} \dfrac{dP_{N+S}}{dt} = K\sigma_{N+S}\, I_M^{N+S} \exp\left(-(N+S)\dfrac{t^2}{\tau^2}\right) P_0(t) \\ \sum P_i(t) = 1 \end{cases} \tag{10}$$

where P_i are the populations at level i, σ_{N+S} the generalised cross-sections, I_M the peak intensity of a Gaussian pulse of duration τ.

The free parameter is σ_{N+S} which is adjusted for the best fit.

The result is, for the relative values of σ_{N+S} :

$$\frac{\sigma_7}{\sigma_6} = 0.7\ 10^{-30}\ \text{cm}^2.\text{s}$$

$$\frac{\sigma_8}{\sigma_7} = 0.38\ 10^{-30}\ \text{cm}^2.\text{s}$$

within a factor of 2. Of course this fit is also subject to the error on the intensity absolute measurement, which is not better, in this case, that another factor 2.

This result shows that

i) the ATI cross sections of successive orders are decreasing functions of the order with a "rate" which is similar to the one observed usually in MPI[1]. Obviously this statement can be only qualitative and must be understood as a rough guide.

ii) the threshold intensities I_{th} (defined in the first part) are

$$I_{th6} = 2.6\ 10^{11}\ \text{w.cm}^{-2},\ I_{th7} = 5.3\ 10^{11}\ \text{w.cm}^{-2}$$

while $I_{sat} = 10^{11}\ \text{w.cm}^{-2}$. These values explain why in this case, perturbation theory is still valid since the process saturates before the I_{th} is reached. Thus the (N+S+1) photon ionization probability is kept lower than that of the (N+S) photon process

iii) the total electron signal (equal to the total ion signal) can be written as :

$$\sum_{N+S} I^{N+S} = \sigma_6\, I^6\ (I + 3.8\ 10^{-3}\ I + 7.10^{-6}\ I^2 + \ldots)$$

with I in G W cm^{-2} (and for I $<$ 100 G W cm^{-2}). If one plots this total signal as a function of I, the slope is found to be 6.1 ± 0.3 whis is not distinguistable from the normal slope of 6 for the six-photon ionization of Xenon. This could be an explanation for the contradiction mentioned in the introduction : namely the ATI contribution to the total ionization is too small to be detected by slopes measurements. Another possible explanation can be found in ref.12 (this book): for intensities such than perturbation theory is no longer valid the measured slopes are equal to N whatever the values of S.

Both explanations stress the fact that slopes measurements are not sufficient to support the lowest order perturbation theory as a model, even when they obey its prediction with a good accuracy . Other measurements, like angular distribution[22] are interesting for identifying the ATI processes, but since they do not lead to quantitative comparisons they will not be commented here.

ATI OF CESIUM AT 1.06 μm

Experiments reported in the previous section have made the ATI process clarly observable and have provided for qualitative tests of the lowest order perturbation theory. However, if quantitative tests are considered, one has to rely a simpler atoms. One of the best candidate and one of the most used for MPI (see ref.1) is Cesium. This is because it is easy, both to calculate the MPI and ATI cross-sections and to handle a Cesium atomic beam.

Two-photon ionization and three-photon ATI of Cesium were calculated[23] for the second harmonic of YAG (λ =0.53μ). However it is well known that, at this wavelength, it is very easy to induce photoemissions from metallic surfaces covered by Cesium atoms (due to a decrease of the metal workfunction). Therefore it seemed safer to consider an experiment at 1.06μ, in order to reduce stray electron emissions. The calculation of 4 and 5-photon ionization cross-sections of cesium at 1.06μm has been performed by M.Crance and M. Aymar [24], to the lowest-order. The generalized cross-sections they obtain are :

$$\sigma_4 = 0.18\ 10^{-107}\ \text{cm}^8.\ \text{s}^3 = 14.57\ \text{s}^{-1}/(\text{GW.cm}^{-2})^4$$

$$\sigma_5 = 0.89\ 10^{-139}\ \text{cm}^{10}.\ \text{s}^4 = 0.39\ \text{s}^{-1}/(\text{GW.cm}^{-2})^5$$

and the saturation intensity is given by :

$$I_{sat} \simeq (\sigma_4\ \tau_4)^{-1/4} \tag{11}$$

From this basis one can predict a ratio of the ATI signal to the MPI signal (at I_{sat}) equal to :

$$R = \frac{\sigma_5 I_{sat}^5}{\sigma_4 I_{sat}^4} = \begin{cases} 0.02 \text{ for } \tau_4 = 25 \text{ ps} \\ 0.004 \text{ for } \tau_4 = 20 \text{ ns} \end{cases}$$

The picosecond case is the most favorable (since I_{sat} is larger), but still the expected ratio is small. From a spectroscopy point of view,the problem is to detect a line about 2% of the main component, 1.17 eV apart from it. We have used a time of flight spectrometer with an extracting electric field and a multichannel analyzer which will be described in details in the near future[25]. The Cesium is brought into the interaction region in an atomic beam and the laser is a mode-locked, 10 pps, 50 ps YAG. The beam is focussed down to a diameter of 75 μm. We have measured the absolute intensity in the experiment with the maximum possible accurracy (30%) by measuring the focal spatial distribution with an image digitizer[26] and the pulse duration with a streak-camera. The counting rate is about 1 Cs^{-1}. The collecting angle is 1.3°, thus allowing angular distribution measurements. The measurements reported here are about the differential cross-section along the laser polarization (θ =0) while the calculation mentioned above are for

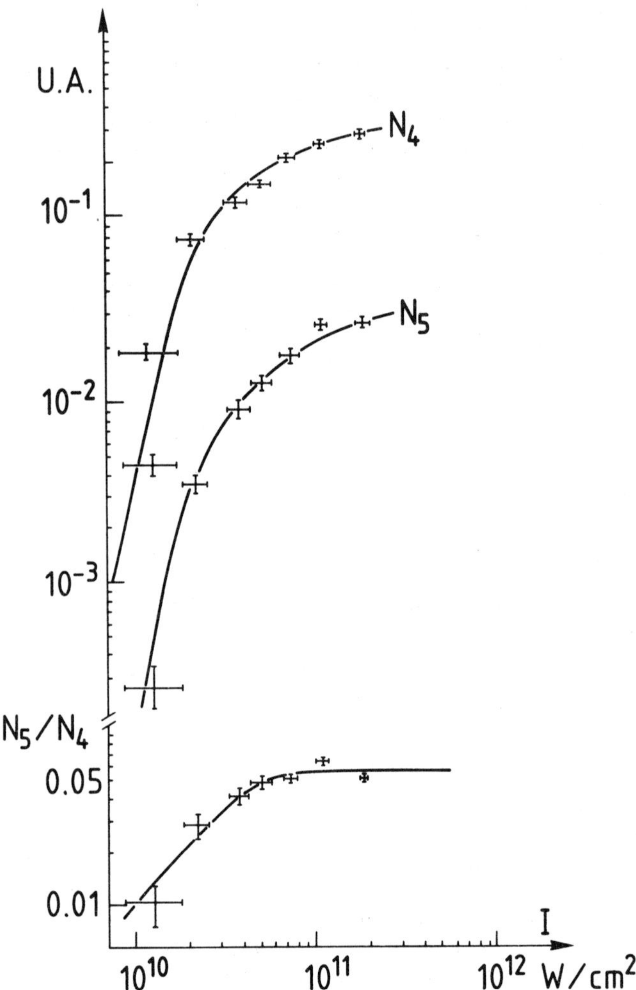

FIGURE 2. ATI of Cesium at 1.06 μm.

The upper part of the figure displays the number of photoelectrons produced by 4 and 5-photon ionization versus intensity. The lower part shows the ratio of the two signals after instrumental corrections.

the total cross-section. It turns out that the theoretical values for the differential cross-sections (for $\theta = 0$) lead to a ratio equal to the one obtained using total cross-sections. Due to the lack of space we can only summarize here the main experimental results:

i) we observe ATI, 5-photon ionization of Cesium.

ii) the intensity dependences (the slopes) are found to be 4.2 ± 0.2 for the 4-photon MPI and 4.8 ± 0.2 for the 5-photon ATI.

iii) the saturation intensity as shown on fig.2 is found to be about 50 GW cm^{-2} (to be compared to the 70 GW cm^{-2} predicted by the calculation).

iiii) the observed ratio is found to be (0.055 ± 0.006) to be compared to the predicted value of 0.02.

To conclude this section it can be said that, for Cesium ATI, lowest order calculation are giving reasonable predictions, but these predictions must be further improved by taking into account the intensity-dependent Stark shifts of the ground states and of some of the excited states, particularly the 6f levels[25]. These shifts are making the cross-sections intensity dependent and modify the angular distribution. There is very little doubt that such corrections will reduce the present discrepancy (about a factor of 2.5) between the experimental and theoretical values of the ratio R.

CONCLUSION

In this paper we have reported experimental results of ATI of Xenon at 0.53μ and ATI of Cesium at 1.06μ. While the Xenon results allow qualitative tests of the theory, the Cesium case is suitable for a full quantitative comparison. Under the conditions where lowest order perturbation theory is pertinent ($I < I_{sat} < I_{th}$), ATI is only a special case of MPI with quite comparable cross-sections. For situations where $I > I_{th}$, the reader is referred to ref.11.

Measurements of differential cross-sections of ATI could give some information about free-free matrix elements which are otherwise very difficult to measure in scattering experiments[5,6].

REFERENCES

1. J. Morellec, D. Normand, G. Petite : Non resonant multiphoton ionization of Atoms - in Adv. in Atom. and Molec. Phys. Acad. Press (1982).
2. E.A. Martin, L. Mandel Appl. Opt. 15, 2378 (1976).
3. M.D. Hollis, Opt. Comm. 25, 395 (1978).
4. M. Gavrila, M.J Van der Wiel : Comments Atom. Mol. Phys. 8, n°1-2 (1978).
5. N.K. Rahman in : Photon-assisted Collisions and related topics.
N.K. Rahman and C. Guidotti Ed., Harwood Acad. Publishers (1982).
6. P. Agostini in : Photon-assisted Collisions and related topics,
N.K. Rahman and C. Guidotti Ed., Harwood Acad. Publishers (1982).
7. F. Fabre, G. Petite, P. Agostini, M. Clement, J. Phys. B : At. Mol. Phys. 15, 1353 (1982).
8. W. Zernike and R.W. Klopfenstein, J. Math. Phys., N.Y. 6, 262 (1965).
9. P. Agostini, F. Fabre, G. Mainfray, G. Petite and N.K. Rahman, Phys. Rev. Lett. 42, 1127 (1979).
10. P. Kruit, J. Kimman, M.J. Van der Wiel in Abstracts of Proceedings of the first European Conference on Atomic Physics, Heidelberg, Federal Republic of Germany,
J. Kowulski, G. Zu Pulitz, H.G. Weber Ed. (1981). (unpublished).
11. M.J. Van der Wiel in: Collisions and Half Collisions with Lasers (This Book)
12. A. L'Huillier, L.A. Lompré, G. Mainfray, C. Manus, Phys. Rev. Lett. 48, 1814 (1982).
13. F. Fabre, P. Agostini, G. Petite, Phys. Rev. A 27, 1682 (1983).
14. E. Karule, J. Phys. B: Atom. Mol. Phys. 11, 441 (1978).
15. S. Klarsfeld anf A. Maquet, Phys. Lett., 78A , 40 (1980).
16. Y. Gontier, M. Trahin, J. Phys. B: Atom. Mol. Phys. 13, 4383 (1980).
17. M.Aymar and M. Crance, J. Phys.B : Atom. Mol. Phys. 12, 3585 (1981).
18. T. Nitzan, J. Jortner, B. Berne, Molecular Physics 26, 281 (1973).
19. W. Reinhardt and S. Chu in Multiphoton Processes (J.H. Eberly and P. Lambropoulos ed.) Wiley and Sons (1977).
20. P. Agostini, M. Clement, F. Fabre, G. Petite, J. Phys. B: Atom. Mol. Phys. 14, L491 (1981).

21. F. Fabre, Thesis, Paris (1983) (in French).
22. F. Fabre, P. Agostini, G. Petite, M. Clement, J. Phys. B: Atom. Phys. Mol. Phys. 14, L677 (1981).
23. M. Aymar and M. Crance, J. Phys. B: Atom. Mol. Phys., 12, 3585 (1981).
24. M. Aymar and M. Crance (unpublished).
25. G. Petite, F. Fabre, P. Agostini, M. Aymar, M. Crance, (to be published).
26. L.A. Lompré, G. Mainfray and J. Thébault, Rev. Phys. Appl. 17, 21 (1982).

ABOVE-THRESHOLD MULTIPHOTON IONIZATION

M.J. VAN DER WIEL
FOM-Institute for Atomic and Molecular Physics,
Kruislaan 407, 1098 SJ Amsterdam, The Netherlands.

INTRODUCTION

In a multiphoton ionization process, the number of photons absorbed in the creation of one ion does not necessarily equal the lowest integer exceeding IP/hω, where IP is the ionization potential. First evidence that 'above threshold' ionization (ATI) - or absorption of additional quanta to higher continuum states - occurs, was presented by Agostini et al.[1] on the basis of an energy analysis of the photo-electrons. This work was followed by similar studies at higher energy resolution by Kruit et al.[2] and Fabre et al.[3].

Further work on this subject has concentrated on two areas: One is that of establishing quantitative agreement with lowest-order perturbation theory, under conditions of light intensity where this theory can be expected to be valid. The main contribution in this area is from the Saclay group and is reported elsewhere in this volume. Our contribution in this area concerns a comparison of above-threshold relative intensities in Xe with calculated continuum-continuum matrix elements. This work was reported in refs. 4-6.

The second area is that of qualitatively understanding the dynamics of above-threshold phenomena for intensities well beyond the point where perturbation theory breaks down. This is the subject of the present contribution. The physics studied is summed up in two questions: How do these intense fields affect the atomic spectroscopy, and what is the behaviour of the transition probabilities into the various continua? We have performed an energy analysis of 'above threshold' photoelectrons from Xe irradiated by 1064 nm light, in a range of intensities extending up to 10^{14} cm^{-2}, i.e. up to the point of saturation.

EXPERIMENTS FOR Xe

Before discussing the results and their interpretation, we briefly mention some essential points of the experiment. Firstly, and very importantly, the determination of the intensity in the laser focus. It is well known that this constitutes a problem even for single-mode pulses. In our experiment, we decided to accept the multimode ($\sim$400 modes) nature of the 10 ns Nd-YAG pulse and to assume a chaotic intensity distribution[7]. Given such a distribution, it is straightforward to show that an N-th order ionization signal

$$S = N \cdot \frac{1}{\bar{I}} e^{-I/\bar{I}}$$

is composed of different contributions from the full range of available intensities. In particular, the maximum contribution to S comes from a narrow intensity range around $I_{eff} = N \cdot \bar{I}$. This definition of I_{eff} is used throughout this article, while $\bar{I}$ is obtained in the customary way, i.e. from the focal diameter and the gross time behaviour of the pulse.

Secondly, the electron spectrometer[8] used accepts an opening angle of 2π sterad. This high efficiency makes it possible to work at conditions well below saturation and still obtain electron spectra in acceptable counting times. Further advantages are the constancy of the transmission down to practically zero eV, and the avoidance of complica-

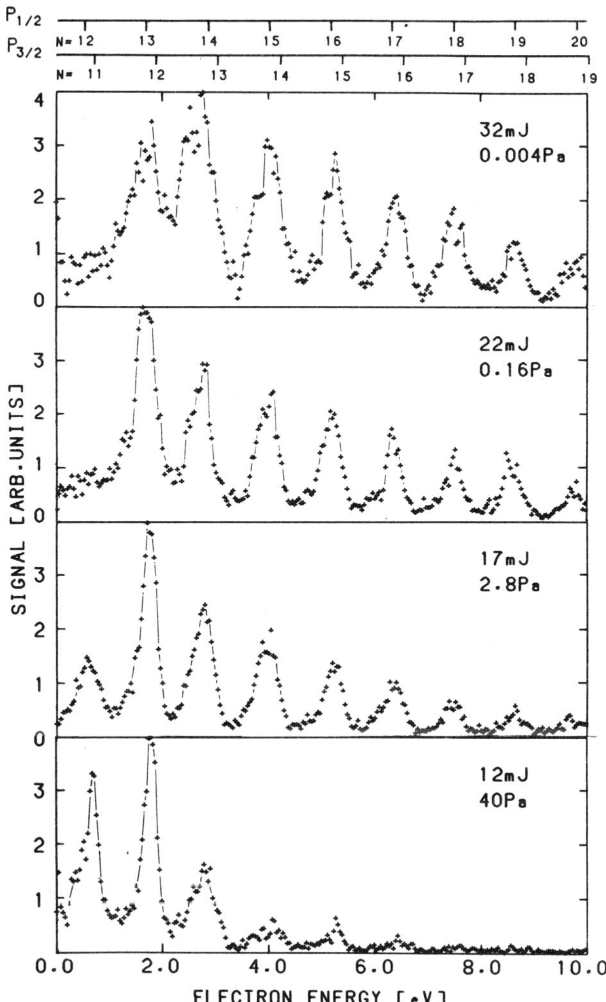

FIGURE 1 Electron spectra from multiphoton ionization of Xenon at 1064 nm. The vertical scales are chosen to make the maximum measured signal equal in each spectrum. The pulse energy F and pressure P at which each spectrum is taken is given in the figure. In the spectrum at 0.004 Pa, the background has been subtracted. The estimated effective intensity is F x 2.10^{12} W/cm^2 (F in mJ).

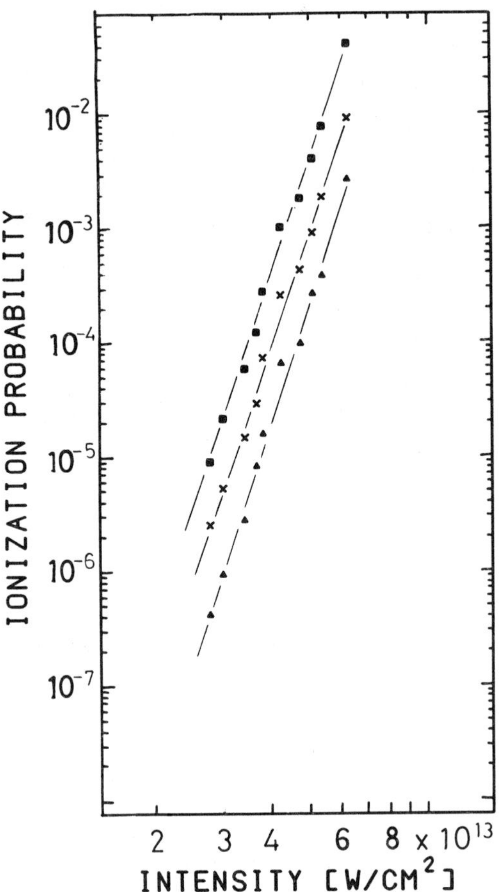

FIGURE 2 Ionization probability for (N + S) photon ionization of xenon at 1064 nm (N = 11).
■ - total number of electrons, x - number of electrons in the 1.8 eV peak, ▲ - number of electrons in the 8.7 eV peak, as a function of pulse energy. The conversion from number of electrons to absolute ionization probability is accurate to within a factor of two.

cations due to the angular dependence of the photoelectrons w.r.t. the laser polarization.

Thirdly, in taking electron spectra at different laser powers, we have made sure that the total number of ionization events is kept constant at a few hundred per laser shot. This is done by changing the target pressure accordingly. Thus, the space charge induced energy broadening is small and the same for all spectra. Moreover, even at the lowest target pressure, the number of neutral atoms in the laser focus exceeded that of the electron-ion pairs created by a factor of at least ten. Depletion of the ground state therefore did not play a role.

Under the above conditions, the spectra shown in fig. 1 have been obtained. Note that each of the peaks is an unresolved doublet arising from e.g. 11-photon ionization leading to ions in the $P_{3/2}$-state and a 12-photon process into the $P_{\frac{1}{2}}$ continuum. The two energies differ by 70 meV, a splitting too small to be observed at the present peak broadening. Before drawing conclusions from the figure, we first present two more results concerning these processes.

Figure 2 gives the orders of non-linearity of a few of the electron peaks. The lowest-energy peak, which seems to disappear at high intensities, behaves as $I^{7.8}$ over the region where it can be observed.

Figure 3 presents the average electron energy, i.e. the weighted average over the spectral range of 0 - 10 eV, as a function of laser intensity. In the low-intensity range the data points show a linear relationship; the deviation from linearity at higher intensities is ascribed to the limited range of our spectra. Inclusion of further peaks at energies beyond 10 eV would yield an upward correction of the data points and thus extend the linear behaviour.

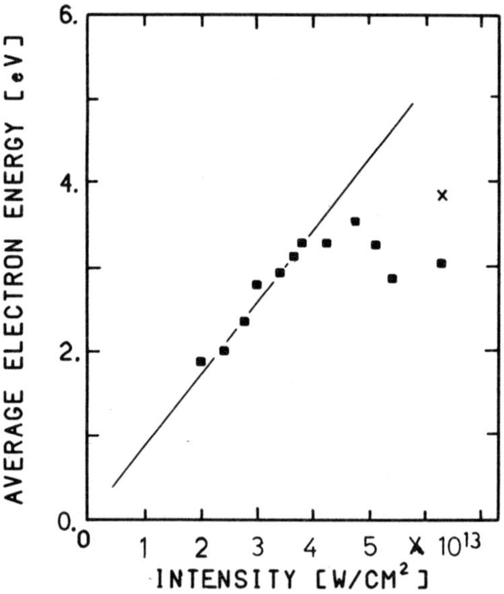

FIGURE 3 Average energy of electrons arising from multiphoton ionization of Xe at 1064 nm as a function of intensity.

■ - data from measurements as in fig. 1; × - background subtracted. Note that the deviation from the straight line at high intensities is a result of the limited energy range (max. 10 eV) over which the spectra of fig. 1 have been measured.

DISCUSSION

The observations in the three figures may be summarized as follows:

i) The peak positions are not affected by the increase in intensity.

ii) The lowest and next-lowest peak seem to be suppressed at high intensities.

iii) A peak broadening is present of the order of 200 meV, i.e. significantly in excess of the instrumental width of 15 meV (proven under the same conditions for a five-photon process).

iv) All 'above-threshold' processes show rates having orders of non-linearity between 10 and 11, with only a slight increase for higher-order processes.

v) The average electron energy behaves approximately as $0.7 * 10^{-13}$ I (I in W cm^{-2}).

With regard to item i), we note that it contradicts predictions based on the usual formulation of the ponderomotive force[9]. This prediction is an energy gain for each individual peak of $\Delta E = 10^{-13}$ I [I in W cm^{-2}] (for the Nd-YAG laser wavelength), which for our intensities would amount to several eV. Since there can be little doubt that the electrons do undergo this acceleration on the way out of the laserfocus, only one conclusion remains: in an intense field, an upward shift occurs of the ionization potential, which exactly cancels the energy gain by the ponderomotive force. Before discussing the theoretical support for this idea, we note that it has far-reaching implications for e.g. the earlier work by Hollis[10] and by Borcham et al.[11,12]. This work involved very high electron energies - hundreds of eV - created in fields of 10^{15} W cm^{-2}. The interpretation in terms of energy gain from the ponderomotive force clearly needs revision; our data shows that it is ATI which is responsible for the high energies.

Returning to the experimental observations, a large upward shift of the IP is consistent with the disappearance of the low energy peaks at high intensity: e.g. for a shift of more than 0.6 eV, an 11-photon process no longer reaches the continuum. The disappearance of the peak obviously is not disrupt, since we never deal with one intensity but with a wide range of intensities. Concerning the observed width (item iii)), it should be realized that the cancellation of 'IP shift' and 'ponderomotive shift' is complete only if the laser field does not change its intensity between the instant of creation of the photoelectron and the instant the electron reaches the zero field outside the laser focus. Typical escape times being in the order of 10 ps, it is quite conceivable that in our multimode pulse variations occur on that time scale. Such variations could then be responsible for at least part of the observed peak broadening. (Below we will see that another part is due to AC Stark shift of the ground state.)

We now turn to the rates of the various ATI peaks. The almost equal orders of non-linearity indicate a kind of saturation of the final steps in the continuum. The idea is that of an 11-th order rate-limiting transition to the lowest continuum state, followed by a redistribution over various higher continuum state with probabilities close to unity. It turns out that matrix elements for continuum-continuum transitions, calculated using a potential model for the Xe atom[13], do indeed lead to probabilities near unity at the field strengths involved here. Moreover, there is other evidence from double-ionization experiments[14] that the 'cross section' per step in a multiphoton process is much larger in the continuum than in the discrete part of the spectrum (for non-resonant cases).

Finally, the linear dependence of average energy on intensity has no simple explanation. However, it is reminescent of the predictions of Jung[15] for free-free scattering in the soft-photon limit: also in this case the average energy gain, weighted over the entire spectrum including absorptions and emissions, is found to be linear with laser intensity.

THEORY

So far we have described the observations in qualitative terms. For completeness, we mention recent developments of the theory[16], which support the above ideas. In the first place, it is rather straightforward to show that the ionization potential in a strong field shifts up by an amount exactly equal to the energy shift of the ponderomotive force. At the same time, it is easy to calculate the AC Stark shift of the ground state from the known polarizability; its value turns out to be about 50 meV at 10^{13} W cm^{-2}. This shift could partly be responsible for the electron peak broadening, given the fact that a distribution over intensities contributes to the ionization signal.

In addition, a non-perturbative approach to the problem of photoionization in arbitrarily strong fields[16] has been developed. So far, the atom is modelled by a single electron in a potential, with a spectrum of only one bound state and a continuum. This simplification was chosen for numerical convenience, but work is in progress on an extension to Coulombic potentials. The laser field is taken to be circularly polarized, which permits removal of the time-dependence. The bound state now becomes a resonance embedded in the various continua, belonging to the various numbers of photons absorbed. An implicit expression for the spectrum of complex energies of this resonance is obtained by means of the complex dilatation method, and computed numerically. The imaginary part of each of the resonance energies represents the rate for production of electrons of the corresponding energies. An example of the results as a function of field strength a

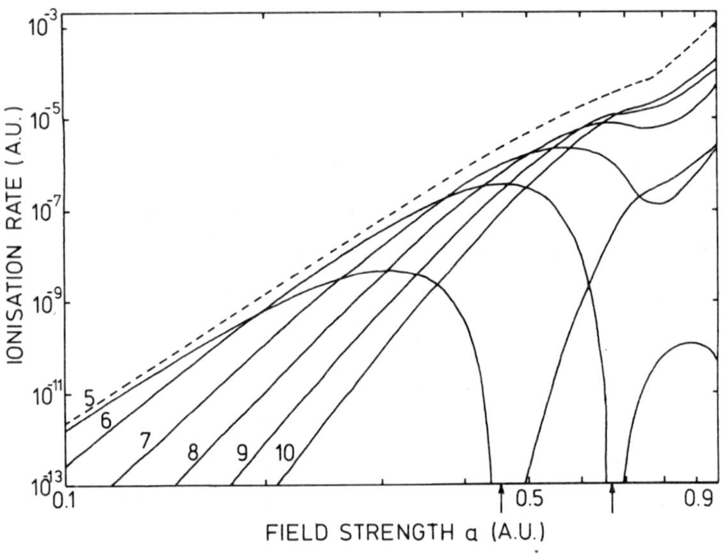

FIGURE 4 Calculated rate of absorption of various numbers of photons (5 to 10) as a function of the field strength a. The photon energy equals 0.24 times the ionization energy in zero field. Dashed curve - total ionization rate. The arrows along the abscissa indicate the field strength at which the AC Stark shift pushes the ionization potential beyond 5, respectively 6 times the photon energy.

(in a.u.) is shown in fig. 4 for the following case: a minimum of five photons is needed to overcome the ionization potential at zero field strength. It is quite clear that this theory correctly describes the disappearance of the 5-photon peak at an intensity where the ionization potential exceeds 5 $\hbar\omega$. At higher intensities, even the 6-photon peak is suppressed, while the 5-photon peak rises again, although it remains smaller than the higher-order peaks. This renewed rise is interpreted as a lifetime broadening, causing part of the resonance profile to exceed the threshold. Note that the total ionization rate (dashed curve in fig. 4) has an order of non-linearity slightly smaller than 5, even in the regime where 6- and more photon-ionization dominates. This is also in accord with our experimental findings for the strong-field process in Xe, which has a total order of non-linearity 10 ± 1 (fig. 2).

Finally, we note that the calculation produces realistic rates, in the sense that even at the highest intensity - at 1 a.u. - the rate still is of the order of the inverse of one hundred light periods.

ACKNOWLEDGEMENTS

This work is part of the research program of the Stichting voor Fundamenteel Onderzoek der Materie (Foundation for Fundamental Research on Matter) and was made possible by financial support from the Nederlandse Organisatie voor Zuiver-Wetenschappelijk Onderzoek (Netherlands Organization for the Advancement of Pure Research).

REFERENCES

1. P. Agostini, F. Fabre, G. Mainfray, G. Petite and N.K. Rahman, Phys.Rev.Letters 42, 1127 (1979).
2. P. Kruit, J. Kimman and M.J. van der Wiel, J.Phys.B. 14, L597 (1981).
3. F. Fabre, G. Petite, P. Agostini and M. Clement, J.Phys. B. 15, 1353 (1982).
4. P. Kruit, J. Kimman, H.G. Muller and M.J. van der Wiel, J.Phys.B. 16, 937 (1983).
5. P. Kruit, Thesis, University of Amsterdam, 1982.
6. P. Kruit, H.G. Muller, J. Kimman and M.J. van der Wiel, J.Phys.B. 1983, to be published.
7. R. London, The Quantum Theory of Light (Clarendon Press, Oxford) 1972.
8. P. Kruit and F.H. Read, J.Phys.E 16, 313 (1983).
9. T.W.B. Kibble, Phys.Rev.A 150, 1060 (1966).

10. M.J. Hollis, Optics Communications 25, 395 (1978).
11. B.W. Boreham and B. Luther-Davies, J.Appl.Phys. 50, 2533 (1979).
12. K.G.H. Baldwin and B.W. Boreham, J.Applied Phys. 52, 2627 (1981).
13. H.G. Muller and A. Tip, to be published.
14. A. l'Huillier, L.A. Lompre, G. Mainfray and C. Manus, Phys.Rev.A 27, 2503 (1983).
15. C. Jung, Phys.Rev. A 21, 408 (1980).
16. H.G. Muller, A. Tip and M.J. van der Wiel, J.Phys.B. to be published.

ON THE CALCULATION OF TWO PHOTON IONIZATION CROSS SECTION

IVO CACELLI, ROBERTO MOCCIA
Istituto di Chimica Fisica dell'Università di Pisa, Via Risorgimento 35, 56100 Pisa, Italy

VINCENZO CARRAVETTA
Istituto di Chimica Quantistica ed Energètica Molecolare del CNR, Via Risorgimento 35, 56100 Pisa, Italy

Abstract The theroretical methods available to compute the two-photon ionization cross sections are briefly reviewed with particular emphasis upon those which exploit the response function and the many-body Green function approaches. Numerical results for He obtained by simple extension of the usual configuration-interaction (C.I.) computer programs are also presented.

INTRODUCTION

The extraordinary progresses of laser technologies in recent years has made available a wealth of data which supply extremely refined informations about the electronic structure of atoms and molecules[1]. Unfortunately the theoretical calculations of the quantities required to interpret these experimental data pose formidable computational problems to the theorist. Without dwelling upon these general and well known aspects we will concentrate here upon a particular quantity which is needed to compute the probabilities of the two photon processes. The second order perturbation theory gives for the probability per unit time $W_{f\leftarrow i}$ that a system, initially in the state $i\rangle$ will end up, by absorbing two photon of angular frequencies ω_1 and ω_2 and of polarization $\hat{e}_1$ and $\hat{e}_2$, to the final state $f\rangle$ the following expression[1]:

$$W_{f\leftarrow i}=(8\pi^3\hbar e^4/c^2)\,I(\omega_1)d\omega_1\,I(\omega_2)d\omega_2\times \qquad (1)$$
$$\left|g_1 g_2\langle f\,\hat{e}_1\cdot\bar{O}_1[\hbar\omega_2-(\mathcal{H}_A-\mathcal{E}_i)+i\eta]^{-1}\hat{e}_2\cdot\bar{O}_2\,i\rangle+(1\leftrightarrow 2)\right|^2$$

In Eq. (1) the $\mathcal{E}$'s are eigenvalues of $\mathcal{H}_A$, the $I(\omega)$ represents the photon fluxs for unit (angular) frequency range and, in the dipole approximation $g_j\bar{O}_j=\sqrt{\omega_j}\,\underline{r}$ if the length expression of the radiation-matter interaction operator is employed while $g_j O_j=(1/m\sqrt{\omega_j})\,\underline{P}$ if the velocity form is adopted. Eq. (1) once integrated upon the density (for unit energy range) of the final states, gives the looked for quantity which allows to correlate observable data with the structure of the system[1]. It is clear that at this level of approximation where a strict second order perturbation treatment is employed and therefore no account is given for the width and shifts of the levels, the individuality of the systems will turn up only through the quantity

$$\sum\!\!\!\!\!\!\int_I \frac{\langle f\,\hat{e}_1\cdot\bar{O}_1 I\rangle\langle I\,\hat{e}_2\cdot\bar{O}_2\,i\rangle}{\hbar\omega_2-(\mathcal{E}_I-\mathcal{E}_i)+i\eta}+(1\leftrightarrow 2) \qquad (2)$$

where the $I\rangle$ are the discrete and continuum eigenstates of $\mathcal{H}_A$ with eigenvalues $\mathcal{E}_I$ which act as intermediate relay states[1]. In the following it will be considered only the simpler case of non resonant intermediate states.It is worthwhile to point out that at this level of approximation the length form of the interaction operator, length gauge (LG), and the velocity form, velocity gauge (VG), should give identical results[2]. This is not true if level shift and width were admitted in some way in the denominators.

The reason for this may be found in the mixing of perturbation orders arising by the inclusion of the higher order terms required to account for the shift and widths of the levels due to the radiation. The computation of (2) requires in principle the knowledge of the complete spectrum and it may be accomplished only for the simplest systems. Of paramount importance, among the solvable cases,is the one-electron hydrogenic system[3] which is a good testing ground of the approximation methods and may represent a resonable starting point for the many-electron atoms[4]. Several approximation methods have been proposed to obtain reasonable estimate of (2) and the interested reader may consult the abundant literature on the subject[1,5]. Here we wish to draw attention to some less popular techniques based upon the response function approach and the many-electron Green function approach. In addition some results are presented for He which will demonstrate that, by making some minor extensions, it is expedient to use the standard computer programs based upon the limited basis set configuration interaction method, to obtain accurate values of the photo-ionization cross sections.

RESPONSE FUNCTION APPROACH[6].

Let $V(t_1,x_1) - f(t_1)h(x_1)$ be a one electron ($x\equiv \underline{r},\sigma$ indicates the space-spin coordinate) time dependent perturbation and $B(x_a)$ a one-electron observable (for the sake of simplicity only local operators are considered, the extension to non-local operators is immediate). If it is assumed that, in the interaction representation for $t\rightarrow -\infty$ $V_I(t,x)\rightarrow 0$ and the density matrix $\rho_I\rightarrow|\varphi_0\rangle\langle\varphi_0|$; then by esploiting the properties of the step function θ the value of the observable B at the time t is given[6] by (atomic unit will be used from now on)

$$\langle B\rangle_t-\langle\varphi_0 B\varphi_0\rangle=-i\int_{-\infty}^{\infty}dt_1\, f(t_1)\,\Theta(t-t_1)\langle\varphi_0[B_I(t),h_I(t_1)]_-\varphi_0\rangle$$

$$+(-i)^2\int_{-\infty}^{\infty}dt_1\int_{-\infty}^{\infty}dt_2\,\Theta(t-t_1)\Theta(t_1-t_2)\langle\varphi_0[[B_I(t),h_I(t_1)]_-,h_I(t_2)]_-\varphi_0\rangle \quad (3)$$

$+\ \cdots\cdots$

The perturbation series (3) allows to write for the different orders of h

$$\Delta\langle B\rangle_t^{(1)}=\int_{-\infty}^{\infty}dt_1 f(t_1)\int dx_a dx_1 V(x_1)\, R^{(1)}(x_a,x_1;t-t_1) \quad (4a)$$

$$\Delta\langle B\rangle_t^{(2)}=\int_{-\infty}^{\infty}dt_1\int_{-\infty}^{\infty}dt_2\, f(t_1)f(t_2)\int dx_a\, dx_1\, dx_2\, V(x_1)V(x_2)$$

$$\times R^{(2)}(x_a,x_1,x_2;t-t_1,t_1-t_2) \quad (4b)$$

$\cdots\cdots$

The comparison of Eq's (4) and (3) affords the precise definition of the functions $R^{(1)}$ $R^{(2)}$ etc. which are labeled as linear response function, quadratic response function etc. Of particular importance are their Fourier Transforms which may be readily seen to be

$$R^{(1)}(x_a,x_1;\omega)=\sum_\ell\left(\frac{\rho_0^\ell(x_a)\rho_\ell^0(x_1)}{\omega-\omega_\ell}-\frac{\rho_\ell^0(x_a)\rho_0^\ell(x_1)}{\omega+\omega_\ell}\right) \quad (5a)$$

and (of the symmetrized form $R_S^{(2)} = \frac{1}{2}\,(R^{(2)}(x_a,x_1,x_2;t-t_1,t_1-t_2) + (1\leftrightarrow 2)\,)$

$$R_S^{(2)}(x_a,x_1,x_2;\omega_1,\omega_2)=\frac{1}{2}\sum_\ell\sum_j\left(\frac{\rho_0^\ell(x_a)\rho_\ell^j(x_1)\rho_j^0(x_2)}{(\omega_1-\omega_\ell)(\omega_2-\omega_j)}+\right.$$

$$-\frac{\rho_0^\ell(x_1)\rho_\ell^j(x_a)\rho_j^0(x_2)}{(\omega_2-\omega_\ell)(\omega_1-\omega_2+\omega_j)}-\frac{\rho_0^\ell(x_2)\rho_\ell^j(x_a)\rho_j^0(x_1)}{(\omega_2+\omega_\ell)(\omega_1-\omega_2-\omega_j)}+\frac{\rho_0^\ell(x_2)\rho_\ell^j(x_1)\rho_j^0(x_a)}{(\omega_1+\omega_j)(\omega_2+\omega_\ell)} \quad (5b)$$

$$\left.+\frac{\rho_0^\ell(x_a)\rho_\ell^j(x_2)\rho_j^0(x_1)}{(\omega_1-\omega_\ell)(\omega_1-\omega_2-\omega_j)}+\frac{\rho_0^\ell(x_1)\rho_\ell^j(x_2)\rho_j^0(x_a)}{(\omega_1+\omega_j)(\omega_1-\omega_2+\omega_\ell)}\right)$$

In the above Eq's $\rho_a^b(x)$ indicated the one-electron transition density matrix between the N-electron states $\varphi_a\rangle$ and $\varphi_b\rangle$ and the $\omega_\ell = \mathcal{E}_\ell - \mathcal{E}_o$ are the excitation energies of the ground state $\varphi_o\rangle$.

Eq. (5a) shows that the residue, at $\omega = \omega_\ell$, of the linear response (in order to avoid the formal difficulties arising form the continuous part of the spectrum we may always imagine to confine the system in a box large enough to ensure physically reliable results. Moreover it may be recalled that the physical correctness of the boundary conditions at infinity is open to questions[7]) is what is needed to compute the probability of one-photon transition between $\varphi_o\rangle$ and $\varphi_\ell\rangle$.

The residue at $\omega_1 = \omega_\ell$ of (5b) is proportional to

$$\rho_o^\ell(x_a)\sum_j\left(\frac{\rho_\ell^j(x_1)\rho_j^o(x_2)}{(\omega_2-\omega_j)} + \frac{\rho_\ell^j(x_2)\rho_j^o(x_1)}{(\omega_\ell-\omega_2-\omega_j)}\right) \tag{6}$$

which, apart the unimportant factor $\rho_o^\ell(x_a)$, is equal to (2) once we recall that $\omega_1 + \omega_2 = \omega_\ell$.

The practical bearing of the relations illustrated so far lies in the possibility to calculate by variational methods directly the response functions. For instance, by exploiting time-dependent variational principles for approximate w.f., which depend upon a set of arbitrary variational parameters, it is possible to obtain the generalized eq. of motion of the amplitudes of the linear response[8]. While this approach only recently has been successfully employed with rather sophisticated variational w.f.[9] it represents a widely utilized method known as Time Dependent Hatree Fock (TDHF) approximation (also labeled Random Phase Approximation (RPA) with exchange!) when the variational w.f. is of the SCF type. Very recently a time-dependent variational principle has been exploited to obtain the quadratic response function for the particular case of a monodeterminantal SCF variational w.f.[10].

As far as we know there are not as yet numerical results available obtained by this method but it certainly deserves to be examined since, apart some peculiar deficiences (it yields residues at some clearly unphysical poles), it, analogously its first order counterpart, the RPA, may ensure the gauge invariance for the individual matrix elements between different excited states. It must be recalled that the case of final states lying in the continuum poses some additional numerical problems with respect to the simpler case of the completely discretized spectrum yielded by the limited basis set expansion generally used in practice. It has been shown however that this problem may be handled successfully[11] and the recent progresses made on this subject are very encouraging[12]. In the following it will be proved that for systems with few electrons this problem may also be solved by simple numerical artifices.

THE MANY-BODY GREEN FUNCTION APPROACH.

For the one-photon transition processes the quantity of interest is the one-electron transition density matrix

$$\rho_f^o(x_j, x_j') = \langle \varphi_f^N \hat{\psi}^+(x_j') \hat{\psi}(x_j) \varphi_o^N \rangle \tag{7}$$

where the $\hat{\psi}(x_j)$ is the field operator which annihilates one electron at the spin-space point x_j ($x_j \equiv \underline{r}_j \sigma_j$). It has been shown by Namiki[13] that

$$\langle \varphi_f^N = \left(\lim_{\eta \to 0^+} \int_{t_0}^{\infty} e^{-\eta t_a} \eta \, dt_a \int dx_a \frac{\langle \varphi_f^{N(-)} \hat{\psi}_H(a) \varphi_0^{N-1} \rangle}{|\langle \ \rangle|^2} \right) \langle \varphi_0^{N-1} \hat{\psi}_H(a) \tag{8}$$

where $\hat{\psi}_H(a)$ is the field operator, in the Heisenberg representation, which annihilate one electron at the spin-space-time point a $(a \equiv x_a, t_a)$ and the denominator $|\langle \ \rangle|^2$ is the square modulus of the Dyson amplitude[14] $\langle \varphi_f^{N(-)} \hat{\psi}_H^+(a) \varphi_0^{N-1} \rangle$.

In addition it may also be proved[15] that

$$\varphi_0^N \rangle = \left(\lim_{\eta \to 0^+} \eta \int_{-\infty}^{t_0'} e^{\eta t_b} dt_b \int dx_b \frac{\langle \varphi_0^{N-1} \hat{\psi}_H(b) \varphi_0^N \rangle}{|\langle \ \rangle|^2} \right) \hat{\psi}_H^+(b) \, \varphi_0^{N-1} \rangle \tag{9}$$

Thus by indicating with $\mathcal{L}^+_{f,to}(a)$ and $\mathcal{L}_{o,to'}(b)$ the operators contained between parentheses in the r.h.s. of Eq.'s (8) and (9) respectively the matrix element (7) may be concisely written as

$$\langle \varphi_f^{N(-)} \hat{\psi}^+(x_j') \hat{\psi}(x_j) \varphi_0^N \rangle = \lim_{\substack{t_j' - t_j \to 0^+ \\ t_j \to 0}} \mathcal{L}^+_{f,t_0}(a) \, \mathcal{L}_{o,t_0'}(b) \, G_2(a, j; b, j') \tag{10}$$

The two-electron Green function G_2 appearing in Eq. (10) complies with the general definition[16] of a p-body Green function G_p^N in the zero temperature limit and for non degenerate ground state

$$G_p^N(1,2\ldots p; 1'2'\ldots p') = (-i)^p \langle \varphi_0^N T[\hat{\psi}_H(1) \hat{\psi}_H(2) \ldots \hat{\psi}_H(p) \hat{\psi}_H^+(p') \ldots \hat{\psi}_H^+(1')] \varphi_0^N \rangle \tag{11}$$

It must be emphasized that Eq. (10) does not depend from to and to'[17]. At this point several of the methods available to approximate the G_2 may be exploited[14,16] in order to calculate the transition density matrix through Eq. (10)[17]. For instance by making recourse to the approximation scheme afforded by the functional derivative technique[16] it is possible to obtain a rather compact expression for (10), which allows to represent graphically the matrix element as

$$\langle \varphi_f \, \bar{O} \, \varphi_o \rangle \simeq \lim_{\substack{t_j' - t_j \to 0^+ \\ t_j \to 0}} \left(\text{[diagram]} + \text{[diagram]} \right) \tag{12}$$

where the Dyson amplitudes $\langle \varphi_o^{N-1} \hat{\psi}_H(j) \varphi_o^N \rangle$ and $\langle \varphi_f^N \hat{\psi}^+(j') \varphi_o^{N-1} \rangle$ have been indicated by (g_j)→ and →($f_{j'}$) respectively, ⊗∿∿ indicates the one-electron-photon interaction, (R) is the linear response function[14] and the dot ● stands for the antisimmetrized two electron interaction. The interpretation of Eq. (12) is rather appealing since while the first term represents the "direct" transition the second one take into account the "indirect" transition induced by the electron sea dynamically polarized by the photon. For the two-photon transition processes it may be verified[18] that for $\omega_1 + \omega_2 = \varepsilon_f^N - \varepsilon_o^N$

$$-\mathcal{F}\left\{ \lim_{t_j \to 0} \mathcal{L}_f^+(a) \mathcal{L}_o(b) G_3^{N-1}(a, i, j; b, i'^+, j'^+) \right\}_{(-\omega_1)} = \tag{13}$$

$$= \sum_{\ell} \left(\frac{\rho_f^{\ell}(x_i x_i') \rho_j^{o}(x_j x_j')}{\omega_1 - \omega_\ell} + \frac{\rho_f^{\ell}(x_j x_j') \rho_\ell^{o}(x_i x_i')}{\omega_2 - \omega_\ell} \right)$$

Eq. (13) means that the matrix elements needed for the two-photon transition processes may be obtained from the Fourier Transform of a properly operated (by $\mathcal{L}_f^+ \mathcal{L}_o$ and the indicated time limit) three electron Green function. By exploiting an approximation scheme based upon the functional derivative approach it is possible to obtain reasonable expressions for the quantity in parentheses of the l.h.s. of Eq. (12).

By performing also the integrations which involve the electron-photon interactions it may be seen[18] that the most important contributions of the above mentioned quantity are

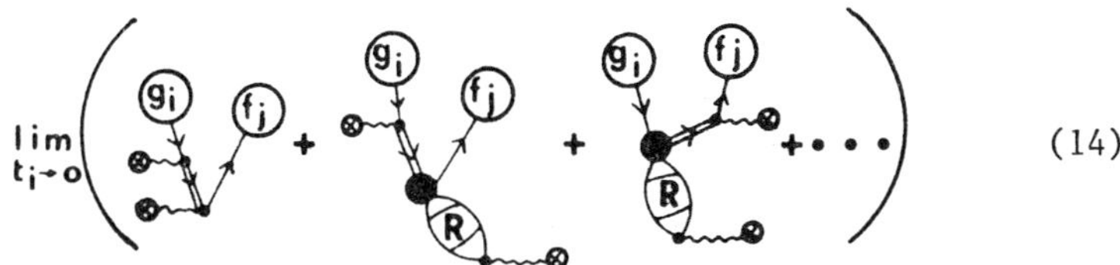

(14)

where $\underset{1' \quad 1}{\Rightarrow} = G_1^{N-1}(1,1')$. The different terms appearing in Eq. (16) may easily be interpreted analogously to what has been done for Eq. (11). It is important to emphasize that the terms of (14) depends upon other quantities which have a easily identificable bearing upon the global property. Thus it should be possible to figure out a computational ways which should remove the most serious numerical artifacts contained in (14).

LIMITED BASIS SET APPROACH

Most of the computer programs developed to calculate the electronic properties of atoms and molecules are generally conceived to perform the computations only for a few of the lowest lying bound states. Particular methods must be employed for the continuous states[19] although several methods have been proposed to extract the properties of the continuum from the completely discretized approximate spectrum yelded by the usual limited basis set expansion[12].

The situation in the case of the two-photon ionization processes where the required matrix element (2) involves, besides the final state lying in the continuum, also an integration upon all the continuous manyfold of the intermediate states, is certainly more critical than for the one-photon ionization processes. We have examined the feasibility of obtaining reliable data of the properties of the continuous part of the spectrum by a slight extension of the standard computer programs based upon the C.I. method with normalizable functions.

The underlying idea is to enrich the basis set with one electron functions which may mimic the asymptotic behaviour of some of the continuous states of the many electron system, approximated by a limited C.I. The basis functions selected to this purpose are

$$f_{\ell,m}(\xi_1,\xi_2\ldots\xi_\nu;K;\underline{r})=N\Big(\sum_{j=1}^{\nu}c_j e^{-\xi_j r}\Big)r^{\ell}Y_{\ell,m}(\hat{r})\cos(Kr) \qquad (15)$$

The linear combination of exponential is selected requiring that it behaves roughly as $r^{-(l+1)}$ for $a \leqslant r \leqslant b$. N is the normalization constant and the values of a and b should be chosen according to the particular system under study. For He, which was the test case examined here, $a \simeq 2$ a.u. and $b \simeq 30$ a.u. The calculations were performed according to the following steps:

a) the "localized" basis sets (i.e. the usual STO functions more or less localized at the origin plus a dozen or so of the appropriate Rydberg hydrogenic orbitals) were chosen in order to obtain a reasonable accurate representation of the discrete part of the spectrum; generally the maximum values of the l's was limited to 5 or 6 and about 70÷80 configurations were included. The calculations were carried out for the $^{1,3}S^e$, $^{1,3}P^o$ and $^{1,3}D^e$ manyfolds.

b) Several basis functions like (15) were added to the previous basis sets which, for each value of l (the first 3 or 4 values of l were admitted), differed only for the values of k. The k's were generally chosen equally spaced between a value close to zero (~ 0.1) up to ~ 2. The configurations added to the previous ones were mostly of the $1S\ f_1>$ type for the manyfold of total angular momentum 1, and a total of $140 \div 150$ configurations were admitted.

c) By diagonalizing the metric of the resulting configuration space the most redundant linear combinations were identified and eliminated by all the subsequent calculations. The diagonalization of the hamiltonian in the resulting configurational space gave a set of $130 \div 140$ discrete states $\varphi_{n,1}(x_1,x_2)$ (normalized to 1).

d) The spinless radial diagonal elements $P_{n,1}(r_1)$ of the one electron density matrix

$$P_{n,\ell}(r_1) = 2\int dx_2\, d\sigma_1\, d\hat{r}_1 \left|\varphi_{n,\ell}(x_1,x_2)\right|^2 \tag{16}$$

were computed for each of the discrete states lying in the continuum (i.e. the corresponding eigenvalue $E_{n,l} > -2.0$ a.u.) for $0 \leqslant r \leqslant 60$ a.u.

It is expected that at large distances the behaviour of $P_{n,1}$ (presuming that the wave packet $\varphi_{n,1}$ yelded by the C.I. calculation are narrow enough) should be close to the true asymptotic behaviour (a normalization in the unity energy range is assumed)

$$r^2 P_{\mathcal{E},\ell}(r) \underset{r\to\infty}{=} (2/\pi\mathcal{E}^{1/2})\left\{\cos\delta_\ell(\mathcal{E})\left[F_{\mathcal{E},\ell}(r) + tg\,\delta_\ell(\mathcal{E})\,G_{\mathcal{E},\ell}(r)\right]\right\}^2 \tag{17}$$

In eq. 16 $F_{\mathcal{E},1}$ and $G_{\mathcal{E},1}$ are the regular and the irregular Coulomb waves[20], $\delta_1(\mathcal{E})$ is the phase shift and $\mathcal{E} = E_{n,1} + 2.0$. By comparing the behaviour of (16) with that of

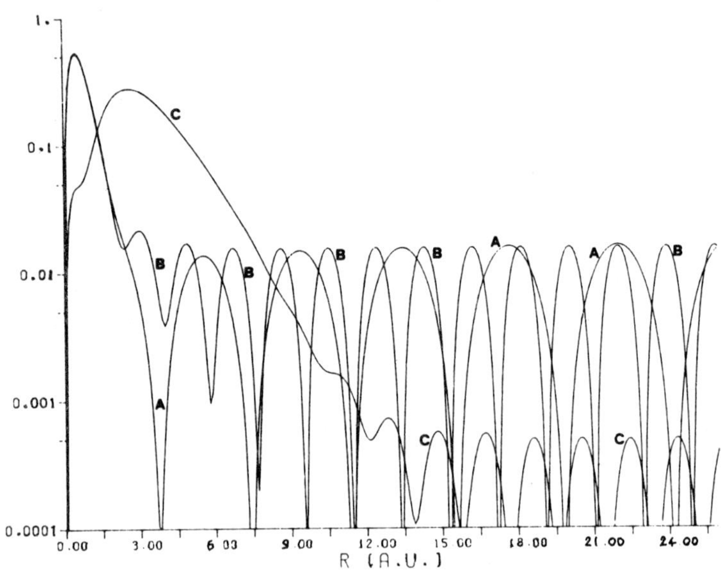

FIGURE 1. Radial spinless densities $P_{n,1}$ times r^2 versus r for three $^1P^o$ states of energies: -1.78426 A; -0.71725 B; -0.68963 C

(17) in the range of r between 20 and 50 a.u. it was possible to obtain the proportionality constant, needed to δ-normalize the states $\varphi_{n,1}$, and the phase shifts. In fig.1 the $P_{n,1}(r)$ of three states of the $^1P^o$ manyfold are plotted in a logaritmic scale, versus r. From the curves displayed it may be easily inferred that the states labeled A and B are normal continuous states (the behaviour of the corresponding $P_{n,1}$, what is true for almost all cases considered, is extremely well described by (17), once the appropriate normalization constants and phase shifts are determined, for r up to 60 a.u.) while the state labeled by

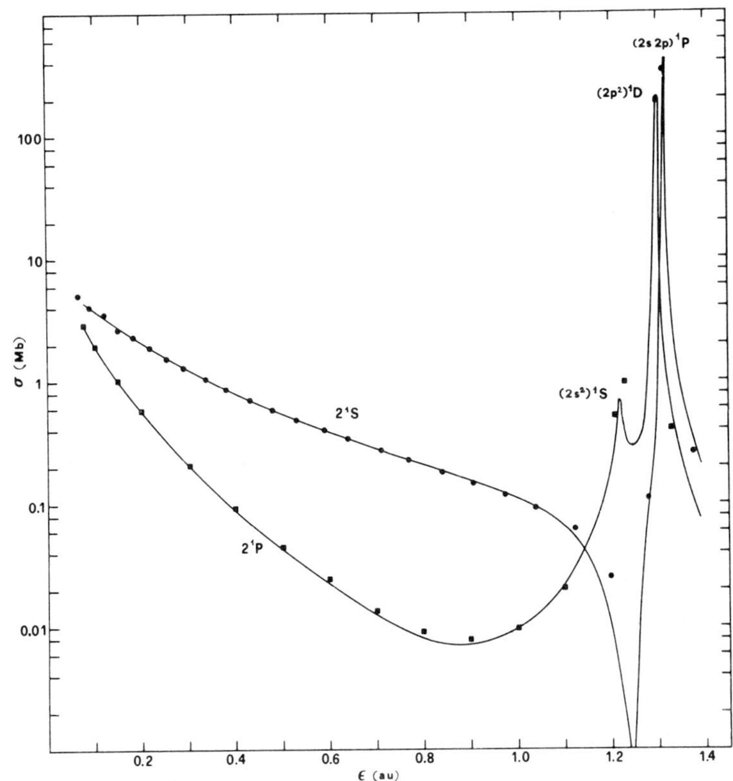

FIGURE 2. σ_{1tot} of the 2^1S^e and 2^1P^o states of He. —— ref. 24; round dots: present results for 2^1S^e; square dots: present results for 2^1P^o.

C, largely localized around the nucleus, is a resonance state.

By this simple numerical artifice it was possible to obtain the looked for properties of the continuous states including those arising by the presence of resonances. The accuracy obtained may be judged by the results obtained, which are presented in the Tables I, II, III and IV and in

TABLE I Bound states Energies of He

n	$^1S^e$	$^1P^o$	$^1D^e$
1	-2.903513	-2.123796	-2.055618
	(-2.903724)[a]	(-2.123843)[a]	(-2.055621)[a]
2	-2.145957	-2.055129	-2.031278
	(-2.145974)[a]	(-2.055146)[a]	(-2.031280)[a]
3	-2.061266	-2.031062	-2.020015
	(-2.061272)[a]	(-2.031069)	(-2.020016)[a]

a) Ref. 21

TABLE II f values for $^1S^e \rightarrow {}^1P^o$ transition in He

		2^1P^o	3^1P^o	4^1P^o	5^1P^o
	L	0.2754	0.0733	0.0298	0.0150
1^1S^e	V	0.2743	0.0729	0.0297	0.0150
		0.2762[a]	0.073[a]	0.030[a]	0.015[a]
	L	0.3772	0.1512	0.0494	0.0227
2^1S^e	V	0.3815	0.1512	0.0492	0.0224
		0.3764[a]	0.1514[a]	0.049[a]	0.02[a]
	L	0.1459	0.6274	0.1441	0.0515
3^1S^e	V	0.1459	0.6345	0.1420	0.0498
		0.1454[a]	0.626[a]	0.144[a]	0.05[a]
	L	0.0261	0.3090	0.8581	0.1498
4^1S^e	V	0.0261	0.3116	0.8657	0.1438
		0.0258[a]	0.306[a]	0.85[a]	0.15[a]

a) Ref. 21

TABLE III Static polarizability of some $^1S^e$ states of He

		$1\ ^1S^e$	$2\ ^1S^e$	$3\ ^1S^e$
α	LG	0.6914	399.8	8428
	VG	0.6904	404.2	8522
		0.6916[a]	401.0[b]	

a) Ref. 23; b) Ref. 24

TABLE IV Autoionization energies of He

$^1S^e$	$^1P^o$	$^1D^e$
-0.7776(-0.7788[a])	-0.6916(-0.6913[a])	-0.7000
-0.6203(-0.6227[a])	-0.5969(-0.5963[b])	
-0.5896(-0.5900[a])	-0.5640(-0.5620[b])	
-0.5476(-0.5482[a])	-0.5468(-0.5463[b])	

a) Ref. 26; b) Ref. 27

the fig. 2, 3 and 4, together with the most accurate results available in the literature. It is very interesting for instance to see that, as shown in fig.2, the resonances of the one-photon photoionization cross section are quite well reproduced. It may also be added that the photoionization cross section of the 2^1S^e state in the region of low photoelectron energies agrees extremely well with the very accurate results of Stewart[25]. In fig. 3 there is also

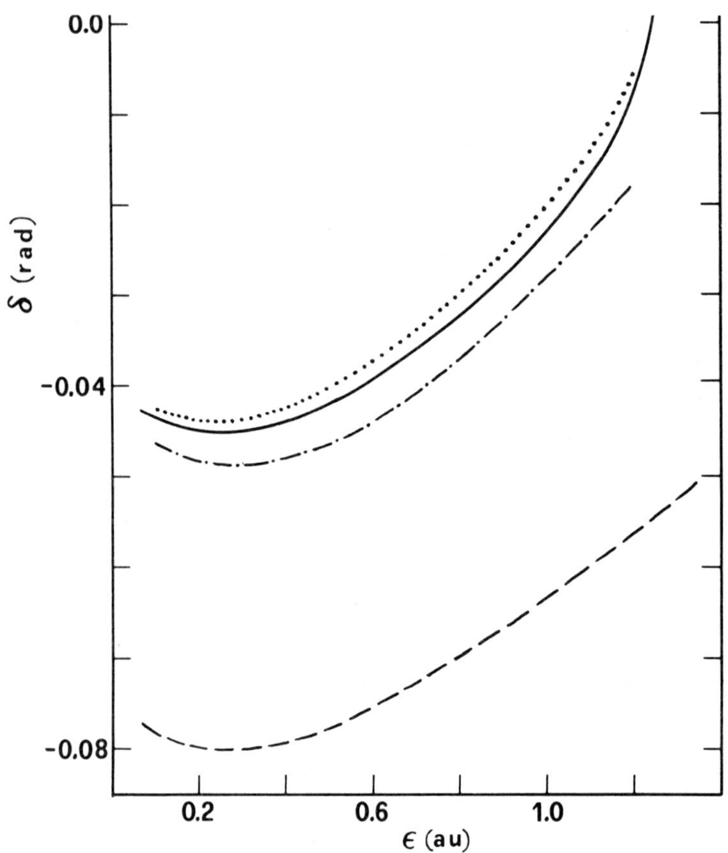

FIGURE 3. Phase shift for the $^1P^o$ [24]. •••• Stewart[25]; —— present results; —·—·— Jacobs[24]; - - - - - SCA 1SεP>

the plot of the results obtained by approximating the final continuous states by a Single Configuration Approximation (SCA) 1s εp>.

The radial p-waves (as well as the radial εs- and εd-waves which will be needed in the following) were determined by solving, by the R-matrix method[28], the SCF

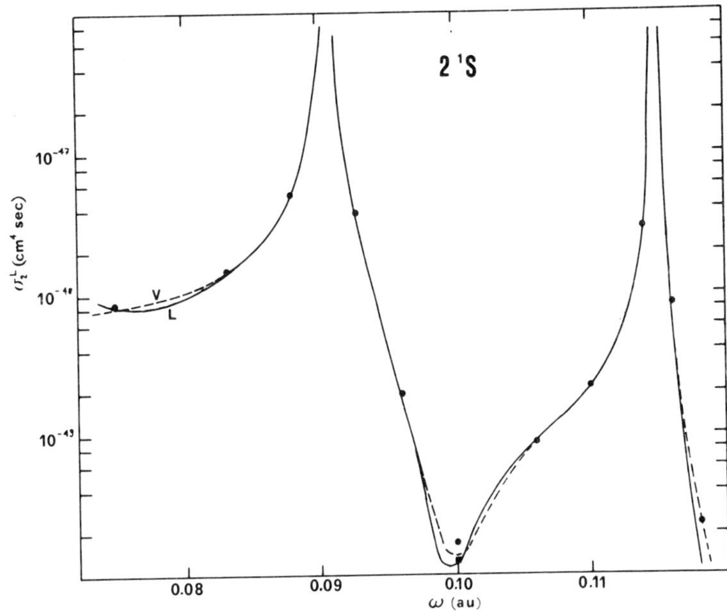

FIGURE 4. σ^{L}_{2tot} of 2 $^{1}S^{e}$ state of He versus the photon energy ($\omega_1 = \omega_2$). ——SCA LG results; ----SCA VG results; dots:wave packet results

variational equation obtained by freezing the 1s orbital to an hydrogenic 1s(Z=2) one.

It should be pointed out that the LG and VG wave packets results are very close to each other. Thus, unless otherwise indicated, the results reported are coincident within the accuracy of the display. The quality of the results shown so far indicates that the intermediate states lying in the continuum may be thought to be rather narrow wave packets. Thus a summation upon these discrete states should represent a reliable approximation for the resolvent

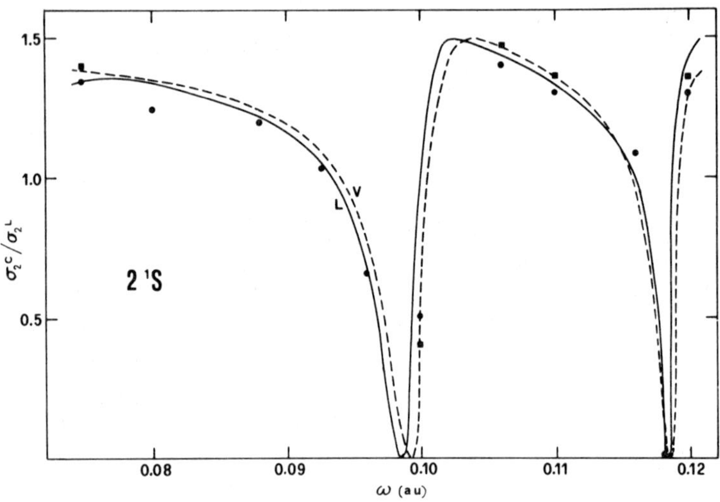

FIGURE 5. $\sigma^C_{2tot}/\sigma^L_{2tot}$ of the 2 $^1S^e$ state of He versus the photon energy ($\omega_1 = \omega_2$). —— SCA LG results; -----SCA VG results; dots:wave packet results. Square dots refer to wave packet VG results when different from LG results.

appearing in (2).

By this way the two photon ionization cross-sections were computed for several states of He, both for linear σ^L_2 and circular σ^C_2 polarized light. Fig.4 and Fig.5 report the results obtained for the 2 $^1S^e$ state of He.

The plots displayed in these figures show that for low photoelectronic energies, therefore far from the resonances, the SCA is quite suitable to describe the final states.

It is obvious however that for higher energies, closer to the resonances, where the correlation effects would strongly undermine the SCA, the wave packets, which take in due consideration the electronic correlation, will give more reliable results.

In our opinion, the numerical method described, may be applied without excessive difficulties to Be and, with some minor modifications, to Mg by exploiting the group-function approach[29]. For these atoms in fact the closed shell cores may be described by a much less sophisticated approximation than that needed for the ns^2 outer shell which, for low photoelectronic energies, should be the main responsible of the two-photon transition processes.

REFERENCES

1. E.Giacobino and B.Cagnac, Progress in Optics XVII, 85 (1980); S.Reynaud, H.Himbert, J.Dalibert, J.Dupont-Roc and C.Cohen Tannoudji, Opt.Comm, 42, 39 (1982); J.Morellec, D.Normand and G.Petite, Adv.in Atom. and Mol.Phys. Vol. 18, 98 (New York Acad. 1982)
2. See for instance G.Wilse Robinson, Phys.Rev. A26, 1482 (1982) and ref. therein.
3. E.Karule, J.Phys.B: Atom.Mol.Phys., 11, 441 (1978).
4. M.Aymar and M.Crance, J.Phys.B: Atom.Mol.Phys, 13, 2527 (1980); ibid 14, 3585 (1981); ibid 15, 719 (1982). E.McGuire, Phys.Rev. A24, 835 (1981).5. R.Moccia, N.K.Rahman and A.Rizzo, J.Phys.B: Atom.Mol.Phys. 16 (1983) (in press).
6. D.N.Zubarev, "Non-equilibrium Statistical Thermodynamics" (Plenum 1974)
   M.Asdente, M.C.Pascucci and A.M.Ricca, Nuovo Cimento 32B, 369 (1976).
7. R.Yaris, R.Lovett and P.Winkler, Chem.Phys. 43, 29 (1979)
8. R.Moccia, Int.J.Quantum Chem. VIII, 293 (1974).
9. M.Jazunski and R.McWeeny, Mol.Phys., 46, 863 (1982).
10. E.Dalgaard, Phys.Rev. A26, 42 (1982).
11. M.Ya.Amusia and N.A.Cherepkov, Case Studies in Atomic Physics, Vol. 5, 50 (North Holland 1975).
    M.Ya.Amusia, N.A.Cherepkov, L.V.Chernysheva, D.M.Davidovic and V.Radojevic, Phys.Rev., A25, 219 (1982)

12. See for instance the several articles which deal with the subject in "Electron-Molecule and Photon-Molecule Collision" Ed. by T.Rescigno, V.McKoy and B.Schneider (Plenum 1979).
13. M.Namiki, Prog.Theor.Phys. 23, 629 (1960).
14. G.Y.Csanak, H.S.Taylor and R.Yaris, Adv.Atom.Mol.Phys. Vol. 7 (Acad. 1971).
15. K.M.Rich, Nucl.Phys, A160, 1 (1971).
16. P.C.Martin and J.Schwinger, Phys.Rev. 115, 1342 (1959).
17. B.Pickup, Chem.Phys., 19, 193 (1979).
18. R.Moccia to be published.
19. See for instance R.F.Barret, B.A.Robson and W.Tobocman Rev.Mod.Phys. 55, 155 (1982).
20. "Handobook of Mathematical Functions" p. 538 ed. by M.Abramovitz and I.Stegun (Dover 1965).
21. Y.Accad, C.L.Pekeris and B.Schiff, Phys.Rev. A4, 516 (1971).
22. Y.Accad, C.L.Pekeris and B.Schiff, Phys.Rev. A4, 885 (1971).
23. A.D.Buckingham and P.G.Hibbard, Symposium of the Faraday Soc. N. 2, 41 (1968).
24. V.Jacobs, Phys.Rev., A3, 289 (1971); ibid A9, 1938 (1974).
25. A.L.Stewart, J.Phys.B.: Atom.Mol.Phys., 11, L431(1978).
26. A.K.Bathia, A.Temkin and J.F.Perkins, Phys.Rev., 153, 177 (1967).
27. T.F.O'Malley and S.Geltman, Phys.Rev. 137,A1344 (1965).
28. P.G.Burke and W.D.Robb, Adv.Atom.Mol.Phys, Vol. 11, 143 (Acad. 1975).
29. R.McWeeny, Rev.Mod.Phys., 32, 335 (1960).

OBSERVABILITY OF OPTICAL COLLISIONS IN THE FREQUENCY AND THE TIME DOMAIN

GERARD NIENHUIS
Fysisch Laboratorium, Rijksuniversiteit Utrecht,
Postbus 80 000, 3508 TA Utrecht, The Netherlands.

Abstract Optical collisions are defined as collisions inducing transitions between the dressed-atom eigenstates, which describe an atom in a strong near-resonant radiation field. We point out that the rate of optical collisions can be directly observed from the intensities of the lines in the three-line fluorescence spectrum of a two-state atom in the field. Another mode of observation is the approach to the steady-state spectrum after switch-on of the field. We demonstrate that this approach is governed by a uniform rate that depends linearly on the rate of optical collisions. The results are valid when the lines are well-separated, which is the case when the precession frequency is large compared to the linewidths. This requires either strong fields, or large detunings from resonance.

INTRODUCTION

An atom in a radiation field with a frequency near resonance, but well outside the natural linewidth, cannot be excited by absorption and spend a natural lifetime in the excited state. However, when collisions perturb the atom, the interaction can momentarily shift the atomic levels into resonance during a collision, and absorptive excitation may occur. The net result is the process

$$A + B + \hbar\omega_L \rightarrow A^* + B.$$

It is this process that enhances the absorption rate in the

far wing of an absorption line, leading to effective collisional line broadening. Weisskopf used the term optical collision for this process, which can be viewed as absorptive excitation during a strong collision. Yeh and Berman[1] introduced the term collisionally aided radiative excitation.

For values of ω_L in the far wing of the absorption line, the absorption coefficient is expected to be simply proportional to the optical-collision cross section[2], at least in the low-intensity limit. For higher intensities, or for frequencies ω_L within the absorption linewidth, the relation between optical collisions and the spectrum deserves more special attention.

FLUORESENCE SPECTRA AT CONSTANT INCIDENT INTENSITY

Recently[3] we have derived a general formal expression for the resonance fluorescence spectrum of an atomic transition, which is valid when the atom suffers binary collisions. For the special case of a two-state, the spectrum takes the form[3]

$$I(\omega) = \frac{A}{\pi} \mathrm{Re}\, \mathrm{Tr}_a d^\dagger \left[\Phi(\omega - \omega_L) - i(\omega - \omega_L - L_d) + \Gamma \right]^{-1} (\bar{\sigma}_o d), \tag{2}$$

where

$$d = | e >< g | \tag{3}$$

is the absorptive part of the atomic-dipole operator. The Liouville operator L_d is the commutator with the dressed-atom Hamiltonian

$$H_d = - \tfrac{1}{2}(\Omega S_x + \Delta S_z), \tag{4}$$

where Ω is the Rabi frequency and $\Delta = \omega_L - \omega_0$ is the detuning of the incident frequency from resonance. The Pauli matrices S_x, S_y and S_z have their usual significance

$$\begin{aligned} S_x &= |e\rangle\langle g| + |g\rangle\langle e| \\ S_y &= -i|e\rangle\langle g| + i|g\rangle\langle e| \\ S_z &= |e\rangle\langle e| - |g\rangle\langle g| \\ S_0 &= |e\rangle\langle e| + |g\rangle\langle g| \,. \end{aligned} \tag{5}$$

The operator Γ, as defined by the relations

$$\begin{aligned} \Gamma\, S_x &= \tfrac{1}{2}AS_x \quad , \quad \Gamma\, S_y = \tfrac{1}{2}AS_y \\ \Gamma\, S_z &= \Gamma\, S_0 = AS_z \end{aligned} \tag{6}$$

describes spontaneous decay, with A the corresponding Einstein coefficient. The collisional relaxation operator Φ has its standard form valid in the unified theory of line broadening[4], but with the dressed- atom Hamiltonian H_d substituted for the atomic Hamiltonian. In a fully quantum-mechanical formalism, Φ is given by its action on an arbitrary atomic density matrix σ_0 according to the relation

$$\Phi(\Lambda)\sigma_0 = N\mathrm{Tr}_1(\Lambda - L_d)(\Lambda - L_d - L_1)^{-1} iL_1\, \rho_p(1)\sigma_0 \quad , \tag{7}$$

with N the number of perturbers, L_1 the Liouville operator for a single perturber, including its interaction with the atom, and ρ_p the equilibrium density matrix of a single perturber. The trace in (7) is taken over the states of this perturber only. An infinitesimal positive imaginary part must be included in Λ for convergence reasons. The steady-state density matrix $\bar{\sigma}_0$ in eq. (2) is the eigenvector of $\Phi(0) + \Gamma + iL_d$ with eigenvalue zero. Initial atom-perturber correlations, which are due to a strong collision at the instant of the fluorescent emission are neglected in eq. (2). This is justified when the frequency difference $\Lambda = \omega - \omega_L$ is separated from the eigenvalues of L_d by no more than an inverse duration of a collision[5].

In the special case of free atoms, where Φ can be ignored, eq. (2) gives a compact notation for Mollow's result[6] for the fluorescence spectrum of a two-state atom. In general eq. (2) relates the spectrum directly to the matrix elements of the collisional relaxation operator Φ.

The collision operator is closely related to the analogue in Liouville space of the transition operator of formal scattering theory[7]. For ordinary-line broadening theory this was demonstrated already by Fano[8]. The only difference of our operator Φ with Fano's case is that we had to include the incident field in the Hamiltonian, which gave rise to the dressed-atom Hamiltonian rather than the Hamiltonian of a free atom.

THE RATE OF OPTICAL COLLISIONS

Optical collisions are defined as collisional transitions between the eigenstates of the dressed-atom Hamiltonian H_d[1,2]. These eigenstates are written as

$$| 1 > = - | e > \sin \tfrac{1}{2}\theta + | g > \cos \tfrac{1}{2}\theta$$

$$| 2 > = | e > \cos \tfrac{1}{2}\theta + | g > \sin \tfrac{1}{2}\theta \, , \qquad (8)$$

where the effective angle θ is defined by

$$\tan \theta = \Omega/\Delta \quad , \quad - \tfrac{1}{2}\pi \leqslant \theta < \tfrac{1}{2}\pi \, . \qquad (9)$$

The eigenvalue of these two states (8) are

$$E_1 = \tfrac{1}{2}\Omega' \qquad E_2 = -\tfrac{1}{2}\Omega' \quad , \tag{10}$$

with

$$\Omega' = \Delta\{1 + \Omega^2/\Delta^2]^{\frac{1}{2}} \tag{11}$$

Then the rate of optical collision, leading to transfer from the dressed state $|1>$ to the dressed state $|2>$ is

$$k(\Omega\ ,\ \Delta) = -\ \mathrm{Tr}\ |2><2|\ \Phi(0)\ |1><1|\ . \tag{12}$$

It will be obvious from eq. (2), that in general the fluorescence spectrum depends on <u>all</u> matrix elements of Φ, and not just on the optical-collision rate (12). This makes it difficult in general to extract this rate from the observed spectrum.

In the special case of large values of Ω' (the secular approximation) the spectrum separates into three non-overlapping lines, each of which has a central part that is basically Lorentzian. The strengths of these lines are then simply related to the rate of optical collisions k. This has been pointed out by several authors[3,9,10]. Furthermore, the expression for the line strengths can then be simply understood in terms of rate equations for the populations of the dressed states[9,10,11]. The widths of the lines depend upon other matrix elements of Φ, but they can be described entirely in terms of the result of completed collisions, which means that knowledge of the S matrix is sufficient to determine the spectrum completely in this secular approximation of well-separated lines. This is best illustrated in the semi-classical picture, where the perturber is assumed to follow a classical rectilinear trajectory. The corresponding semiclassical expression for the collisional relaxation operator Φ is then[3]

$$\Phi(\Lambda)\sigma_0 = \{n \int_0^\infty 2\pi b db \, v \int_{t_2>t_1} dt_2 dt_1 (\Lambda - L_d) \exp[i(\Lambda - L_d)t_2]$$

$$(U_1(t_2,t_1) - 1) \exp[-i(\Lambda - L_d)t_1](\Lambda - L_d)\}_{av}. \quad (13)$$

An integration is performed over the impact parameter b, and an average is taken over the Maxwell distribution of relative velocities v. The relaxation operator is proportional to the perturber density n. The semiclassical evolution operator U_1 has its usual definition

$$U_1(t, t') = \Theta \exp [-i \int_{t'}^{t} ds \, \hat{V}_1(s)] \quad , \qquad (14)$$

with

$$\hat{V}_1(s) = \exp(iL_d s) \, V_1(s) \exp(-iL_d s) \qquad (15)$$

the atom-perturber interaction Liouvillian in the interaction picture. The time dependence of $V_1(s)$ is imposed by the classical trajectory, with the time of closest approach at $t = 0$.

THE THREE-LINE SPECTRUM

The three lines in the fluorescence spectrum in the secular approximation are centered at about the values $\Lambda = \omega - \omega_L = 0, \pm \Omega'$. These positions correspond to the eigenvalues of the dressed-atom Liouvillian occurring in eq.(2). The eigenvalue Ω' corresponds to the eigenvector $|1\rangle\langle 2|$, and gives rise to the three-photon line[12] at $\omega = \omega_L + \Omega'$. The eigenvalue $-\Omega'$, corresponding to the eigenvector $|2\rangle\langle 1|$ produces the fluorescence line at $\omega = \omega_L - \Omega'$.

Finally the eigenvector 0 is twofold degemerate, and has the two eigenvectors $|1><1|$ and $|2><2|$. This eigenvalue yields the Rayleigh line, at $\omega = \omega_L$. An explicit expression for the profiles of these three lines is found by expanding $\overline{\sigma}_0 d$ in terms of these eigenvectors, and accounting only for diagonal parts of Φ and Γ in these eigenvalue subspaces. In this way we simply ignore line coupling between the components. Furthermore, the central part of each line (within a region of the order of the inverse duration of a collision) is well described by setting Λ equal to the corresponding eigenvalue. The relevant matrix elements of Γ and Φ for the three-photon line are

$$\Phi_T = \mathrm{Tr}|2><1|\Phi(\Omega')|1><2| \quad , \quad \Gamma_T = \mathrm{Tr}|2><1|\Gamma|1><2| \ . \tag{16}$$

If we use eq. (13) for Φ, the prescribed value $\Lambda = \Omega'$ (apart from the infinitesimal positive imaginary part) effectively amounts to the limits $t_2 \to \infty$ and $t_1 \to -\infty$ for this matrix element. By using the identity

$$U_1(\infty,-\infty)\sigma_o = S\sigma_o S^{\dagger} \tag{17}$$

with S the semiclassical S matrix, we obtain

$$\Phi_T = \{n \int_o^{\infty} 2\pi b db \ v \ (1 - <1|S|1><2|S^{\dagger}|2>\}_{av} . \tag{18}$$

The radiative contribution Γ_T to the linewidth of the three-photon line can be directly found from (6), (8) and (16), with the result

$$\Gamma_T = A(3\Omega^2 + 2\Delta^2)/4\Omega'^2 \ . \tag{19}$$

The analogous matrix element relevant to the fluorescence

line are

$$\Phi_F = \{n \int_0^\infty 2\pi b db\ v\ (1 - \langle 2|S|2\rangle\langle 1|S^\dagger|1\rangle)\}_{av} = \Phi_T^* \qquad (20)$$

and

$$\Gamma_F = \Gamma_T \ . \qquad (21)$$

Evaluation of the Rayleigh line requires knowledge of $\Phi(0) + \Gamma$, projected on the two-dimensional subspace $|1\rangle\langle 1|$ and $|2\rangle\langle 2|$ for which L_d has eigenvalue zero. On this basis the projected part of $\Phi(0) + \Gamma$ has the matrix representation

$$M = \begin{pmatrix} q & -p \\ -q & p \end{pmatrix} \qquad (22)$$

with

$$p = Ag_F^2 + k(\Omega,\Delta)$$
$$q = Ag_T^2 + k(\Omega,\Delta) \ , \qquad (23)$$

where we introduce the abbreviated notation

$$g_F = \cos^2\tfrac{1}{2}\theta = (\Omega' + \Delta)\ /2\Omega' \qquad (24)$$
$$g_T = \sin^2\tfrac{1}{2}\theta = (\Omega' - \Delta)\ /2\Omega'$$

The rate of optical collisions k in the semiclassical limit has the explicit form

$$k(\Omega,\Delta) = \{n \int_0^\infty 2\pi\ b db\ v\ |\langle 1|S|2\rangle|^2\}_{av} \ . \qquad (25)$$

(Note that it follows from the unitarity of the S matrix that

$$|\langle 1|S|2\rangle|^2 = 1 - |\langle 1|S|1\rangle|^2 = 1 - |\langle 2|S|2\rangle|^2 . \qquad (26)$$)

The steady-state density matrix $\bar{\sigma}_o$ in the secular approximation is diagonal in the dressed states, and it is the eigenvector of M with eigenvalue zero. Hence we find

$$\bar{\sigma}_o = |1\rangle f_1 \langle 1| + |2\rangle f_2 \langle 2| \qquad (27)$$

with the dressed-state populations

$$f_1 = p/(p + q) , \qquad f_2 = q/(p + q) . \qquad (28)$$

The three-line emission spectrum in the secular approximation can now be expressed as the sum of three Lorentzians and a delta peak. The final result is

$$I(\omega) = I_F(\omega) + I_T(\omega) + I_{Ri}(\omega) + I_{Rc}(\omega) \qquad (29a)$$

with

$$I_F(\omega) = (S_F/\pi)\ \mathrm{Re}[\, \Phi_F + \Gamma_F - i\ (\omega - \omega_L + \Omega')]^{-1}$$

$$I_T(\omega) = (S_T/\pi)\ \mathrm{Re}[\, \Phi_T + \Gamma_T - i\ (\omega - \omega_L - \Omega')]^{-1}$$

$$I_{Ri}(\omega) = (S_{Ri}/\pi)\ \mathrm{Re}\ [\, p + q - i(\omega - \omega_L)]^{-1} \qquad (29b)$$

$$I_{Rc}(\omega) = S_{Rc}\ \delta(\omega - \omega_L) .$$

The strengths of the fluorescence line and the three-photon line are

$$S_F = Ag_F^2\, f_2 \qquad , \qquad S_T = Ag_T^2\, f_1 . \qquad (30)$$

The Rayleigh line consists of an incoherent and a coherent component with strengths

$$S_{Ri} = Ag_R^2\ 2(pf_2 + qf_1)/(p + q)$$

$$S_{Rc} = Ag_R^2\ (p - q)(f_1 - f_2)/(p + q), \qquad (31)$$

when

$$g_R = \cos\tfrac{1}{2}\theta \, \sin\tfrac{1}{2}\theta = \Omega/2\Omega' \qquad (32)$$

is the geometrical means of g_F and g_T. The strengths of the lines can be used as a probe for the determination of the rate of optical collisions, for instance by measuring these strengths as a function of the perturber density for a fixed value of Ω and Δ. A plot of these strengths as a function of n is includedin figs.1 and 2. The fluorescence line has a strength that is most sensitive to the value of k. The sum of the two Rayleigh strengths is independent of k

$$S_R = S_{Ri} + S_{Rc} = Ag_R^2 \qquad (33)$$

From the definition (12) of k one can derive that the low-intensity limit of k is closely related to the frequency-dependent linewidth $\gamma(\omega_L)$ of the low-intensity absorption profile[3]. One obtains

$$k(\Omega,\Delta) = (\Omega^2/2\Omega'^2)\, k_o(\Omega,\Delta) \qquad (34)$$

with

$$\lim_{\Omega\downarrow o} k_o(\Omega,\Delta) = \gamma(\omega_L) \; . \qquad (35)$$

Hence for not too strong intensities, the line strengths are fully determined by the low-intensity absorption profile. The widths of the side-bands are equal, and depend upon the phases of the S matrix elements.

TIME-DEPENDENT EMISSION SPECTRUM

The high intensities needed for the actual observation of the three-line spectrum often require the use of pulsed lasers, and some understanding is needed of the way in which the strengths of the lines follow the variations in the incident intensity. This problem has been discussed in the case of adiabatic intensity variations[12,13], when the rate of change of the incident intensity is small compared to the frequency separation Ω' of the lines. These studies refer directly to the physical time-dependent spectrum[14], which is the total time-dependent intensity passing the spectrometer as a function of its setting frequency. This spectrum is not a property of the fluorescent system alone, but of the system and the spectrometer combined. It can be demonstrated[15], however, that this physical spectrum $\overline{I}(\omega,t)$ is equivalent to a double convolution over frequency and time of a filter function $s(\omega,\tau)$, which contains the properties of the spectrometer, and a quasi-spectrum $I(\omega,t)$, which is the time derivative of the spectral distribution of the total fluorescence energy emitted up to a given time. Hence we write

$$\overline{I}(\omega,t) = \int d\omega' \int_0^\infty d\tau \; s(\omega',\tau) \; I(\omega - \omega', t - \tau). \quad (36)$$

The smoothing function $s(\omega,\tau)$ has a frequency width $\Delta\omega$ and a time spread $\Delta\tau$ obeying the uncertainty relation $\Delta\omega \, \Delta\tau > 1$, so that a very good spectral resolution is bound to give rise to a poor resolution in time, as it should be. In general the quasi-spectrum I (which is equivalent to the Page-Lampard definition[16,17] of a time-dependent spectrum) can be negative. However, for a stationary incident intensity, I is equal to the steady-state fluorescence spectrum.

A major advantage of the separation (36) of the properties of the spectrometer and of the emitting system is that the same techniques that led to eq. (2) for the stationary spectrum can likewise be applied to the evaluation of the time-dependent quasi-spectrum I. The main difference is that now the Rabi-frequency Ω, and thereby the dressed-atom Hamiltonian H_d, and the relaxation operation Φ become functions of time. If we adopt the assumption that the incident intensity varies negligibly during a collision, then we can demonstrate that the quasi-spectrum $I(\omega,t)$ can be put in the form

$$I(\omega,t) = \frac{A}{\pi} \mathrm{Tr}_a d^{\dagger} D_o(\Lambda,t) \tag{37}$$

with D_o the solution of the inhomogeneous first-order differential equation

$$\frac{d}{dt} D_o(\Lambda,t) = [i\Lambda - iL_d(t) - \Gamma - \Phi(\Lambda,t)]D_o(\Lambda,t) + \sigma_o(t)d \ . \tag{38}$$

The density matrix σ_o of the fluorescent atom is determined by the homogeneous differential equation

$$\frac{d}{dt} \sigma_o(t) = [- iL_d(t) - \Gamma - \Phi(o,t)]\sigma_o(t) \ . \tag{39}$$

The relaxation operator Φ depends on time through the dressed-atom Liouvillian L_d, but otherwise eq.(7) is unchanged.

Eqs.(37)-(39) completely determine the emitted spectrum for an arbitrary given time-dependent Rabi frequency $\Omega(t)$. When Ω has been constant for several radiative lifetimes, the spectrum approaches steady-state spectrum (2), as can be checked immediately.

APPROACH TO STEADY STATE

A problem of practical interest is the time-dependent spectrum immediately following the switch-on of the incident radiation.[18] We adopt the secular approximation that $\Omega' \gg A, \Phi$, so that the lines are well-separated, and we assume that the Rabi frequency Ω is constant from some initial time $t = o$. Initially the density matrix σ_o can have off-diagonal elements in the dressed state $|1\rangle$ and $|2\rangle$. However, these off-diagonal terms produce rapidly oscillating contributions to the operator D_o, with a frequency of the order of Ω'. When the spectral resolution of the spectrometer is better than Ω', which is certainly needed for the separate observation of the three lines, the contribution of these off-diagonal terms can be ignored. The diagonal part of σ_o takes the form (27), where the time-dependent populations f_1 and f_2 obey the matrix equation

$$\frac{d}{dt}\begin{pmatrix} f_1(t) \\ \\ f_2(t) \end{pmatrix} = -M \begin{pmatrix} f_1(t) \\ \\ f_2(t) \end{pmatrix} \tag{40}$$

with M given by (22). The solution of (40) for given initial values $f_1(o)$ and $f_2(o)$ is determined by

$$\begin{aligned} pf_2(t) - qf_1(t) &= (pf_2(o) - qf_1(o)) \exp(-(p+q)t) \\ f_1(t) + f_2(t) &= 1 \ . \end{aligned} \tag{41}$$

The time-dependent quasi-spectrum can be expressed as the sum of four components, as in (29a), where the components obey the uncoupled differential equations

$$\frac{d}{dt}I_F(\omega,t) = [i(\omega-\omega_L+\Omega')-\Gamma_F-\Phi_F]\, I_F(\omega,t)+\frac{1}{\pi}Ag_F^2 f_2(t)$$

$$\frac{d}{dt}I_T(\omega,t) = [i(\omega-\omega_L-\Omega')-\Gamma_T-\Phi_T]\, I_T(\omega,t)+\frac{1}{\pi}Ag_T^2 f_1(t)$$

$$\frac{d}{dt}I_{Ri}(\omega,t) = [i(\omega-\omega_L)-p-q]\, I_{Ri}(\omega,t)+\frac{2}{\pi}Ag_R^2(pf_2(t)+qf_1(t))/(p+q)$$

$$\frac{d}{dt}I_{Rc}(\omega,t) = i(\omega-\omega_L)I_{Rc}(\omega,t)+\frac{1}{\pi}Ag_R^2(p-q)(f_1(t)-f_2(t))/(p+q) \; . \tag{42}$$

Eqs. (41) and (42) determine the quasi-spectrum completely, for given initial populations $f_1(o)$ and $f_2(o)$, and for given initial profiles. The time-dependent strengths of the four components is still given by eqs. (30) and (31), where we have to substitute the time-dependent values of the populations f_1 and f_2. An important general conclusion is that each of the components has a strength that approaches its steady-state value at the same rate p+q. This opens a way of determining p+q, and thereby the rate of optical collisions in the time domain. The precise values of the initial strengths at time zero are irrelevant for this method.

We wish to illustrate these general results in two complementary cases of practical interest. When the switch-on of the incident intensity occurs adiabatically, during a rise time long compared to Ω'^{-1}, but still short compared to the inverse linewidths, then at time zero the atom is in the dressed state $|1\rangle$, so that we may write

$$f_1(o) = 1 \quad , \quad f_2(o) = 0 \; . \tag{43}$$

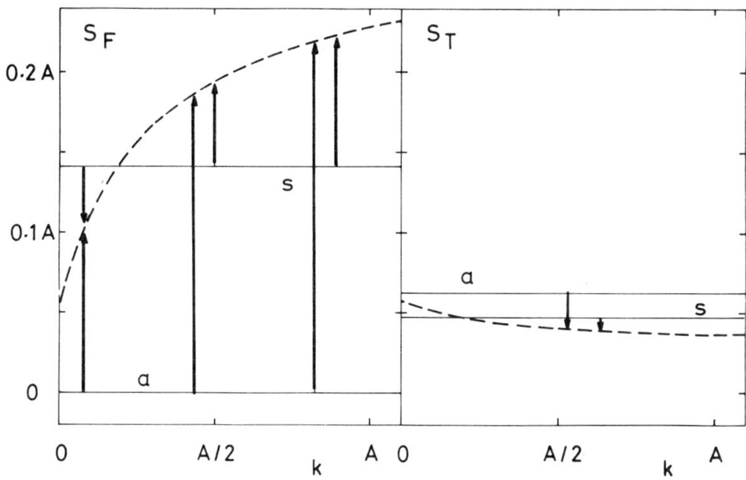

FIGURE 1. Initial and steady-state values of the strengths of the fluorescence (S_F) line and the three-photon (S_T) line as a function of the rate of optical collisions, in the case that $\Omega^2 = 3\Delta^2$. Solid lines indicate the initial strengths in the case of adiabatic (a) and sudden (s) switch-on. The broken curves represent the steady-state values. The arrows denote the transition from the initial values to the steady-state values at the uniform rate p + q.

Then the fluorescence component has initially a strength zero, whereas the three-photon line has an initial strength above its stationary value.

When the rise time is short compared to Ω'^{-1}, the atom is still in its unperturbed ground state $|g\rangle$ at t=o. Then the initial dressed-state populations are

$$f_1(o) = g_F \qquad f_2(o) = g_T. \tag{44}$$

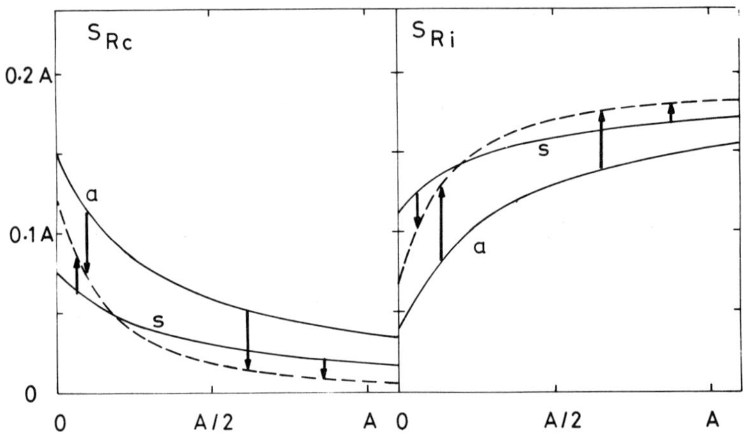

FIGURE 2. Initial and steady-state values of the strengths of the incoherent (S_{Ri}) and the coherent (S_{Rc}) Rayleigh line. The significance of the curves is the same as in fig.1.

The population of state $|1\rangle$ is higher than in the steady state for low values of the perturber density, for which $k < Ag_R^2$. Then the strengths of the fluorescence line and the incoherent Rayleigh line decrease towards their steady-state value, and the other two components have increasing strengths. The opposite is true for higher perturber densities, where $k > Ag_R^2$.

The behaviour of the initial strengths and the steady-state strengths of the four components is illustrated in figs. 1 and 2, as a function of the rate of optical collisions k. The approach to the steady-state strengths occurs at the uniform rate p + q, which increases linearly with k.

CONCLUSIONS

The rate of optical collisions and thereby the optical-collision cross section is a dominant quantity determining the strengths of the four components in the fluorescence spectrum of a two-state atom, as well as their approach to their steady-state values. This offers various ways of measuring this quantity directly from experimental observations. At the same time these observations can verify the predicted behaviour of the ratios of these strengths, thereby testing the validity of the theoretical derivations.

REFERENCES

1. S. Yeh and P. R. Berman, Phys. Rev. A19, 1106 (1979).
2. V. S. Lisitsa and S. I. Yakovlenko, Sov. Phys.-JETP 39, 759 (1974).
3. G. Nienhuis, J. Phys. B 15, 535 (1982).
4. E. W. Smith, J. Cooper and L. J. Roszman, J. Quant. Spectrosc. Radiat. Transfer 13, 1523 (1973).
5. K. Burnett and J. Cooper, Phys. Rev. A22, 2044 (1980).
6. B. R. Mollow, Phys. Rev. 188, 1969 (1969).
7. J. R. Taylor, Scattering Theory (Wiley, New York, 1972).
8. U. Fano, Phys. Rev. 131, 259 (1963).
9. K. Burnett, J. Cooper, P. D. Kleiber and A. Ben-Reuven, Phys. Rev. A25, 1345 (1982).
10. S. Reynaud and C. Cohen-Tannoudji, J. Physique 43, 1021 (1982).
11. G. Nienhuis, Acta Phys. Pol. A61, 235 (1982).
12. E. Courtens and A. Szöke, Phys. Rev. A15, 1588 (1977).
13. P. D. Kleiber, J. Cooper, K. Burnett and C. Kunasz, Phys. Rev. A27, 291 (1983).
14. J. H. Eberly and K. Wódkiewicz, J. Opt. Soc. Am. 67, 1252 (1977).
15. G. Nienhuis, Physica C95, 391 (1979).
16. C. H. Page, J. Appl. Phys. 23, 103 (1952).
17. D. G. Lampard, J. Appl. Phys. 25, 803 (1954).
18. J. H. Eberly, C. V. Kunasz and K. Wódkiewicz, J. Phys. B 13, 217 (1980).

COLLISIONAL REDISTRIBUTION BY TWO-ATOM COHERENCE IN WEAK FIELDS

ABRAHAM BEN-REUVEN
Chemistry Department, Tel Aviv University, 69978
Tel Aviv, Israel

KEITH BURNETT AND JOHN COOPER
Joint Institute for Laboratory Astrophysics,
University of Colorado and the National Bureau
of Standards, Boulder, Colorado 80309, U.S.A.

Abstract Conventional description of coherence phenomena by the use of Bloch-type linear master equations may become inadequate in pure gases in which collisions occur between similar atoms (or molecules). Significant deviations from linear-response behavior are expected in one-photon spectra only when the driving coherent radiation is sufficiently strong. It is however argued that in two-step resonances (such as resonance fluorescence) linear response may become inadequate even in the limit of weak fields, as result of collisions in which both participants are coherently driven before (or after) their encounter. An experiment is suggested for the observation of such effects in the collisional redistribution spectrum.

INTRODUCTION

Whenever radiation-assisted collisions are discussed, one usually refers to processes occurring in very strong radiation fields. The strength of the coupling between atom and field Ω (the Rabi frequency), is generally required to obey

$$\Omega\tau_c \gtrsim 1 \tag{1}$$

where τ_c is the duration of a single collision. In this region, to be called here the strong-coupling region, one usually treats individual collision events as occurring between two "dressed" atoms (or molecules) modified by the presence of the strong radiation field. These atom-plus-radiation states serve as a basis for describing the collision dynamics.[1] Such collisions may lead to processes that are almost totally forbidden in the absence of the radiation.

A special case in this category is the case of "optical collisions", or radiation-dissipating collisions, involving similar radiation-modified atoms (or molecules).[2] Andreev and Lisitsa[3] count this type of collisions as the major source of resonance broadened atomic line absorption in the strong-coupling limit defined by Eq. (1). However, as atomic absorption exists near resonance even in relatively weak fields where optical collisions are negligible, it would require measuring the absorption intensity in the far wings (outside the so-called impact domain) to see the effects of optical collisions. A more direct way to measure the optical-collision rate in strong fields by means of the resonance-fluorescence spectral line shape was suggested recently.[4]

As one reaches down below the strong-coupling region, i.e., when

$$\Omega\tau_c \ll 1 \tag{2}$$

it is claimed by Lisitsa and Andreev that ordinary linear-response-type collision broadening theories should prevail. This class of theories[5] usually originates in the weak coupling region where, in addition to (2) also

$$\Omega T_2 \ll 1. \tag{3}$$

holds, T_2 being the coherence damping time.

The underlying assumption to all these weak-field theories is what one may call the linear-response (or thermal-bath) paradigm. According to this assumption, each atom (or molecule) driven by the applied radiation field is surrounded by other molecules acting as a thermal bath, characterized by an equilibrium distribution of states. This assumption is undoubtedly plausible in the case of foreign-gas broadening, where the surrounding-medium molecules are not in resonance with the applied radiation. However, this paradigm has been frequently carried over to the treatment of resonance broadening involving a pure gas of similar atoms. The excuse behind such practices is the claim that in sufficiently weak fields only a small fraction of the atoms is excited and therefore neighboring atoms still act as thermal-bath particles.

The linear-response paradigm has been even extended[6] to the medium-coupling region in which (2) still holds, but

$$\Omega T_2 \gtrsim 1. \tag{4}$$

This region is characterized by saturation phenomena and a variety of dynamic Stark effects resulting from a strong modification of the atomic states by the field in between collisions. The collision-damping rates are retained in the equations of motion in their linear-response form, under the pretence that the collision kernels describing the outcome of the encounter have no time to "see" the effect of the radiation owing to inequality (2). However, the collision-damping rates depend not only on the collision kernels but also on the state of the perturbing atom. And in a strongly-saturating field there is no excuse to treat the distribution of perturbing atoms as thermal equilibrium, all collision partners being alike.

This inconsistency may be removed by replacing the linear-response paradigm with the self-consistency paradigm.[7] According to this paradigm the actual coherently-driven distribution of perturbing atoms should be used in calculating the damping rates. However, the latter are needed in order to calculate the distributions. Thus the damping rates and the distributions mutually depend on each other in a self-consistent manner. Some of the consequences of this self-consistency paradigm will be briefly described below.[8] They are generally of significance in the saturation dominated medium-coupling region. In the weak-coupling region, one may plausibly suspect, the linear-response paradigm should take over, as it does indeed in one-photon resonances. The major conjecture raised here, however, is that when dealing with two-step resonances, with an intermediate resonant level (e.g., double-resonance absorption, resonance fluorescence, or resonance Raman scattering) the linear-response paradigm may fail even in weak fields.

Deviations from linear response, resulting from collisions in which both collision partners are coherently driven before (or after) their encounter (i.e., collision-induced two-atom coherence[7]) may show up in the response of the system even to lowest order in the applied field strength. More particularly, we argue that such collisions should lead to a measurable modification of the intensity of the diffuse collisional-redistribution peak observed in resonance fluorescence spectra[9,10] alongside the sharp elastic-scattering Rayleigh peak.

In linear-response theories (assuming binary-collision homogeneous impact broadening of nondegenerate two-level atoms), the ratio of redistribution to Rayleigh line intensities is equal to $2\gamma^{(g)}/\gamma_1$, twice the ratio of collisional dephasing rate to upper-level spontaneous decay rate, and hence is proportional to the gas pressure. The most significant correction resulting from two-atom coherence is the addition of a term $\zeta/(2\gamma^{(g)} + \gamma_1)$ to the intensity ratio, in which ζ is a collision rate representing the scattering of the spatial phase coherence (rather than its damping) by the encounter of two driven atoms. In the pressure range where $\gamma^{(g)} >> \gamma_1$, this term is independent of the pressure, and may therefore be measured by linear extrapolation to zero pressure.

THE SELF-CONSISTENCY PARADIGM

Consider now more specifically the meaning and consequences of the self-consistency paradigm. The discussion is confined here to the medium-coupling region. In addition we demand that the atoms be excited sufficiently close to resonance so that

$$|\Delta\omega|\tau_c << 1 \tag{5}$$

where $\Delta\omega = \omega_L - \omega_0$ is the detuning of the beam frequency from the atomic resonance frequency. Condition (5) is often called the impact condition (in weak fields).[5] It allows the treatment of collisions as instantaneous events. It is also consistent with the binary-collision condition

$$\tau_c << T_2 \tag{6}$$

if $\Delta\omega$ is allowed to scan the core of the line profile. For simplicity's sake we shall further assume that velocity effects (such as Doppler broadening or velocity-dependent collisions) are negligible, i.e.,the line broadening is homogeneous. Also, we shall assume two-level atoms, neglecting space-degeneracy and neighboring-level effects.

The linear-response paradigm leads under these conditions to the well known set of linear master equations (the Bloch equations) for the coherently-driven one-atom density matrix (defined on the atomic internal states),

$$\frac{\partial}{\partial t}\rho^1(t) = -i[\mathcal{H}_0^{\ 1} + \mathcal{V}^{1,R}]\rho^1(t) - \Gamma^1\rho^1(t). \tag{7}$$

Here $\mathcal{H}_0^{\ 1}$ is the Hamiltonian super-operator (or commutator-operator) for the isolated atom "1"; $\mathcal{V}^{1,R}$ is the corresponding super-operator describing the coupling to the applied radiation; Γ^1 is a super-matrix of damping rates. It can incorporate additively[11] collision-broadening and spontaneous-emission decay rates. In the dressed-atom Floquet representation, in the rotating-wave approximation, $\mathcal{H}_0^{\ 1}$ incorporates the radiative energy transfers $\Delta\omega$, and $\mathcal{V}^{1,R}$ is time independent[7] in a coherent monochromatic radiation.

The binary-collision contributions to Γ^1 are obtained in linear-response theory by averaging a two-atom collision kernel $\Phi^{1,2}$ over an equilibrium distribution of perturber-atom "2" states, ρ^2_{eq}, and summing over all N perturbers:

$$(\Gamma^1)_{coll.} = \Gamma^{\ell.r.} = N\ tr_2(\Phi^{1,2}\ \rho^1_{\ eq}). \tag{8}$$

The self-consistency paradigm asserts that neither atom "1" nor "2" are thermal-bath particles. Actually, being identical, their instantaneous distributions at every moment must be identical. Therefore the equilibrium distribution in (8) ought to be replaced by the actual distribution $\rho^2(t)$ at the moment the collision takes place. A bilinear rate equation is thus obtained[8] instead of (8):

$$\frac{\partial}{\partial t}\rho^1(t) = -i[\mathcal{H}_0^{\,1} + \mathcal{V}^{1,2}]\rho^1(t)$$

$$-N\ tr_2[\Phi^{1,2}\rho^2(t)]\rho^1(t). \qquad (9)$$

Here ρ^1 and ρ^2 pertain to different degrees of freedom. Yet their functional forms must be identical.

The derivation of Eq. (9) is based implicitly on the assumption that the motion of the two atoms is uncorrelated prior to their encounter. Otherwise the product $\rho^1\rho^2$ should be replaced by a correlated two-atom distribution. However, this approximation is consistent with the binary-collision approximation, as encounters between two atoms correlated by past interactions amount to a recollision event (e.g., a triple collision). This is reminiscent of the assumptions underlying the derivation of the Boltzmann equation,[12] except that here only internal states are concerned, and the system is driven by a coherent radiation field.

A major implication of Eq. (9) is the modification of the apparent line widths by the applied field. Spectral line shapes can be obtained from steady-state solutions of (9). Therefore, whereas (8) produces the unmodified linear-response rates, a self-consistent set of modified line-width parameters is produced by

$$\Gamma^{s.c.} = N\ tr_2(\Phi^{1,2}\rho^2_{s.c.}) \qquad (10)$$

where $\rho_{s.c.}$ is the steady-state solution of (9).

Consider an <u>optical two-level model</u> for the atomic transition from the ground state g to the excited state e. By "optical" we imply that $\hbar\omega_0 >> k_BT$, where ω_0 is the transition frequency, k_B - the Boltzmann constant, and T - the gas temperature. Then

$$(\rho^2_{eq})_{gg} = 1 \; ; \qquad (\rho^2_{eq})_{ee} = 0 \; . \tag{11}$$

The linear-response collisional-dephasing rate, describing the damping of ρ^1_{eg} while atom 2 is in the ground state, can be denoted as

$$(\Gamma^{\ell.r.})_{eg;eg} \equiv \gamma^{(g)} \tag{12}$$

the superscript (g) specifying the perturber state.

In the self-consistent theory, $\rho^2_{s,c}$ completely differs from the equilibrium distribution, especially when saturation sets in. It can have both diagonal and non-diagonal elements that mix various elements of the supermatrix $\Gamma^{\ell.r.}$. Suppose, however, that the T_2 (dephasing) rate is the only significant collisional rate. We then can simply write for the self-consistent dephasing rate

$$\gamma^{s.c.} = \gamma^{(g)}(\rho^2_{s.c.})_{gg} + \gamma^{(e)}(\rho^2_{s.c.})_{ee} \tag{13}$$

Here $\gamma^{(e)}$ is defined like $\gamma^{(g)}$, but with all perturber atoms in the excited state e before the collision, and $(\rho^2_{s.c.})_{aa}$ is the steady-state self-consistent population of level a = g or e.

A general solution of the nonlinear equations is not an easy task. However, since we are only interested in the steady state, we can make the ansatz that the steady-state populations are solutions of the linear Bloch equations (7) but with $\gamma^{\ell.r.}$ replaced by $\gamma^{s.c.}$. Using the well-known solution for the two-level problem,

$$(\rho^2_{s.c.})_{ee} = \frac{1}{2}\,\frac{\Omega^2(\gamma^{s.c.}/\gamma_1)}{(\Delta\omega)^2 + (\gamma^{s.c.})^2 + \Omega^2(\gamma^{s.c.}/\gamma_1)} \tag{14}$$

(γ_1 being the spontaneous decay rate of the excited state, assumed here $\ll\gamma^{(g)}$), and inserting it in (13), we obtain a closed algebraic equation for $\gamma^{s,c.}$. On resonance ($\Delta\omega=0$), this equation becomes quadratic, with the stable solution,

$$\gamma^{s.c.} = \frac{1}{2}\,\gamma^{(g)}\,\{(1-D) + [(1+D)^2 + 2AD]^{\frac{1}{2}}\} \qquad (15)$$

where

$$A = (\gamma^{(e)} - \gamma^{(g)})/\gamma^{(g)}; \quad D = \Omega^2/\gamma^{(g)}\gamma_1 \; . \qquad (16)$$

Obviously, a deviation from linear response requires $A \neq 0$; i.e., that the damping by collisions be sensitive to the internal state of the perturber atom. The dimensionless parameter A (dependent only on the interatomic collision dynamics) is a measure of this sensitivity.

The most suitable place to look for such phenomena is a gas in which the excited state has a much larger collision radius than the ground state (e.g., high Rydberg states in atomic vapors). In such cases the apparent line width must become a nonlinear function of the gas pressure and dependent on the radiation beam power (through D). Off resonance, it must also depend on $\Delta\omega$.

The variation of the self-consistent line width with the dimensionless parameter D (representing the ratio of beam power to gas density) is shown in Figure 1 for A = 2. The plot of $\gamma^{s.c.}/\gamma^{(g)}$ vs. D displays the line width as a monotonously increasing function of the beam power (keeping the gas density fixed). A similar plot of $\gamma^{s.c.}$ vs. D^{-1} (keeping the beam power fixed) shows the line width as a monotonously-increasing (but not linear) function of the gas density. At relatively high densities (or weak coupling) the linear-response width is approached:

$$\gamma^{s.c.} \sim \gamma^{(g)} \qquad (D << 1) \qquad (17)$$

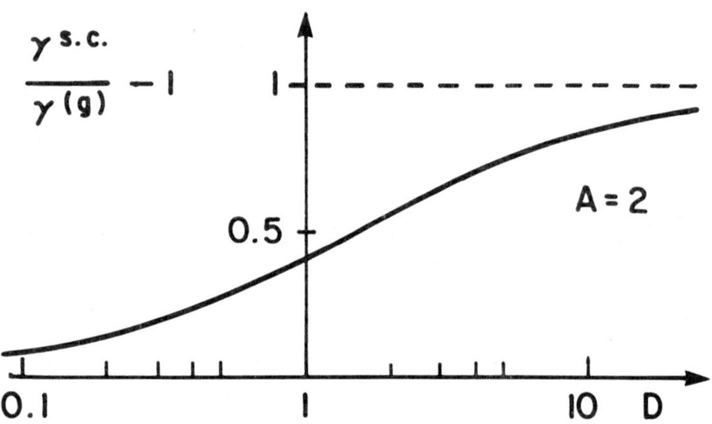

FIGURE 1. The variation of the apparent line width with the ratio of radiation beam power to gas density (defining the parameter D) for one value (A=2) of the collisional state-sensitivity parameter.

while at low densities (or relatively strong coupling), saturation brings the two perturber levels to bear equally in determining the apparent line width:

$$\gamma^{s.c.} \sim \frac{1}{2}\,(\gamma^{(g)} + \gamma^{(e)}) \qquad (D >> 1). \tag{18}$$

TWO-ATOM COHERENCE IN COLLISIONAL REDISTRIBUTION

The nonlinear behavior described above clearly requires a moderately-strong radiation, in the medium-coupling region. A lowest-order-perturbation expression of the steady-state solution to the Bloch equation for a single-photon resonance does not allow for two-atom coherence effects. However, this conjecture is not true anymore when one deals with two-step resonances such as double-resonance absorption, resonance fluorescence, or resonance Raman scattering. In the first instance, the process usually requires two applied field modes (the "pump" and the "probe"). In the latter two instances the "probe" role is supplied by the scattered field mode specified by the detector. It is our major claim here that in these cases, even to lowest-order perturbation, one may get nonvanishing contributions from collisions involving _both_ collision partners in a coherently-driven state before (or after) their encounter.

In resonance scattering, unlike the case of response to an applied radiation, correlations created by the collision between the two atoms do matter. As a result the single-atom-distribution equation (9) is inadequate and one generally should resort to the next equation in a hierarchy of n-atom correlations.[7] However in perturbative solutions a simple diagrammatic expansion can be alternatively used, incorporating in it also the possibility of two-atom coherence extended after the collision, as well as before it.

To demonstrate our major conjecture, let us consider first a simple diagrammatic expression of the perturbative solution to the Bloch equations (i.e., in the absence of two-atom coherence effects) for resonance fluorescence in the optical two-level model described earlier. Figure 2 represents the three typical diagrams that (together with their complex conjugate) lead in the impact-limit rotating-wave approximation to the well known expression[9,10] for the low-field collisionally-redistributed resonance-fluorescence spectrum. The vertical bars in Figure 2 represent propagators (or resolvent operators in steady-state solutions) for the field-free single particle density matrix[7] _including_ here collision broadening of the linear response type. The

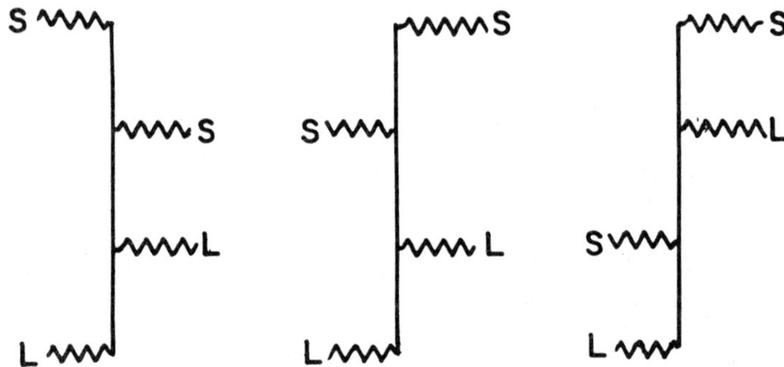

FIGURE 2. Diagrammatic lowest-order-perturbation solution to resonance fluorescence in the density-matrix formalism (see text for detail).

horizontal wavy lines represent couplings to the radiation modes L (the "pump") and S (the "probe", here the scattered mode) as indicated. The orientation of the wavy lines specifies whether the coupling super-operator $\mathcal{V}^{1,R}$ in Eq. (7) operates to the left or right of the density matrix.

As we deal here with the density-matrix formalism, i.e., with scattering <u>rates</u> rather than amplitudes, the two-photon process is represented by fourth-order perturbation, to lowest order. Consider, however, the type of diagrams represented by Figure 3. The cross-hatched rectangle in Figure 3 represents a collision in which the state of both participating atoms is a coherent-superposition state of levels g and e, before as well as after the collision. Provided such a collision kernel does not vanish, the diagram is clearly of the same perturbation order as Figure 2, and therefore should contribute in the weak-field limit. There are other diagrams of the same order, but analysis shows that the one depicted in Figure 3 (with all its time-ordering variations) is the only nonvanishing one for our model case.[8]

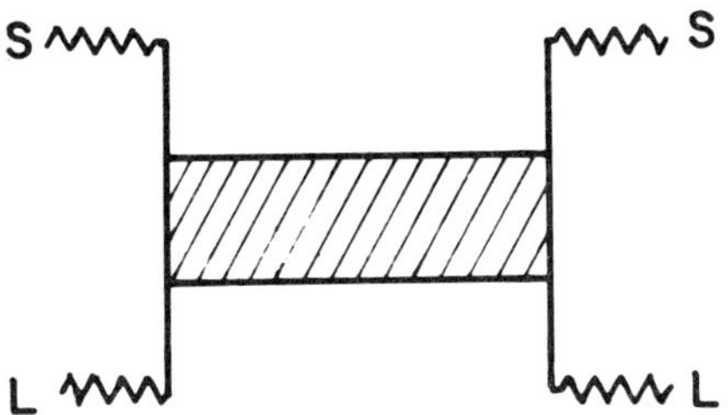

FIGURE 3. A two-atom-coherence contribution to resonance fluorescence. The hatched rectangle represents a collision kernel involving both atoms in a coherence state before and after their encounter.

Before presenting the outcome of such terms, let us discuss further the nature of the collision involved here. A single-atom coherence element of the density matrix, e.g., ρ^1_{eg}, produced by the driving field, is characterized by (among other things) the driving frequency of the radiation (ω_L) and its wave-vector ($\underline{k}_L$). If we consider an atom specified initially in the momentum state $\underline{p}$, its excited state will have $\underline{p}+\underline{k}_L$ for momentum. Although in homogeneously-broadened spectra we eventually average the collision kernel over all momenta, the momentum <u>excess</u> $\underline{k}_L$ remains fixed, and produces a spatially-harmonic dependence of the density matrix on the position of the atom. Ordinary collision-damping processes are "diagonal" in $\underline{k}_L$, in the sense that this momentum excess remains unaltered as result of the collision. A different situation is encountered here in which the collision kernel transforms $\underline{k}_L$ into $\underline{k}_S$ (the scattered-mode wave-vector) in order to facilitate the scattering of the radiation. In other words, we deal here with "coherence scattering" rather than coherence damping.

Still within the optical two-level model, using a classical-trajectory S-matrix description of collisions in the homogeneous-broadening impact limit,[14,15] we may write

$$\gamma^{(g)}(\underline{k}_L) = \int d\nu_c [1 - S_{eg;eg}(\underline{q}) S^*_{gg;gg}(\underline{q}) - S_{ge;eg}(\underline{q} + \tfrac{1}{2}\underline{k}_L) S^*_{gg;gg}(\underline{q})] \tag{19}$$

for the linear-response coherence-damping rate. Here $S(\underline{q})$ is an S-matrix element corresponding to a collision trajectory characterized by the momentum transfer $\underline{q}$ (in center-of-mass frame of reference). The last term in the integrand is an addition to the Anderson-type expression, needed in resonance broadening in order to take into account resonance exchange between the "driven" and "bath" atoms.[16] The integration in (19) symbolizes a summation over all collision trejectories with the differential rate

$$\int d\nu_c = \frac{1}{2} n \int b db \int v f_{eq}(v) dv \int d\Omega_{\underline{q}} \tag{20}$$

n being the gas density, b the impact parameter of the trajectory, v the relative initial velocity, $f_{eq}(v)$ the Maxwell equilibrium distribution, and $d\Omega_{\underline{q}}$ a solid angle extended over orientations of the collision frame relative to the laboratory (or $\underline{k}_L$-fixed) frame.

Using the same notation, one can show that the coherence-scattering collisions depicted in Figure 3 should occur at the rate

$$\zeta(\underline{k}_L \to \underline{k}_S) = \int d\nu_c [S_{eg;eg}(\underline{q} + \tfrac{1}{2}\underline{k}_L - \tfrac{1}{2}\underline{k}_S) S^*_{ge;ge}(\underline{q} - \tfrac{1}{2}\underline{k}_L + \tfrac{1}{2}\underline{k}_S) + S_{ge;eg}(\underline{q} + \tfrac{1}{2}\underline{k}_L + \tfrac{1}{2}\underline{k}_S) S^*_{eg;ge}(\underline{q} - \tfrac{1}{2}\underline{k}_L - \tfrac{1}{2}\underline{k}_S)]. \tag{21}$$

Here, again, the last term was added so as to take into account the exchange of roles between the identical collision partners. Consider for a moment the first term. The momentum transfer here at large impact parameters must be bounded from below by $\frac{1}{2}\,|\underline{k}_L - \underline{k}_S|$ and therefore the corresponding impact parameter defines an upper bound for the scattering cross section incorporated in (21) which, all else considered, is generally finite.[8]

We are now in a position to study the modification due to such collisions of the linear-response spectral distribution of resonance fluorescence in two-level atoms. As mentioned above, the diagrams shown in Figure 2 involving single-atom coherence produce the linear-response expression for the scattered light spectrum. In addition to an infinitesimally sharp Rayleigh-scattering peak (assuming an ideal monochromatic incident beam) with frequency of scattered mode ω_S equal to ω_L, they accommodate a collision-induced diffuse line (as a function of ω_S) centered around the atomic resonance frequency ω_0. The spectral distribution in the impact-limit rotating-wave approximation is[9,10]

$$F_1(\omega_S) = B[L(\omega_S-\omega_L;0) + (\frac{2\gamma_2}{\gamma_1} - 1)L(\omega_S-\omega_0;\gamma_2)]. \tag{22}$$

Here B is a common factor dependent on the geometry of the experiment and characteristics of the incident radiation (e.g., its absorption rate); $L(x;\gamma)$ is the Lorentz line profile

$$L(x;\gamma) = \frac{\gamma/\pi}{\gamma^2 + x^2} \tag{23}$$

γ_1, as earlier, is the spontaneous (radiative) decay rate, and γ_2 is the total T_2-damping rate (collisional and radiative),

$$\gamma_2 = \gamma^{(g)} + \frac{1}{2}\gamma_1. \tag{24}$$

The subscript 1 in the left-hand side of (22) refers to the single-atom coherence.

A straightforward fourth-order-perturbation expansion for diagrams such as Figure 3 and its various time-ordering variations leads to an additive correction to (22) owing to two-atom coherence,

$$F_2(\omega_S) = B\,\frac{\zeta}{2\gamma_2}\,L(\omega_S - \omega_0;\,\gamma_2). \tag{25}$$

I.e., the redistribution peak for which the second term in (22) stands is augmented by a ζ-dependent term.

As both ζ and $\gamma^{(g)}$ are proportional to the gas density in dilute gases, while γ_1 is not, one should be able to measure the coherence-scattering rate ζ by plotting the intensity of the redistribution peak (relative to that of the Rayleigh peak) vs. pressure. The linear-response contribution to the relative intensity is linearly dependent on the pressure whereas the corresponding contribution of two-atom coherence is independent of the pressure at gas densities in which $\gamma^{(g)} >> \gamma_1$. Linear extrapolation from that region to zero pressure, as demonstrated schematically in Figure 4, should intercept the ordinate at the pressure-independent value of $\zeta/2\gamma^{(g)}$.

Of course, this result should be modified so as to take account of space degeneracy in real gases, as in linear-response theories. Nevertheless, provided coherence scattering has a non-negligible cross section, an experiment of the kind proposed here should serve as a demonstration that collision-induced two-atom coherence effects exist not only in medium-coupling radiation fields, but even in weak fields where linear response is commonly believed to prevail.

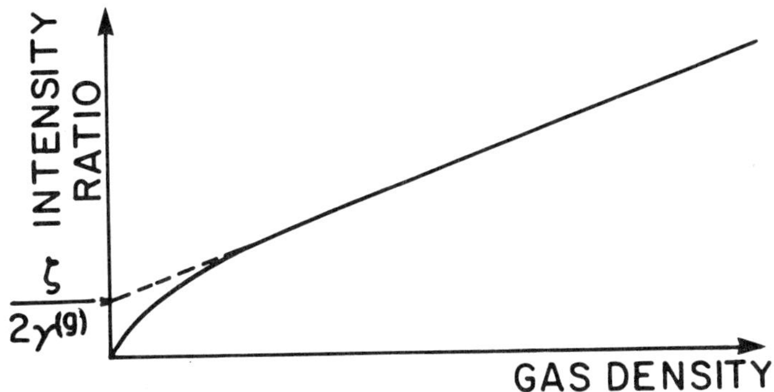

FIGURE 4. Linear extrapolation to zero pressure of the ratio of redistribution to Rayleigh line intensities as a proposed measurement of the coherence-scattering rate ζ.

REFERENCES

1. S. I. Yakovlenko, Kvantovaya Elektron. (Moscow), 5, 259 (1978) [Sov. Phys. J. Quantum Electron., 8, 151 (1978)].
2. V. S. Lisitsa and S.I. Yakovlenko, Zh. Exp. Teor. Fiz., 68, 479 (1975) [Sov. Phys. JETP, 41, 233 (1975)].
3. S. P. Andreev and V. S. Lisitsa, Zh. Eksp. Teor. Fiz., 72, 73 (1977) [Sov. Phys. JETP, 45, 38 (1977)].
4. Y. Rabin and A. Ben-Reuven, J. Phys. B: Atom. Molec. Phys., 13, 2011 (1980); K. Burnett, J. Cooper, P. D. Kleiber and A. Ben-Reuven, Phys. Rev. A, 25, 1345 (1982).
5. R. G. Breene, Jr., The Shift and Shape of Spectral Lines (Pergamon, Oxford, 1961); Theories of Spectral Line Shape (Wiley, New York, 1981).
6. B. R. Mollow, Phys. Rev. A, 2, 76 (1970); 15, 1023 (1977).
7. A. Ben-Reuven, Phys. Rev. A, 22, 2572 and 2585 (1980).
8. A. Ben-Reuven, K. Burnett and J. Cooper, to be published.

9. D. L. Huber, Phys. Rev., 178, 93 (1969); 187, 392 (1969).
10. A. Omont, E. W. Smith and J. Cooper, Ap. J., 175, 185 (1972); 182, 283 (1973).
11. K. Burnett, J. Cooper, R. J. Ballagh and E. W. Smith, Phys. Rev. A, 22, 2005 (1980); K. Burnett and J. Cooper, Phys. Rev. A, 22, 2027 and 2044 (1980).
12. See, e.g., P. Résibois and M. De Leener, Classical Kinetic Theory of Fluids (Wiley, New York, 1977).
13. J. M. Raimond, G. Vitrant and S. Haroche, J. Phys. B: Atom. Molec. Phys., 14, L655 (1981), have recently observed power-dependent line widths in Cs-vapor beams excited to high Rydberg states. Although the collision broadening in their experiments pertains to the quasi-static rather than impact domain, it essentially has the same origin, namely, excited-atom-excited-atom collisions.
14. P. W. Anderson, Phys. Rev., 76, 647 (1949).
15. M. Baranger, Phys. Rev., 111, 481 and 494 (1958); 112, 855 (1958).
16. A. Ben-Reuven, Advan. Chem. Phys., 33, 235 (1975).

LASER SWITCHED COLLISIONS - A DIAGRAMMATICAL APPROACH

M.C. Gower
Laser Division
SERC Rutherford Appleton Laboratory
Chilton, Didcot,
Oxon.

T.K. Yee
Research Laboratory of Electronics
Massachusettes Institute of Technology, Cambridge,
Massachusettes 02139
USA

Abstract We have used a diagrammatic approach first developed for nonlinear optical calculations of the density operator to obtain expressions for the transition probability in various laser-switched collision processes. Not only do these diagrams considerably simplify the algebra involved in such calculations, but they also afford a clear visualization of the interaction. In particular, pair absorption, laser-switched Raman scattering, and collisionally enhanced hot luminescence processes are considered.

INTRODUCTION

Laser-induced colisional processes of the type:-

$$A^* + B + \hbar\omega \rightarrow A + B^* \qquad \text{..........(1)}$$

in which the energy of atom A* is transferred to atom B in the presence of a photon field at frequency ω, were first discussed by Gudzenko and Yakovlenko[1] and Harris and Lidow,[2] and have been reviewed by Harris and White[3] and Yakovlenko.[4] In this paper we discuss an extension of the diagrammatic technique recently developed for density matrix operator calculations in non-linear optics[5] to the case of laser-switched collisions.[6] This diagrammatical approach has the distinct advantage of being able to more clearly visualise the interaction than would otherwise be possible using simple energy level diagrams. In addition, one is able to immediately write integral expressions for the density operator describing the interaction which involve a considerable simplification of the algebra. These expressions may then be integrated either numerically or analytically to obtain transition probabilities. Furthermore, dampings can be more rigorously included in the calculation.

LASER-SWITCHED COLLISION

As a first example, we consider the process described by Eq (1) and the energy level diagram of Fig 1 (a). The system is initially in the state $|1> \equiv |a_1 b_o>$(atom A in level a_1 and B in b_o). During the course of a collision between A* and B, the system is transferred to state $|2> \equiv |a_o b_1>$, whereafter the absorption of a photon at frequency ω puts the system into the final state $|3> \equiv |a_o b_2>$. The diagrams which explicitly describe this interaction are shown in Figs. 1 (b) - (d) and depict the time-ordered evolution (vertically) of the system wave function (L.H.S. of the figure), and its complex conjugate (R.H.S), according to the general perturbation solution of the density operator, ρ_{33}. The complete expression for ρ_{33} is the sum of all three diagrams and their complex conjugates (reflection in the time axis). The contribution of Fig 1 (b) is given by (Eq (8), Ref. 5, $\rho_{33}^{(b)}$ =

$$\iiint\int_0^\infty A_{33}(\tau_4)H'_{23}(t-\tau_4)A_{32}(\tau_3)H'_{12}(t-\tau_4-\tau_3)A_{31}(\tau_2)H'_{32}(t-\tau_4-\tau_3-\tau_2) \times$$

$$A_{21}(\tau_1)H'_{21}(t-\tau_4-\tau_3-\tau_2-\tau_1)d\tau_4 d\tau_3 d\tau_2 d\tau_1 \ldots\ldots \quad (2)$$

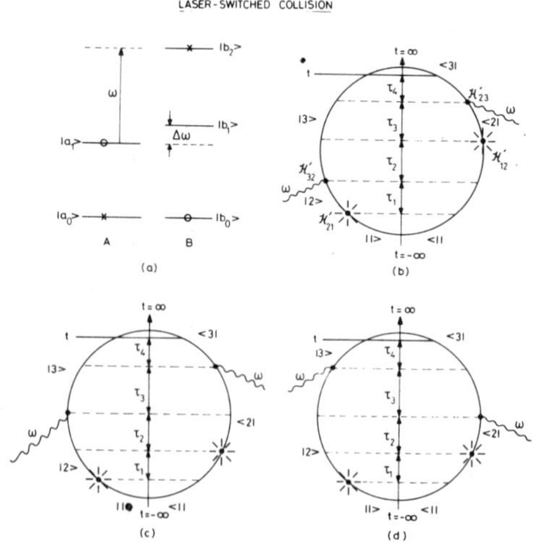

Fig.1

(a) Energy level diagram Atoms are initially in O and finally in X.

(b) Diagram of time evolution (vertically) of the wave-function (LHS) and its complex conjugate (RHS). Photon absorptions (and emissions) are denoted by rising (and descending) wiggley lines, while collisions are depicted as ✱

(c) and (d): Other time orderings.

where the τ's are the times between the collisions and the photon absorptions as shown in Fig. 1 (b). The perturbational Hamiltonians,

$$H'_{32}(t) = H'^{*}_{23}(t)',$$

are the dipole absorption operators between states $|b_1\rangle$ and $|b_2\rangle$ of atom B from the classical photon field $\tilde{E}=E\cos\omega t$, and

$$H'_{21}(t) = H'^{*}_{12}(t) = \langle 2| - \frac{\mu^{A}_{10}\mu^{B}_{01}}{4\pi\varepsilon_0 i\hbar} |1\rangle \frac{1}{R^3(t)} \quad \ldots\ldots.(3)$$

for dipole-dipole collisions between atoms A and B with an inter-nuclear separation R (t). The dipole transition moments μ^A_{10} and μ^B_{01} refer to transitions $|a_1\rangle \rightarrow |a_o\rangle$ in A and $|b_o\rangle \rightarrow |b_1\rangle$ in B respectively. The diagonal operators (in the isolated - line approximation), A_{ij} (t), in Eq (2) are statistically averaged operators given by:-

$$A_{ij}(t) = \langle\langle ij|\ A(t)\,|ij\rangle\rangle_{ave} = e^{-(i\omega_{ij} + \Phi_{ij})t} \quad \ldots\ldots.(4)$$

where $\hbar\omega_{ij} = \hbar(\omega_i - \omega_j)$ is the system energy change associated with the transition from the ket state $|i\rangle$ to the bra state $\langle j|$. Φ_{ij} is the radiative and collisional damping parameter. Once one is familiar with these diagrams, following the time evolution of the system wave function and its complex conjugate enables one to immediately write down expressions similar to Eq (2) by writing down the operators H′ and propagators A_{ij} as they occur in the diagram. Assuming a dipole-dipole collision, from Eq (2), $\rho^{(b)}_{33} =$

$$C\iiiint_0^{\infty} \frac{e^{-\Phi_{33}\tau_4+i\omega(t-\tau_4)-(i\omega_{32}+\Phi_{32})\tau_3-(i\omega_{31}+\Phi_{31})\tau_2-i\omega(t-\tau_4-\tau_3-\tau_2)-(i\omega_{21}+\Phi_{21})\tau_1}}{R^3(t-\tau_4-\tau_3)\ R^3(t-\tau_4-\tau_3-\tau_2-\tau_1)}\, d\tau_4 d\tau_3 d\tau_2 d\tau_1$$

...... (5)

where $C = \left|\frac{\mu^A_{10}\ \mu^B_{01}}{4\pi\varepsilon_o\hbar}\right|^2 \left|\frac{\mu^B_{12}E}{2\hbar}\right|^2$

For the on-resonance case ($\omega=\omega_{31}, \Delta\omega=\omega_{21} = \omega-\omega_{32}$) shown in Fig.1 (a), then Eq. (5) becomes:-

$$\rho_{33}^{(b)} = \frac{C}{(\Delta\omega-i\Phi_{21})(\Delta\omega+i\Phi_{32})} D_1(\Phi_{31},t) \qquad \ldots\ldots (6)$$

where

$$D_1(\Phi_{31},t) = \int_{-\infty}^{t}\int_{-\infty}^{x} \frac{e^{-\Phi_{31}(t-y)}}{R^3(x)R^3(y)} dxdy. \text{ and } x=(t-\tau_4) \text{ and } y=(t-\tau_4-\tau_3-\tau_2).$$

In deriving Eq.(6) we have assumed that $\Phi_{31} \simeq \Phi_{33}$ and that changes in $1/R^3(t)$ occur much slower than the oscillations in $e^{-i\Delta\omega t}$. In a similar fashion, the contributions from the diagrams shown in Fig. 1 (c) and (d) are given by the following expression:

$$\rho_{33}^{(c)} + \rho_{33}^{(d)} = \frac{-iC(\Phi_{32}+\Phi_{23})}{(\Delta\omega-i\Phi_{21})(\Delta\omega+i\Phi_{32})(\Delta\omega-i\Phi_{23})} D_2(\Phi_{22},t) \qquad \ldots\ldots\ldots\ldots (7)$$

where $D_2(\Phi_{22},t) = \int_{-\infty}^{t}\int_{-\infty}^{x} \frac{e^{-\Phi_{22}(t-y)}}{R^6(y)} dxdy.$

For $\Delta\omega \neq 0$ and all Φ's = 0, Eq (7) is zero. This implies that the contribution of Fig. 1(c) exactly cancels the contribution of Fig. 1 (d) when dampings are ignored. Diagrams which are complex conjugates of Figs. 1(b)-(d) also contribute to ρ_{33} and must be included. Thus for a real Φ_{32} (i.e. no frequncy shift of transition $|b_2\rangle \rightarrow |b_1\rangle$, so $\Phi_{32} = \Phi_{23}$)then:-

$$\rho_{33} = (\rho_{33}^{(b)} + \rho_{33}^{(c)} + \rho_{33}^{(d)}) + \text{c.c.}$$

$$= \frac{2C}{(\Delta\omega^2+\Phi_{21}^2)(\Delta\omega^2+\Phi_{32}^2)} \left[(\Delta\omega^2+\Phi_{21}\Phi_{32})D_1 + 2\Phi_{21}\Phi_{32}D_2 \right] \qquad \ldots.(8)$$

In general, the double integrals, $D(\Phi,t)$, are difficult to solve. However, if we assume a straight line classical trajectory such that $R^2(t) = \rho^2 + \bar{v}^2 t^2$ ($\bar{v}$ is the mean collision velocity and ρ the impact parameter) and ignore the dampings, the integrals in D, can be performed analytically and the result is given by:-

$$\rho_{33}(\Phi=0, t\to\infty) = 4C\left(\frac{1}{\rho^2 \bar{v}\, \Delta\omega}\right)^2 \qquad \text{.......(9)}$$

which is identical to the expression for the transition probability derived by Harris et al(7) , using standard perturbation theory.

PAIR ABSORPTION

By interchanging the photon absorptions with the collisions in Fig. 1 (b)-(d) one obtains the process of pair-absorption(8) shown in Fig.2.

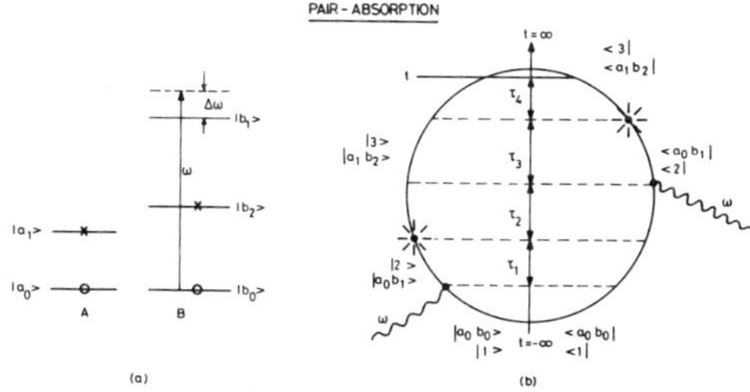

Fig.2
Diagrams describing the process of pair absorption. Only one of the three possible time orderings is shown in (b).

Keeping the propagators A_{ij} identical to those for Fig 1, it is easily shown that, with the same assumptions discussed above, the expression for the density matrix for on-resonance ($\omega=\omega_{31}$) pair absorption (PA) is ρ_{33} (PA) =

$$\frac{2C}{(\Delta\omega+i\Phi_{21})(\Delta\omega-i\Phi_{32})}\left[D_1(\Phi_{31},t) + \frac{2\Phi_{21}\Phi_{32}}{\Phi_{22}(\Delta\omega-i\Phi_{21})(\Delta\omega+i\Phi_{32})} D_3(\Phi_{33},t)\right] \qquad \text{......(10)}$$

$$\text{where} \quad D_3(\Phi_{33},t) = \int_{-\infty}^{t} \frac{e^{-\Phi_{33}(t-x)}}{R^6(x)}\, dx$$

It can be seen that, when dampings are ignored, the transition probability for pair-absorption is identical to that for the switched collision of Fig. 1 (Eq 9).

Rotation of the diagram shown in Fig 2 (b) through 180°, which is equivalent to time reversal of pair absorption, leads to the process of radiative collisional fluoresence as observed by White et al[9]. Since these two processes are the time reversal of each other they have identical transition probabilities.

LASER-SWITCHED RAMAN SCATTERING

Using these diagrams, not only is it much easier to include damping and off-resonance cases in the calculations, it is also very straightforward to write down expressions for higher-order processes than have been considered thus far. Indeed, most nonlinear optical interactions occurring in gases may in principle be enhanced by laser-switched collisions. Examples of such higher-order cases are laser-

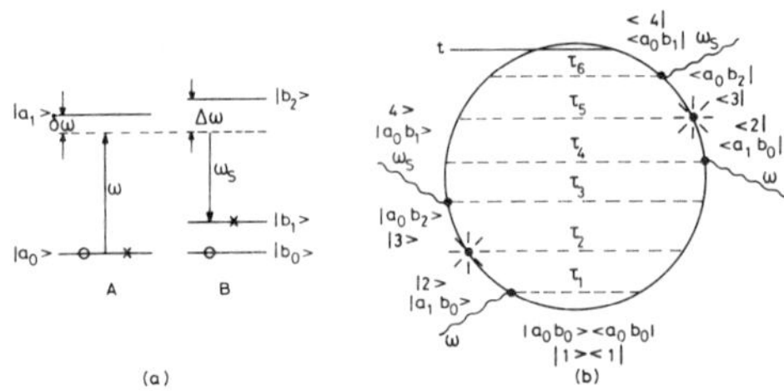

Fig 3.
One time-ordered diagram of the laser-switched Raman effect with ω_s the scattered frequency.

switched Raman scattering (Fig.3) and collisionally enhanced hot luminescence (Fig.4). By following the interactions of the wave function and its complex conjugate as a function of time, integral equations for ρ, similar to (2) and (6) may be written down. With similar approximations to those discussed previously ($\Delta\omega$ and $\delta\omega > \bar{v}/\rho$), the laser-switched Raman scattering process (LSRS) shown in Fig. 3 can be written as

$$\rho_{44} = \frac{C}{(\Delta\omega - i\Phi_{31})(\Delta\omega + i\Phi_{43})(\delta\omega - i\Phi_{21})(\delta\omega + i\Phi_{42})} D_1(\Phi_{41}, t) \quad \ldots\ldots (11)$$

where now

$$C = \left|\frac{\mu^A_{10}\mu^B_{02}}{4\pi\varepsilon_0 \hbar}\right|^2 \left|\frac{\mu^A_{10} E}{2\hbar}\right|^2 \left|\frac{\mu^B_{21} E_s}{2\hbar}\right|^2$$

and in D_1 as given in (6).

$x = (t - \tau_6)$ and $y = (t - \tau_6 - \tau_3)$.

The total contribution to the transition probability is then ρ_{44} plus its complex conjugate. If dampings are neglected, in the limit of $t \to \infty$ then

$$\rho_{44} + c.c. = 4C\left(\frac{1}{\rho^2 \bar{v}\, \Delta\omega\delta\omega}\right)^2 \quad \ldots\ldots\ldots\ldots (12)$$

This equation, which characterizes the time ordering of Fig.3, is now identical to the collision-induced Raman expression obtained by Green(10) using standard perturbation theory.

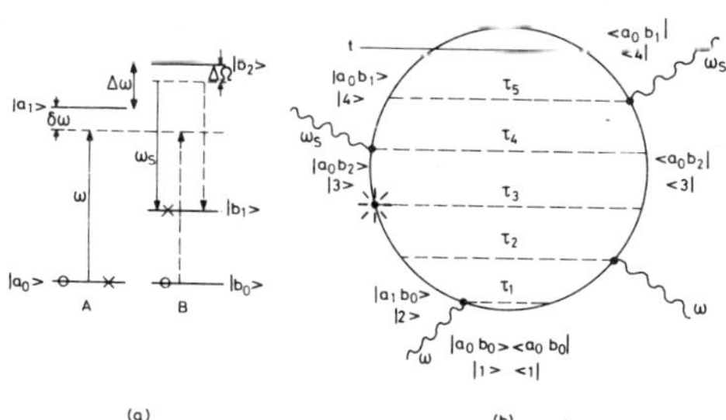

Fig 4

Collisional-enhanced hot luminescence, (a) Energy level diagram, solid (dotted) arrows represent photon interaction on ket (bra) side (b) Diagrammatic representation. Only the ket side undergoes a collisional interaction.

COLLISIONALLY-ENHANCED HOT LUNINESCENCE

One of the other possible time orderings of Fig.3 (with the collisional interaction on the bra side removed) is given in Fig.4. This diagram also contributes to the LSRS transition probability and describes a process called collisionally-enhanced hot luminescence (CEHL)[5]. With the identical assumptions and approximations as before, the contribution to ρ_{44} (CEHL) due to Fig.4 is given by

$$\rho_{44} = \frac{-C}{(\Delta\omega+i\Phi_{23})(\delta\omega-i\Phi_{21})(\Delta\Omega+i\Phi_{43})} D_4(\Phi_{33},\Phi_{44},t) \quad \text{.........} (13)$$

$$\text{where} \quad D_4(\Phi_{33},\Phi_{44},t) = \int_{-\infty}^{t}\int_{-\infty}^{x} \frac{e^{-\Phi_{33}(x-y)}\, e^{-\Phi_{44}(t-x)}}{R^3(y)}\, dydx \ .$$

with $x = (t-\tau_5)$, $y = (t-\tau_5-\tau_3)$, and

$$C = \left(\frac{\mu^A_{10}\mu^B_{02}}{4\pi\varepsilon_0\hbar}\right)\left|\frac{\mu^A_{10}E}{2\hbar}\right|^2\left|\frac{\mu^B_{21}E_s}{2\hbar}\right|^2$$

In contrast to the LSRS transition probability of Fig.3 and (11), as expected the probability for CEHL as given by (13) predicts a resonant enhancement for emission at a frequency $\omega_s = \omega^B_{21}$ (i.e., $\Delta\Omega = 0$). Also, it can be seen from (11) and (13) that as the states $|b_2\rangle$ and $|a_1\rangle$ become farther apart, then so the contribution of CEHL to the overall transition probability for LSRS increases.

CONCLUSION

In conclusion, we have extended the use of the diagrammatical analysis used in nonlinear optical calculations of the density operator to include the case of various laser-switched collision processes. Not only do the diagrams greatly simplify the relevant algebra for calculating transition probabilities, but they also enable laser-switched higher order nonlinear optical processes to be envisaged and calculated. Furthermore, inclusion of dampings in the calculations ensure the convergences of the integrals involved. Indeed, for some processes these dampings cannot be neglected in the calculations.

REFERENCES

1. L.I. Gudzenko and S.I. Yakovlenko, Sov-Phys.JETP 35, 877. 1972.
2. S.E. Harris and D.B. Lidow, Phys Rev Letts 33, 674 1974
3. S.E. Harris and J.C. White, IEEE J.Quant. Electr. QE-13, 972. 1977.
4. S.I. Yakovlenko, Sov. J.Quant. Electr. 8, 151. 1978.
5. T.K. Yee and T.K. Gustafson, Phys. Rev A 18, 1597 1978.
6. T.K. Yee and M.C. Gower, IEEE J.Quant. Electr. QE-18, 437 (1982)
7. S.E. Harris, R.W. Falcone, W.R. Green, D.B. Lidow, J.C. White and J.F. Young, 'Tunable Lasers and Applications' Springer-Verlag. 1976.
8. L.I. Gudzenko and S.I. Yakovlenko, Phys, Lett. A 46, 475. 1974.
9. J.C. White, G.A. Zdasiuk, J.F. Young and S.E. Harris, Phys. Rev. Letts. 41, 1709 (1978)
10. W.R. Green, 'Laser-Induced Interactions', Stanford Univ G.L. Rep 2936 (1979)

COLLISION INDUCED COHERENT PHENOMENA

YEHIAM PRIOR†
Department of Chemical Physics, Weizmann Institute of Science, Rehovot, Israel 76100

Abstract The newly discovered family of Pressure Induced Extra Resonances (PIER) in coherent four wave mixing is discussed. These resonances occur when collisions induce a coherent contribution to the third order polarizability, and are therefore a result of the interaction between the input laser beams and the short lived "collision complex". The experiments are analyzed in the framework of non-linear optics, and their possible extension to the study of the interaction potential between the colliding atoms (molecules) is discussed.

INTRODUCTION

Four wave mixing (FWM) is a process where four laser beams (three input and one generated) interact coherently with a "non linear" medium. At the end of such an interaction, energy has been exchanged between the laser fields, but due to the coherent, parametric, nature of the interaction, the atom remains in the state it had occupied before the interaction. Many examples of FWM are known and widely used - CARS (Coherent Anti Stokes Raman Scattering), SRS (Stimulated Raman Scattering) CSRS (Coherent Stokes Raman Scattering), THG (Third Harmonic Generation) are a few examples. The motivation for performing FWM experiments has been either to obtain coherent emission at new frequencies, or to develop new spectroscopic techniques and perform spectroscopic studies of non linear molecular properties (eg. Raman or two photon or other $\chi^{(3)}$ properties). In all these experiments relaxation process are detrimental to the coherent effect and since most theoretical treatments have been semi classical in nature, relaxation (T_1) and dephasing (T_2) phenomenological rates have been included in the calculation. This approach has usually been enough to account for most of the experiments.

Recently a new family of pressure induced resonances has been predicted[1] and later observed in FWM in sodium vapor.[2-5] These resonances depend on dephasing collisions for their very existence. In a FWM experiment, two input frequencies ω_1,ω_2 are tuned near the 3S-3P resonances in sodium, but neither one, nor any of their combinations is resonant with any material levels. Under these conditions a non resonant FWM signal is expected and observed. In addition, in the presence of collisions with a buffer gas, for a fixed ω_1 and a scanned ω_2, the following collision induced resonances are observed in the intensity of the coherent signal generated at $\omega_p(=2\omega_1-\omega_2)$:

1. For $\omega_1-\omega_2=17cm^{-1}$, namely when the difference between the input frequencies matches the difference between the upper (unpopulated) doublet levels.

2. For $\omega_1=\omega_2$, a degenerate FWM signal for a frequency which is not resonant with the atomic transition.

3. For $\omega_1\simeq\omega_2$, a Raman transition between hyperfine or Zeeman components of the ground state which are equally populated and noramlly would not give rise to a Raman Signal.

In what follows the experimental observation of the resonance in point 1 above will be described, followed by theoretical expressions derived from density matrix calculations. The degenerate four wave mixing collision induced signal is discussed next, and preliminary attempts to use it in the study of lineshapes are described. The last section of this paper describes more recent theoretical studies and attempts to broaden the scope of the phenomenological treatment of collisional processes in non linear optics.

EXPERIMENTAL

In these FWM experiments, three beams (two at frequency ω_1 and one at ω_2) are obtained from two tunable dye lasers (both pulsed and CW lasers have been used) and interact in a heated cell containing sodium vapor and a variable pressure of a buffer gas (Helium). The frequency ω_1 is set close to the $3S$-$3P_{1/2}$ (or $3S$-$3P_{3/2}$) resonance of sodium, and the frequency ω_2 is scanned around $\omega_1 \pm 17cm^{-1}$ such that neither frequency nor any of their combinations is resonant with a Na transition. A FWM signal is generated at $\omega_p = 2\omega_1 - \omega_2$ and detected. Since all frequencies are very close, we had to use a new three dimensional phase matching arrangement[6] that enables the observation of degenerate (and nearly degenerate) four wave mixing with spatial separation of all four beams. In this configuration, also known as folded boxcars[7] the three input beams are directed to be on the three corner of a square when they are at the plane of the lens focussing them into the cell, and the signal is phase matched in a direction that completes the square at the plane of their exit from the cell.

The intensity of the FWM signal at ω_p is measured as a function of $\omega_1 - \omega_2$, and Figure 1 depicts the results of such measurements for several different buffer gas pressures.

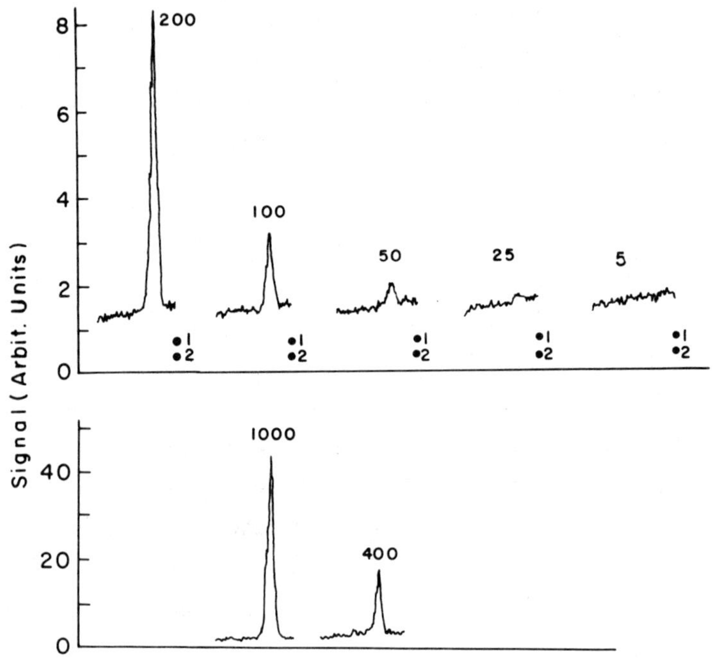

FIGURE 1. Pressure induced extra resonance for various helium pressures. The points marked 1,2 are the scattered light from lasers 1,2 respectively. There is a FWM present even in the absence of pressure (the non resonant signal), but a sharp peak appears with the increase of pressure. Note the change of scale for the higher pressures.

For low Helium pressure, a non resonant FWM is expected and observed, but as the helium pressure is increased, a strong resonant feature appears at $\omega_1-\omega_2=17\ cm^{-1}$, which grows as the square of the pressure.

This collision induced signal should not be confused with other signals that can be obtained from population building up in the excited state,[8] as is discussed in Ref. 7. The reader is referred to the original papers[2-5] for full description of the experimental conditions.

THEORY

Four wave mixing and many other non linear phenomena have been described by a complex third order susceptibility $\chi^{(3)}$. Expressions for $\chi^{(3)}$ can be derived from the Liouville equation of motion for the density matrix, with damping of atomic states included explicitly,[1] or from double sided Feynman diagrams, where the bra and ket vectors evolve separately.[9] When dephasing processes, collisions, are present, proper time ordering of the evolution of the density matrix is extremely important, and one is not allowed to add the dephasing formally at the end of the calculation.[10] For the purpose of the current discussion we can represent the Na atom as a three level system with levels $|g\rangle$ corresponding to the ground 3S state and $|k\rangle$, $|j\rangle$ to the upper $3P_{1/2}$ and $3P_{3/2}$ states. Degeneracy and hyperfine levels will be discussed below. Each transition from state m to state l is assigned a width Γ_{ml} which could be due to spontaneous emission only or due to spontaneous emission as well as pressure broadening.

The complete expression for $\chi^{(3)}$ will not be reproduced here, but one finds that it contains a term proportional to the coherence induced between the $3P_{1/2}$ and $3P_{3/2}$ states:

$$\rho^{(2)}_{kj}(\omega_1-\omega_2)\sim\frac{P_{gg}(0)}{(\omega_{kg}-\omega_1-i\Gamma_{kg})(\omega_{jg}-\omega_2+i\Gamma_{jg})}\left[1+\frac{\Gamma_{kj}-\Gamma_{kg}-\Gamma_{jg}}{\omega_{kj}-(\omega_1-\omega_2)-i\Gamma_{kj}}\right] \tag{1}$$

This equation predicts a resonance for $\omega_1-\omega_2=\omega_{kj}$. The daming terms Γ_{ml} are due to spontaneous emission and collisions and can be written as

$$\Gamma_{ml} = \Gamma_{ml}^{sp} + \Gamma_{ml}^{collision} \tag{2}$$

where Γ_{ml}^{sp} is due to natural lifetime, and $\Gamma_{ml}^{collision}$ is due to collisional broadening (and is proportional to buffer gas pressure in first approximation). Since g is the ground state, a well known relationship exists between the damping terms if they are due to spontaneous emission only:

$$\Gamma_{kj}^{sp} - \Gamma_{kg}^{sp} - \Gamma_{jg}^{sp} = 0 \tag{3}$$

Thus, in Eq. (1) the resonance at $\omega_1-\omega_2=\omega_{kj}$ disappears in the absence of proper dephasing. A detailed quantitative study of this PIER 4 signal has been reported,[5] and it confirms the prediction of the $\chi^{(3)}$ theory.

A different approach to the derivation of these resonances has been taken by Grynberg.[11] Using the dressed atom model he calculates the phase change induced by the colli sion in the wave function of the atom, and shows that such a phase change in two excited states can induce a coherent superposition between them in spite of the random nature of the collision.

The physical origin of these pressure induced resonances can be explained as follows: in the absence of collision, the FWM signal is enhanced by the proximity to the one photon resonance intermediate states k,j. The detuning from each state is different, but a careful examination[1,2] of the full expression for $\chi^{(3)}$ reveals that the enhancement due to these two states exactly cancels. The function of the collision, or any other dephasing process, is to destroy this destructive interference between the large contributions from these two states.

OTHER PRESSURE INDUCED RESONANCES

The discussion so far concentrated on the resonance between two upper levels which are initially unpopulated. However, in the course of these studies, several other resonances have been observed and analyzed:

a. Resonance between equally populated levels. When the ground state is not a single state but rather a collection of several level (hyperfine or Zeeman sub-levels) one may look for Raman type resonances between such states. Normally one does not expect a contribution between equally populated levels, but under the condition of collisional enahncement such a contribution is expected and observed. Recently, Bloembergen, Downer and Rothberg[12] demonstrated collisional Doppler narrowing of these resonances.

b. Pressure induced degenerate four wave mixing (DFWM). When $\omega_2=\omega_1$, one can see from Eq. (1) that only two atomic levels are needed in the equation (j=k) and one may observe a DFWM signal at a given detuning $\Delta\omega$ from the atomic line. In these experiments, only one frequency is used, and we measure the generated DFWM signal at a given intensity and pressure as a function of laser detuning. This signal gives information on the interaction potential between the colliding partners. Attempts to utilize this effect for the study of this interaction potential have been started in our laboratory by E. Yarkoni,[13] but are very preliminary in nature and have not yet extended outside the impact limit. One such signal is depicted in Fig. 2. The DFWM obtained near the $3S\text{-}3P_{3/2}$ Na line is plotted as a function of laser detuning in GHz. The laser is a single mode ring dye laser

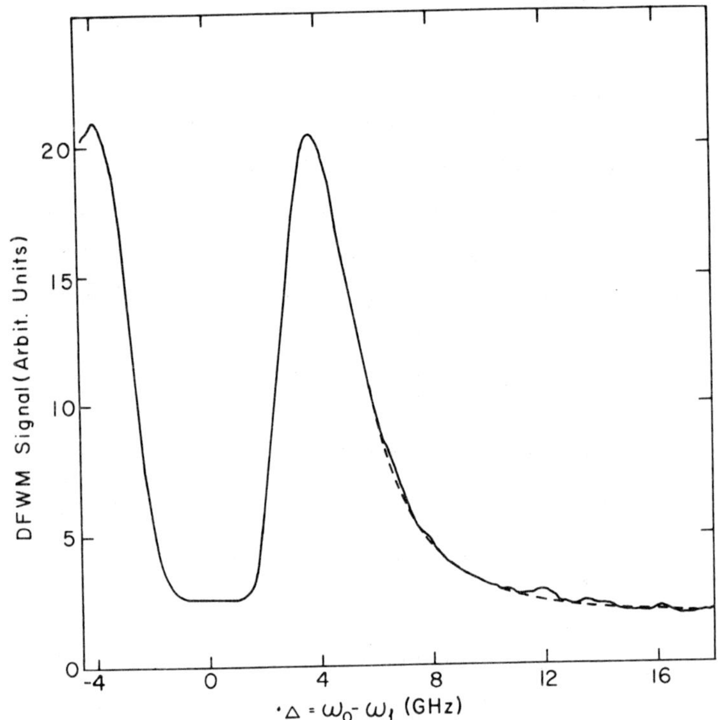

FIGURE 2. DFWM as a function of laser detuning Δ. At line center the absorption is complete. The theory was fitted from Eq. (5) for Δ>6GHz.

(linewidth ∿10 MHz). At line center (Δ=0) the absorption is complete. The signal is fitted to the theoretical prediction (dashed line) in the range Δ>6GHz, where the absorption can be neglected. For Na:He collision this experiment is well within the impact limit, and the use of Lorentzian lineshape functions is justified. For only two levels g,k Eq. (1) gives:

$$\text{signal} \sim \left|\frac{1}{(\omega_{kg}-\omega)^2+\Gamma_{kg}^2}\ \{1+\frac{\Gamma_{kg}+\Gamma_{gk}-\Gamma_{kk}}{\Gamma_{kk}}\}\right|^2 \tag{4}$$

with the notation

$$\Delta=\omega_{kg}-\omega$$
$$\Gamma_{kk}=T_1^{-1}$$
$$2\Gamma_{kg}=2T_2^{-1}=T_1^{-1}+\gamma p_{He}$$

where p_{He} is the helium pressure, one obtains

$$\text{singal} \sim \left(\frac{1+PT_1}{\Delta^2+T_2^2}\right)^2 \tag{5}$$

and as is seen from the figure, for the known γ, p_{He} and T_1, the fit is very good. The best signal to noise ratio obtained so far at the peak of this signal was 1000:1, which is not yet good enough to perform measurements further away from line center. Experiments are under way to study this line shape outside the impact limit.

Related Theoretical Work

The formalism of the third order suceptibility has been very successful in explaining almost all the phenomena observed in NL optics. However, by its very nature it assumes a Lorentzian shape for all transitions. For gases undergoing collisions this is a reasonable assumption within the impact limit, namely for $\Delta\omega\tau_c<1$ where $\Delta\omega$ is the detuning and τ_c is the collision duration. Outside this limit lineshapes may be different from Lorentzians - but the theory was not constructed to handle such lineshapes. The dressed atom model used by Grynberg[11] is more adequate for this task, and indeed sveral recent attempts in that direction have been reported. We are currently engaged in expanding the theory to include non Lorentzian lineshapes[14] but a full theory for an arbitrary lineshape is not yet available. The semiclassical nature of the present treatment, and the phenomenological description of the damping here and in Bloch equation makes it difficult to treat damping more generally. It seems that in a fully quantum mechanical context this task may be easier.[11,14,15]

Even within the semi classical limit, it seems that these PIER resonances open the way to the study of a new kind of phenomena that are similar but stem from the nature of the exciting fields and spontaneous emission and do not require collisions for their existence. Friedman and Wilson-Gordon[16] discussed extra resonances which are due to the three photon process, and Agarwal and Cooper[17] discussed extra resonances due to fluctuation in the exciting fields. No experimental observation of these new effects has been reported to date.

CONCLUSIONS

In this paper a brief outline of the theory and experiments related to the pressure induced extra resonances in four wave mixing (PIER 4) has been presented. Within the framework of non linear optics the theory is well founded and can account for all observable experimental facts. It seems however that these effects may find a wider application and indeed they have been observed in molecular crystals where interaction with the host lattice is responsible for the dephasing.[18]

If these affects are to be used for the study of interaction potentials, one needs to adopt a picture different from the one described above. The exciting lasers are tuned away from the atomic resonance, but during the collision the relevant energy levels are not those of the atom, but rather they are the levels of the "collision complex". As such, they dpend on the internuclear distance. Thus one may expect that for a given detuning of the laser in a degenerate FWM experiment one is probing a particular internuclear distance. Such a theory for the non-linear interaction has not been worked out in detail yet, but a measurement of the FWM signal as a function of detuning $\Delta\omega$ has a strong connection to the details of the interaction potential. The signal is contributed mostly by complexes at a given distance, and only to a lesser extent by complexes at other internuclear distances. The possibility of probing the interaction potential by means of the non-linear optical interaction, is being investigated, and is related to many other topics of current interest.[19]

Parts of this work were done in collaboration with Prof. N. Bloembergen and his group at Harvard. This research was supported in part by a grant from the U.S.-Israel Binational Science Foundation (BSF), Jerusalem, Israel.

† Incumbent of the Jacob and Alphonse Laniado Career Development Chair in perpetuity, established by bequest of Jacob Laniado, Montreal, Canada.

REFERENCES

1. N. Bloembergen, H. Lotem and R.T. Lynch, Jr., Ind.J.Pure and Appl.Phys., 16, 151 (1978).
2. Y. Prior, A.R. Bogdan, M. Dagenais and N. Bloembergen, Phys.Rev.Lett., 46, 111 (1981).
3. A.R. Bogdan, Y. Prior and N. Bloembergen, Opt.Lett., 6, 82 (1981).
4. A.R. Bogdan, M.W. Downer and N. Bloembergen, Opt.Lett., 6, 348 (1981).
5. A.R. Bogdan, M.W. Downer and N. Bloembergen, Phys.Rev.A, 24, 623 (1981).
6. Y. Prior, Appl.Opt., 19, 1741 (1980).
7. J.A. Shirley, R.J. Hall and A.C. Eckbreth, Opt.Lett., 5, 380 (1980).
8. M. Dagenais, Phys.Rev.A, 24, 1404 (1981).
9. J.A. Druet, B. Attal, T.K. Gustafson and J-P.E. Taran, Phys.Rev.A, 18, 1529 (1978).
10. S.A.J. Druet and J-P.E. Taran, Progress in Quant.Elect., 7, 1 (1981).
11. G. Grynberg, J.Phys.B, 14, 2089 (1981).
12. N. Bloembergen, M.W. Downer and L.J. Rothberg, to be published.
13. E. Yarkoni, M.Sc. Thesis, Weizmann Institute of Science, 1983 (unpublished).
14. V. Mizrahi, Y. Prior and S. Mukamel, Opt.Lett., 8, 145 (1983).
15. A. Weiszman and Y. Prior, to be published.
16. H. Friedman and A.D. Wilson-Gordon, Phys.Rev.A, 26, 2768 (1982).
17. G.S. Agarwel and J. Cooper, Phys.Rev.A, 26, 2761 (1982).
18. J.R. Andrews, R.M. Hochstrasser and R.M. Trommsdroff, Chem.Phys., 62, 87 (1981).
19. See for example the contribution of T. George to this volume.

POLARIZATION OF ATOMIC FLUORESCENCE EXCITED BY MOLECULAR PHOTODISSOCIATION

Jacques VIGUÉ
Laboratoire de Spectroscopie Hertzienne de l'E.N.S.
Associé au CNRS LAn°18- 24 Rue Lhomond 75005 PARIS

with the collaboration of :
A. ASPECT, P. GRANGIER, G. ROGER
Institut d'Optique - Université PARIS SUD 91405 ORSAY

M. BROYER
Laboratoire de Spectrométrie Ionique et Moléculaire
Université de LYON I Villeurbanne

J.A. BESWICK
Laboratoire de Photophysique Moléculaire
Université de PARIS SUD 91405 ORSAY

Abstract The principle of this polarization effect is presented and the available experimental results are recalled. We present a new theory in which a strong coherence effect increases the polarization rate and the various quantum interferences are analysed. Comparison with a recent experiment is made.

1. PRINCIPLE

The idea seems to be due to MITCHELL [1] and is quite simple. The molecule AB is photodissociated by a light beam, along the following reaction :

$$AB + h\nu \rightarrow A + B^{*} \qquad (1a)$$

$$\text{or} \rightarrow A + B \qquad (1b)$$

(where B^* designate an excited atom). The internal variables (orbital or spin angular momenta) of A and B^* (or of A and B) should keep some memory of the anisotropy of the excitation process. This anisotropy may be due to the polarization (linear or circular) or to the directivity (when no polarizing element is available) of the light beam. It seems very attractive to obtain information on the dissociation process thanks to the polarization of the light emitted by the excited atom B^* as this polarization is directly sensitive to the angular momenta of B^*. This idea was considered further by Van Brunt and Zare [2] who made a quantitative theory and proposed the following generalizations

-the light beam can be replaced by beams of electrons or of ions.

-the molecule could be polyatomic and the fragments could be atoms, molecules, atomic or molecular ions and electrons.

In the following, we limit our study to photodissociation experiments producing an excited species (reaction 1a). Moreover, we consider that the light beam is linearly polarized along Oz. We defined the polarization rate P of the atomic fluorescence emitted the excited atom B^* as :

$$P = \frac{I_{/\!/} - I_{\perp}}{I_{/\!/} + I_{\perp}} \qquad (2)$$

where $I_{/\!/}$ is the fluorescence intensity polarized along Oz and $I_{\perp}$ the fluorescence intensity polarized along an axis Ox perpendicular to Oz. Figure 1 presents a schematic drawing of this type of experiment :

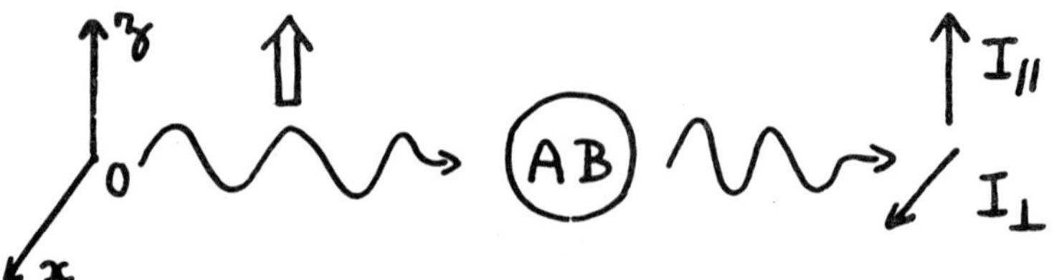

Figure 1. The excitation beam photodissociates the molecule AB ; the atomic fluorescence of B^* is detected and its polarization is measured.

2. EXPERIMENTAL RESULTS

2.1 Various experiments

We present in table 1 all the experiments that we are aware of corresponding the reaction (even when AB is a triatomic molecule)

$$AB + h\nu \rightarrow A + B^*$$

We give for every experiment the nature of AB, A and B, the identification of the excited state of B^*, the range of wavelengths used for the excitation beam $h\nu$ (in nanometers) the range of polarization rates observed and the reference of the publication :

TABLE 1 :

AB	A	B	B^* state	$h\nu$ nanometers	P	Reference
NaI	I	Na	^2P	210-245	~ 0	1,3
H_2O	H	OH	$\tilde{A}^2\Sigma^+$	130	0.01-0.06	4
HCN	H	CN	$B^2\Sigma^+$	125-145	0.05	5
BrCN	Br	CN	$B^2\Sigma^+$	130-155	0 to 0.09	5
ICN	I	CN	$B^2\Sigma^+$	110-170	0 to 0.08	6
Na_2	Na	Na	$^2P_{1/2}$	458	-0.05	7
N_2	e	N_2^+	$B^2\Sigma_u^+$	54 - 66	0.02 to 0.05	8
Ca_2	Ca	Ca	^1P	406	0.64	9

Photodissociation of CsI has been shown [10] to give strongly oriented ground state Cesium atoms, by the reaction :

$$CsI + h\nu \ (\lambda < 360nm) \rightarrow Cs(^2S_{1/2}) + I$$

No absolute measurement of the spin orientation of $Cs(^2S_{1/2})$ has been given.

We are going to describe the Ca_2 experiment with some more details for the following reasons :

-it is the first experiment in which a large polarization rate has been observed.

-the theory developped by Van Brunt and Zare predicts a polarization rate which is smaller than the observed value.

2.2 Ca_2 experiment.

This experiment has been described briefly in [9] and some detailsconcerning the apparatus can be found in a previous publication [11]. The Ca_2 molecules were produced in an oven (Temperature close to 1000K) and formed an effusive molecular beam, (containing mainly Ca atom). These molecules were irradiated by the violet lines (406 or 413 nanometer) of a Krypton laser, and the Ca ^1P- ^1S line at 422,7nm was observed. The signal was proportionnal to the calculated Ca_2 density. The laser beam is linearly polarized and the polarization rate P of the atomic fluorescence is :

$$P = 0.64 \pm 0.01 \quad \text{for } \lambda = 406 \text{ nm}$$
$$P = 0.68 \quad \text{for } \lambda = 413 \text{ nm}$$

(Some polarized stray light made the second measurement more uncertain). As the detection apparatus uses a wide aperture lens (f/0.8) this reduces slightly the polarization rate P. This reduction amounts to 0.02 - 0.03 according to Zinsli [12]. The fraction of dissociated molecules was measured thanks to an absolute calibration of the detection apparatus. This fraction F was :

$$F \simeq 0.04 \quad \text{for } \lambda = 406 \text{ nm}$$
$$F \simeq 0.02 \quad \text{for } \lambda = 413 \text{ nm}$$

with a laser beam waist radius $w_0 = 28 \times 10^{-6}$ m and a laser power

P = 0.2 Watt. The fluorescence signal S (measured in detected photons per second) and the corresponding dissociation fraction F (in %) are plotted as a function of the laser power in figure 2.

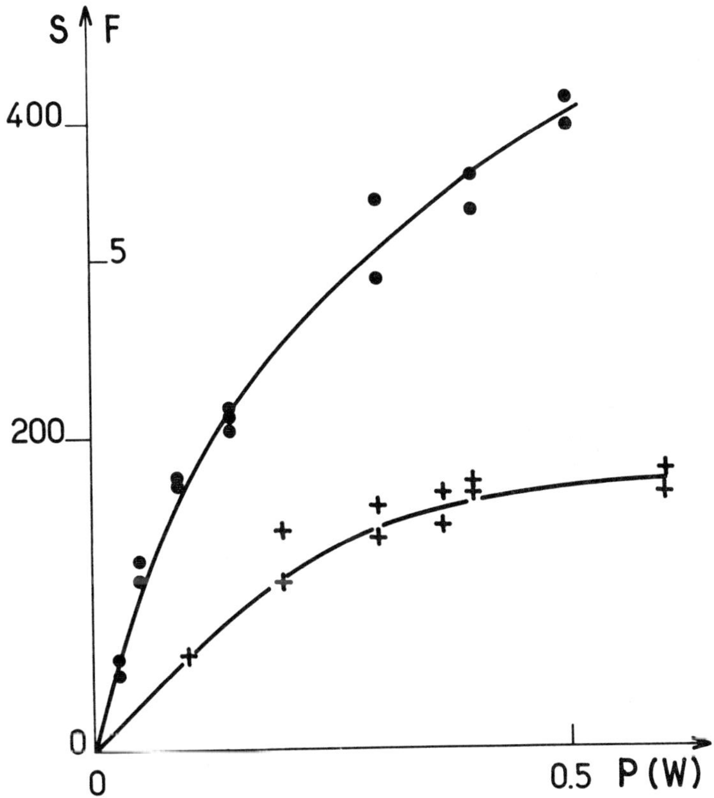

3. THE THEORY OF VAN BRUNT AND ZARE

We refer the reader to the original paper [2] for the complete presentation and we discuss here only the particular case of a diatomic molecule AB without electronic spins. Figure 3 represents the states important for the process.

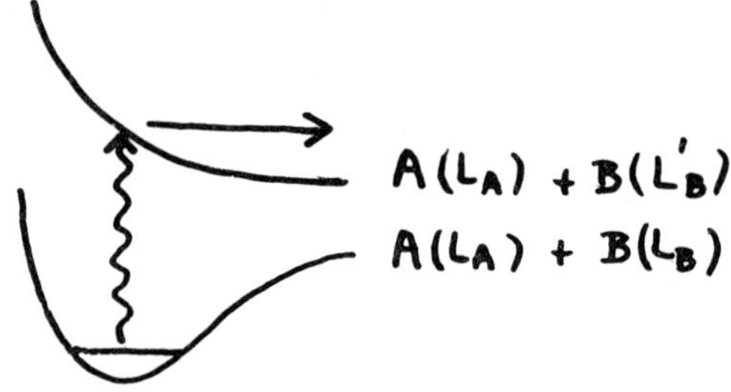

figure 3

The ground state arises from the atomic states L_A, L_B (and where L designates the value of the orbital angular momentum), and the state excited during photodissociation arises from the atomic states L_A, L'_B. The Λ value of the excited state, Λ' is the modulus of the sum of the projections of $\underline{L}_A$ and $\underline{L}_B$ on the internuclear axis.

$$\Lambda' = M_A + M'_B$$

In the absence of non adiabatic interactions, the atom A and B^* are left in the states $L_A M_A$, $L'_B M'_B$. M_A and M'_B being measured on the final direction of the internuclear axis after dissociation. In particular if $L_A = 0$ (this is the case of Ca_2) the value of M'_B is obviously given by :

$$M'_B = \pm \Lambda'$$

The experiment is a type of Stern-Gerlach experiment as emphasized by Van Brunt and Zare, but the quantization axis is not a fixed axis but the internuclear axis after dissociation and therefore varies from one molecule to the next. The intensity of the fluorescence polarized parallel to a vector $\underline{e}$ is evaluated by Van Brunt and Zare as :

$$I(\underline{e}) = \sum_{M_B=-L_B}^{+L_B} \sum_{M'_B=\pm\Lambda'} \left| \langle L_B M_B | \underline{e}.\underline{D} | L'_B M'_B \rangle \right|^2$$

where $\underline{D}$ is the dipôle moment of atom B, and the quantization axis (defined by angle θ ,ϕ) is the final position after dissociation of the internuclear axis AB. (an usual approximation is to neglect rotation during dissociation ; this is the axial recoil approximation). We must average this intensity which depends on θ ,ϕ over the orientations of the internuclear axis ; the probability of finding the axis AB into the infinitesimal solid angle $d\Omega = \sin\theta d\theta d\phi$ being $f(\theta,\phi)d\Omega$ The general form of $f(\theta,\phi)$ is [2,13]

$$f(\theta,\phi) = 1 + \beta \; P_2(\cos\theta)$$

where $P_2(\cos\theta) = (3\cos^2\theta - 1)/2$ is the 2nd Legendre polynomial and β ranges from -1 to +2.

Applied to the case of Ca_2 ($L_B=0, L'_B=1$), this calculation gives the polarization rate P :

$$\text{if } \Lambda = 1 \quad P = \frac{-3\beta}{20-\beta} \qquad -\frac{1}{3} \leqslant P \leqslant \frac{1}{7}$$

$$\text{if } \Lambda = 0 \quad P = \frac{3\beta}{10+\beta} \qquad -\frac{1}{3} \leqslant P \leqslant \frac{1}{2}$$

In this particular case, one finds a general result of Van Brunt and Zare's calculation i.e. that P vanishes always with β . An other important remark is that the measured P values on Ca_2 exceeds noticeably the upper bounds of this calculation. This discrepancy has been explained [9] by the omission of an important coherence effect between the states $M'_B = +\Lambda'$ and $M'_B = -\Lambda'$ ($\Lambda' = 1$).

4. THEORY INCLUDING THE COHERENCE EFFECT

The basic idea [9] is that the two states $M'_B = \pm\Lambda'$ are produced in a completely coherent superposition. In the case of Ca_2 $L'_B = 1$, and this superposition of states is the orbital p_X where X is the vector perpendicular to AB, in the plane containing AB and the polarization of the laser.

To prove this result, we must neglect completely molecular rotation. This means that one may consider the internuclear axis AB fixed during the dissociation process, pointing in the direction θ,ϕ . The ground state of Ca_2 is $X^1\Sigma_g^+$ and the state excited during photodissociation is assumed to be $^1\Pi_u$ (if it was $^1\Sigma_u^+$, there is no coherence effect as $\Lambda'=0$ and Van Brunt and Zare's theory is exact). The transition is a perpendicular transition ; only the component of the laser field perpendicular to the internuclear axis AB can induce the excitation . The excitation probability is proportional to the square of this component of the electric field, i.e. to $\sin^2\theta$. As we neglect all the effects of rotation, this is also the spatial distribution of the axes of dissociated molecules $f(\theta,\phi)\ \alpha\ \sin^2\theta$ (i.e. $\beta = -1$). Among the various combinations of the states $|M'_B=1\rangle$ and $|M'_B=-1\rangle$ only the state even with respect to σ_v (defined as the reflection by the plane containing AB and the laser electric field) can be excited as the initial state and the transition operator are both even with respect to σ_v. This state is precisely the orbital $|P_X\rangle$. Then, using the same calculation as in § 3, we get the fluorescence intensity :

$$I(\underline{e}) = |\langle L_B=0, M_B=0 \mid \underline{e}.\underline{D} \mid P_X\rangle|^2$$

where :

$$|P_X\rangle = \frac{1}{\sqrt{2}}\left(|L'_B=1, M'_B=1\rangle - |L'_B=1, M'_B=-1\rangle\right)$$

This exhibits clearly the coherence between the two states. After averaging over the distribution of internuclear axis ; we get the polarization rate :

$$P = 7/9 \simeq 0.78$$

This calculation has completely neglected molecular rotation ; in order to treat its effects, we are going to present a fully quantal calculation. However, if we replace $f(\theta,\phi) \propto \sin^2\theta$ by an isotropic distribution $f(\theta,\phi) = 1$ the polarization rate becomes P =0.6. This proves that in the presence of the coherence effect P does not vanish with β . An important consequence is that now the polarization rate predicted by the present calculation exceeds the experimental value. This reconciles experiment and theory, the difference being due to various depolarization effects that we are going to investigate now.

5. FULLY QUANTAL THEORY INCLUDING THE EFFECT OF ROTATION IN THE CASE OF Ca

5.1 Principle :

The process that we want to describe is represented on figure 4

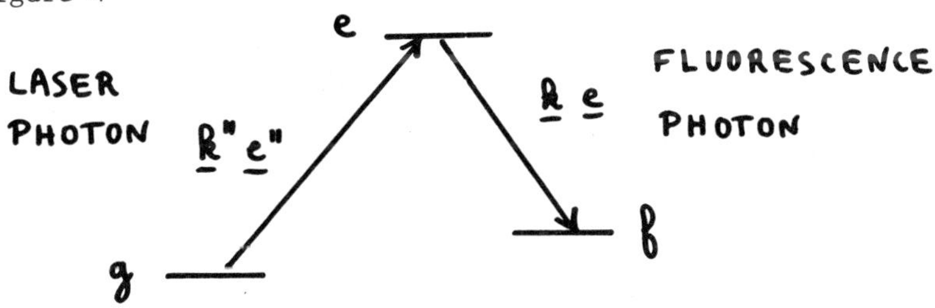

The first step is the absorption of a photon between a rovibrational level of the ground state ($X^1\Sigma_g^+$, v" J" M") and an excited state e lying in the continuum ($^1\Pi_u$, ε', J',M') followed by an emission of a fluorescence photon down to the final state f ($X^1\Sigma_g^+$ ε_f J M). It is important to introduce parity defined states.The cross section for this process is given by [14] :

$$\frac{d\sigma}{d\Omega_{\vec{k}}} (v'' J'' M'' \rightarrow \varepsilon_f J M)$$

$$\left| \sum_{\substack{\alpha=\pm 1 \\ \varepsilon' J' M'}} \frac{\langle {}^1\Sigma_g^+ v''J''M'' | (\underline{D}\underline{e}) | {}^1\pi^\alpha \varepsilon' J' M' \rangle \langle {}^1\pi^\alpha \varepsilon' J' M' | (\underline{D}.\underline{e})^+ | {}^1\Sigma_g^+ \varepsilon_f JM \rangle}{E - \varepsilon' + i\Gamma'} \right|^2$$

where $E = E(v''J''M'') + hck''$. Γ' is the natural radiative width of the level ${}^1\pi_u^\alpha\ \varepsilon'\ J'\ M'$.

$$|{}^1\pi^\alpha\rangle = \frac{1}{\sqrt{2}}(|\Lambda=1\rangle + \alpha\ |\Lambda=-1\rangle)$$

Using sum rule relations, we get :

$$\sum_{\alpha=\pm 1} |\Lambda^\alpha\rangle\langle\Lambda^\alpha| = \sum_{\Lambda=\pm 1} |\Lambda\rangle\langle\Lambda|$$

The physical signal is obtained by summing the cross section over the final state $\varepsilon_f JM$ because this final state is not detected and also on M'' because the initial molecules are not prepared. The fluorescence intensity is finally proportionnal to :

$$I(\vec{e}) = \sum_{M''\varepsilon_f JM} \left| \sum_{\Lambda=\pm 1\ \varepsilon\ J'M'} \frac{\langle {}^1\Sigma_g^+ v''J''M'' | \underline{D}.\underline{e}'' | \Lambda\varepsilon J'M' \rangle}{E - \varepsilon + i\Gamma'} \times \right.$$

$$\left. \langle \Lambda\ \varepsilon\ J'M' | (\underline{D}.\underline{e})^+ | {}^1\Sigma_g^+\ \varepsilon_f\ J\ M \rangle \right|^2$$

In this formula two types of interference effects appear :

a) $\Lambda = 0 \nearrow \Lambda = 1 \searrow \Lambda = 0$; $\Lambda = 0 \searrow \Lambda = -1 \nearrow \Lambda = 0$

This interference is the source of the coherence effect described in part 4 of this paper.

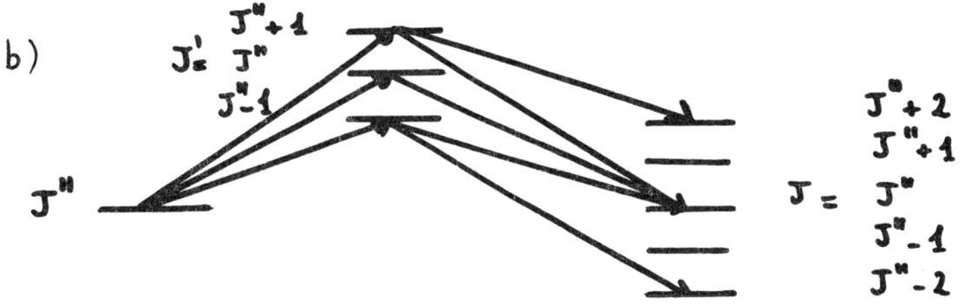

This interference describes the effect of molecular rotation. We are going to investigate these two effects in detail. In order to do the calculation we use the matrix element of $\underline{D}.\underline{e}$ [15,16] and summations are made thanks to various relations of Racah algebra. The details appear in an other paper [14].

5.2 Axial recoil approximation

We are going to show that this case is characterized by the fact that the vibrational wavefunction of the continuum ε' is not dependent on J'. Then we get :

$$I(\vec{e}'') = (2J''+1) \sum_K \mathcal{F}_K \left| \sum_{\Lambda'=\pm 1} \begin{pmatrix} 1 & 1 & K \\ \Lambda' & -\Lambda' & 0 \end{pmatrix} \right|^2$$

where $\mathcal{F}_K$ depends on $\underline{e}$ and $\underline{e}''$ [14]. Using explicit values of $\mathcal{F}_K$ for the geometry considered in figure 1, we get immediately :

$$P = \frac{7}{9} = 0.78$$

in perfect agreement with the calculation of part 4. It is also possible, on this expression to suppress the interference effect between $\Lambda'=1$ and $\Lambda'=-1$ just by taking $\sum_{\Lambda'=\pm 1}$ out of the modulus square. Then the destructive interference for K=1 disappears and the polarization rate becomes :

$$P = \frac{1}{7} = 0.14$$

in agreement with Van Brunt and Zare's result.

5.3 Rotation during dissociation

The effect of rotation is to modify the vibrational wave function of the continuum because the effective potential is $V_{eff}(r) = V_{^1\pi_u}(r) + h^2/2\mu r^2\, J'(J'+1)$

The asymptotic (large internuclear distance r) approximation of this wavefunction is : $\langle r|\varepsilon'_{J'}\rangle = 1/r \sin(Kr - \phi_{\varepsilon' J'})$

There is a direct relation between the phase $\phi_{\varepsilon' J'}$ calculated by WKB approximation and the classical angle of rotation during dissociation α:

$$\alpha = \left.\frac{\partial\phi}{\partial J'}\right)_{\varepsilon'}$$

Then, one can show [14] that the fluorescence intensity is given by :

$$I(\vec{\varepsilon}'') = \sum_J \sum_K \mathcal{F}_K \left| \sum_{J',\Lambda'=\pm 1} (2J+1)^{1/2}(2J'+1) \begin{Bmatrix} 1 & 1 & K \\ J & J'' & J' \end{Bmatrix} \begin{pmatrix} J'' & 1 & J' \\ 0 & -\Lambda' & \Lambda' \end{pmatrix} \begin{pmatrix} J' & 1 & J \\ -\Lambda' & \Lambda' & 0 \end{pmatrix} e^{i\phi_{J'E}} \right|^2$$

This expression may be calculated in the high J" limit. We assume that $\phi_{J''+1,E} - \phi_{J'',E} = \phi_{J'',E} - \phi_{J''-1,E} = \alpha$

Then one gets a closed form result for the polarization rate P

$$P = \frac{1+3(\cos\alpha + \cos^2\alpha)}{7+\cos\alpha + \cos^2\alpha}$$

A classical relation gives [17] the anisotropy parameter ; in the present case :

$$\beta = (1-3\cos^2\alpha)/2$$

One can see that P does not vanish with β ; however, P cannot be calculated with the knowledge of β because β does not depend on the sign of $\cos\alpha$ while P does. This proves that a measurement of P carries more information than a measurement of the anisotropy parameter β of the fragment distribution $f(\theta,\phi)$.

Finally, if we supress the $\Lambda=1$, $\Lambda=-1$ interference in equation (by taking $\sum_{\Lambda'=\pm1}$ out of the modulus square), we find back :

$$P = \frac{3\cos^2\alpha - 1}{13 + \cos^2\alpha} = -\frac{3\beta}{20-\beta}$$

as predicted by Van Brunt and Zare [2].

5.4 Discussion of Ca_2 experimental results

The potential curves of Ca_2 states are represented in figure 5.

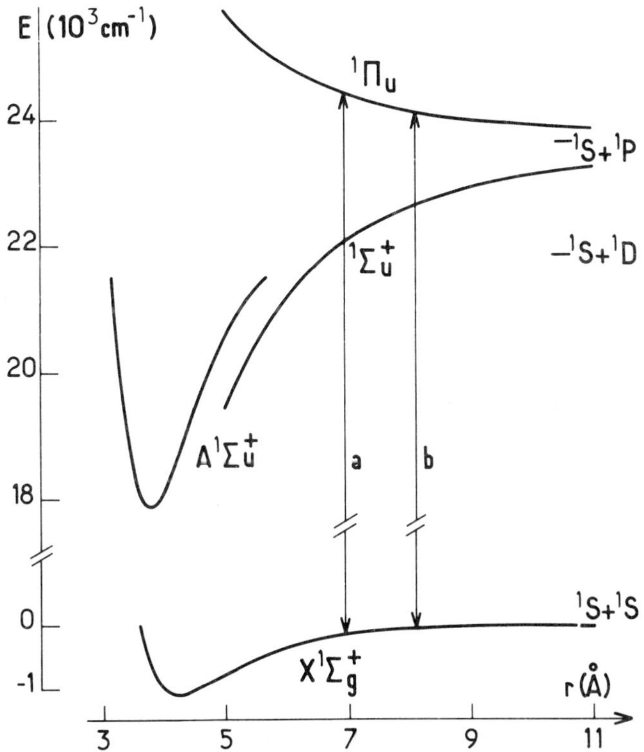

Figure 5. Only the X and $A^1\Sigma^+_u$ potentials curves are known from experiment [18]. The long range $^1\Pi_u$ and $^1\Sigma^+_u$ states dissociating in ^1S + ^1P are calculated by taking into account

only the dipole dipole resonance interaction.

To calculate the polarization P, we need the value of α which depends strongly on the initial level v"J". However only a small fraction of the ground state levels $X^1\Sigma_g^+$ v"J" can be dissociated, because of the Franck Condon principle. The Franck Condon density $< v'', _{J''} | \varepsilon' = E(v'',J'') + h\nu, _{J'} >$ is large only if the speudo crossing point r_0 defined by :

$$V_X(r_0) + h\nu = V_{^1\pi_u}(r_0)$$

is in the classically allowed region of the vibrational motion for the level Xv"J". This condition predicts the dissociation fraction F (at saturation) :

$F \sim 0.08 \qquad \lambda = 406$ nm

$F \sim 0.02 \qquad \lambda = 413$ nm

in agreement with observed values (see figure 2).
We have calculated for each wavelength, the maximum value α_m of the rotation angle and the root mean square value α_{RMS} for the population of molecule that can be dissociated (P depends quadratically on α for low α values). We present these values of α and the corresponding polarization rate .

λ (nm)	α_m (rad)	P	α_{RMS} (rad)	P
406	0.55	0.67	0.31	0.74
413	0.43	0.71	0.24	0.76

The depolarization due to rotation remains small, even for noticeable α values close to 0.5 radians. This effect seems to be too small to explain the observed values (even when corrected for instrumental depolarization). We think that the remaining difference is due to a saturation effect. The depletion rate of the level v"J"M" is given by the Fermi Golden rule

$$\gamma\ (v''J''M'') = \frac{2\pi}{\hbar} \sum_{J'} \left| < X\ v''J''M'' \mid D_z\ E_z \right| {}^1\pi_u\ E\ J'M'\rangle \Big|^2$$

E_z is the laser electric field. The validity of this formula requires[20] that the width $\Delta\varepsilon$ of the Franck-Condon density (considered as a function of ε') $|< v''_{J''}| \varepsilon'_{J'}>|^2$ is large compared to the Rabi Frequency

$$\Delta\varepsilon >> \hbar\ \Omega_R = D_z\ E_z$$

with the experimental values $\hbar\Omega_R \lesssim 1 cm^{-1} << \Delta\varepsilon \sim 10^2\ cm^{-1}$ and the depletion rate $\gamma \sim 10^9\ s^{-1}$. This insures saturation as a molecule crosses the laser beam in 10^{-7}s.

The consequence of saturation is to reduce the polarization rate. In the limit of strong saturation, all the M'' sublevels of the level v''J'' are transferred to the excited state. Classically this means that the excited molecule axis are isotropically distributed. This case leads to a polarization rate (see part 4) : P = 0.60

The signal observed comes from various levels with various states of saturation. It is not possible to evaluate more precisely the polarization rate P, but saturation is surely important here.

6. CONCLUSION

We have presented a new theory for the polarization of atomic fluorescence excited by photodissociation. Strong interference effects occur if the dissociative state has a Λ value equal to 1. This theory explains well a recent experiment made on the Ca_2 molecule.

We want to stress the following points :

- very important consequences may result from symmetry properties of the wavefunction. These symmetry properties may

not be associated to any large energy term.
-the theory presented here shows that the polarization rate carries more information than the distribution of fragments.

ACKNOWLEDGEMENTS

We want to thank J.C. LEHMANN and J. DURUP for helpful discussions.

REFERENCES

1. A.C.G. MITCHELL Z. Fur Physik 49 p.228 (1928)
2. R.J. VAN BRUNT ,R.N. ZARE J. Chem. Phys. 48 p.4304 (1968)
3. H.G. HANSON J. Chem. Phys. 47 p.4773 (1967)
4. G.A. CHAMBERLAIN, J.P. SIMONS Chem.Phys.Lett.32 p.355(1975)
5. G.A. CHAMBERLAIN, J.P. SIMONS J. Chem.Soc. Faraday Transac. II 71 p.2043 (1975)
6. E.D. POLIAKOFF, S.H. SOUTHWORTH, D.A. SHIRLEY, K.H. JACKSON, R.N. ZARE Chem. Phys. Lett. 65 p.407(1979)
7. E.W. ROTHE, U. KRAUSE, R. DÜREN Chem. Phys. Lett. 72 p.100 (1980)
8. E.D. POLIAKOFF, J.L. DEHMER, Dan DILL, A.C. PARR, K.H. JACKSON, R.N. ZARE Phys. Rev. Lett. 46 p.907 (1981)
9. J. VIGUE, P. GRANIER, G. ROGER, A. ASPECT J. Physique Lettre 42 , L 531 (1981), Comptes Rendus Acad. Sci. 293 p.805(1981)
10. O.S. VASYUNTINSKIĨ J.E.T.P. Lett. 31 p.428 (1980)
11. A. ASPECT, C. IMBERT, G. ROGER Optics Comm. 34 p.46 (1980)
12. P.E. ZINSLI J. Phys.E 11 p.17 (1978)
13. S. MUKAMEL, J. JORTNER J. Chem. Phys 61 p.5348 (1074)
14. J. VIGUE, J.A. BESWICK, M. BROYER J. de Physique (to be published)
15. J.M. BROWN , B.J. HOWARD Mol. Phys. 31 1517 (1976)

16. M. BROYER, M. LARZILLIERE, M. CARRE, M.L. GAILLARD M. VELGUE, J.B. OZENNE Chem. Phys. 63 445 (1981)

17. J.A. BESWICK, J. DURUP Chemical Photophysics p. 385 (1979) Editions du CNRS PARIS

18. C.R. VIDAL J. Chem. Phys. 72 p.1864 (1980)

19. M.S. CHILD J. Mol. Spectrosc. 33 p.487 (1970)
J. VIGUE Annales de Phys. 3 p. 155 (1982)

20. C. COHEN-TANNOUDJI , P. AVAN Colloque International du CNRS n°273 p. 93 (Edition du CNRS Paris) (1977).

INTERPRETATION OF THIRD-HARMONIC GENERATION IN MULTIPHOTON IONIZATION EXPERIMENTS BY A COLLECTIVE MODEL

MICHEL POIRIER
Service de Physique des Atomes et des Surfaces, Centre d'Etudes Nucléaires de Saclay, 91191 Gif-sur-Yvette Cedex (France).

Abstract Pressure effects in nearly resonant multiphoton ionization have been investigated by several recent experiments which are briefly reviewed. Two main results have been reported : above several 10^{-3} torr or so, three-photon resonances cen be so affected as to vanish; simultaneously a coherent third-harmonic emission is detected in the forward direction. A collective model describing the system by an average single-atom equation is proposed here. Its mean-field character is emphasized and sample numerical solutions are described. A comparison with experimental results and other theoretical approaches concludes this study.

INTRODUCTION

Resonant multiphoton ionization of atomic vapors is now a familiar topic[1], owing to the tunable high power lasers available today. But the effect of pressure upon the ionization rate near atomic resonances have been investigated much more recently. However, it is worth noticing that the first experimental approaches to multiphoton ionization of gases were performed on dense vapors where mainly plasma effects occur[2]. In this paper the attention will be focused on collective radiation effects in vapors in the pressure range of 10^{-5} torr to 10^{-1} torr, where collisions are assumed to have a smaller effect on the ionization rate than radiative phenomena.

When the pressure increases in a quasi resonant multiphoton ionization experiment, one can observe a strong decrease of the ionization yield and the generation of the third harmonic of the laser frequency. The radiated wavelength generally lies in the region of vacuum ultraviolet (VUV) and can be used, for example, in molecular spectroscopy. This is a way to produce a coherent and tunable source of light in a region where laser sources are not always available. Third harmonic generation in gases is not a new subject since it was reported as early as 1967[3]. But its observation together with photoelectron detection is much more recent. Up to now, four groups have published results on VUV generation in multiphoton ionization experiments.

The group of J.C. MILLER and R.N. COMPTON in Oak Ridge has first observed a VUV generation in nearly resonant multiphoton ionization of rare gases[4]. The excitation and fluorescence spectra have been extensively studied. They have shown that a drastic reduction of the resonant ion yield is concomitant to this radiative effect.

At Los Alamos, J.H. GLOWNIA and R.K. SANDER have reported results on molecules (carbon monoxide) and atoms (rare gases and mercury)[5]. They notice among other things the cancellation of VUV signal with one circularly polarized laser beam, and with two such beams propagating in opposite directions.

This last experiment has been repeated with linearly polarized laser beams in IBM laboratories (Yorktown Heights, N.Y.) by D.J. JACKSON and J.J. WYNNE[6]. They show that the interference between the probability amplitudes induced by the laser field and by the third harmonic field cancels the ionization probability with traveling wave but not with standing waves.

Recently, D. NORMAND, J.MORELLEC and J. REIF in Saclay (France) have observed third-harmonic generation in Mercury[7] They have shown that the generated field can exhibit a higher than third order dependence versus the incident laser field.

As it can be seen when reading this short review, VUV generation in multiphoton experiments is a vaste field of investigation. A theoretical effort has been done to generalize non-linear optics theory to quasi resonant situations. A new approach based on a collective dipole formalism and a mean field hypothesis is exposed here. It tries to answer the question : how can third harmonic generation lower the ionization rate near a multiphoton resonance ?

Sec II is devoted to a brief description of the Miller et al. experiment. A mean-field model is proposed in Sec III and its results are then discussed (Sec.IV). Comparisons with other theoretical models and conclusions are exposed in the last two sections.

EXPERIMENTAL ASPECT

The original work that motivated those experimental studies was the observation by ARON and JOHNSON[8] of multiphoton resonances in Xenon at a pressure of several torr : the ionization signal near the three-photon 6s resonance was absent, and four-photon resonances had an anomalous pressure-dependence. Miller et al have investigated this pressure dependent behaviour with a simultaneous detection of forward emitted VUV radiation. Only a brief introductory account on their Xenon experiments is given here, for more details the reader should refer to the authors papers[4-8].

Three-photon resonances have been studied in the case of a five-photon ionization (resonance on the 6s(3/2)°, J=1 state, laser wavelength near 440 nm) and of a four-photon ionization (resonance on the 6s'(1/2)°, J=1 state, laser wavelength near 389 nm).

The main results of the 6s resonance study are twofold. First, the ionization signal decreases above 0.05 torr and disappears at 0.6 torr. On the other hand, a forward-directed VUV emission is detected. Its wavelength is the third of the laser wavelength, within the resolution of the spectrometer. The line shape of the third-harmonic generation is shifted in the blue wing of the resonance and broadened approximately linearly with pressure. An order of magnitude of this width and this shift is several Å per torr, and reaches 150 Å at 300 torr. A careful study shows that the weak ionization signal, as long as measurable, exhibits a similar spectrum, and suggests that, off resonance, the two effects are linked. Moreover, the tunability of third harmonic generation allows to observe it together with four photon resonances. Such resonances produce an enhancement of ionization signal and a sharp dip in the third harmonic line shape. The excited level is likely reached by absorption of one VUV photon plus one laser photon, rather than by four laser photon absorption.

In the same way, the 6s' resonant ion signal disappears and third harmonic generation increases with increasing pressure. The excitation spectrum of third harmonic is also broadened and shifted to the blue wing. But here, a broad ion signal is observed in the blue wing too. The authors attribute it, rather than to dimer ionization, to a non resonant atomic ionization by absorption of one VUV photon plus one laser photon. Once again, processes involving a third harmonic photon plus additional laser photons produce ions more efficiently than laser photons alone.

In summary, at exact resonance pressure effects results in a cancellation of ion signal. Third harmonic emission is intense in the blue wing and can significantly contribute to ionization.

A MEAN-FIELD MODEL

Since the first experimental papers invoked a collective emission from the populated resonant state, Dicke's formalism of collective dipole operators should be extended in order to include the ionization phenomenon. The present model, closely related to the superradiance treatments in an optically thin medium, has been developped in a recent publication[9], in which the reader will find the details of the formalism.

The atom is described by a two-level density matrix. The population of the ground state n_g, of the excited state n_e and the atomic coherence σ_{ge} are treated non perturbatively. Continuum and non resonant states effect is included in the dynamical Stark shift, three-photon Rabi frequency and ionization width. The hamiltonian of the third-harmonic field includes a strong damping term: it was introduced first in laser theory and then in the statistical study of superradiance by Bonifacio et al[10]. Physically this term means that generated photons leave the cell without interacting once more with the medium. Therefore the field variables follow adiabatically the atomic dipole. This enlights the basic approximation of the model : the medium is optically thin to the third-harmonic photons. In other words, the atomic state is uniform in the interaction region and the radiated field can be approximated by its mean value. We make another rather equivalent hypothesis by assuming that the phase-matching condition is satisfied : the two fields propagate at the same velocity. Lastly, this treatment neglects molecular and collisional effects.

All these hypothesis lead to a Schroedinger equation where the atom-field interaction transforms into non linear terms. It is written as :

$$\dot{n}_g = -\Omega r_2 + \frac{1}{T_{SR}}(r_1^2 + r_2^2) \quad (1a)$$

$$\dot{n}_e = \Omega r_2 - \Gamma n_e - \frac{1}{T_{SR}}(r_1^2 + r_2^2) \quad (1b)$$

$$\dot{r}_1 = \Delta r_2 - \frac{\Gamma}{2} r_1 + \frac{1}{T_{SR}} r_1 \frac{n_e - n_g}{2} \quad (1c)$$

$$\dot{r}_2 = -\Delta r_1 - \Omega \frac{n_e - n_g}{2} - \frac{\Gamma}{2} r_2 + \frac{1}{T_{SR}} r_2 \frac{n_e - n_g}{2} \quad (1d)$$

$r_1 + ir_2$ is the atomic coherence, and the initial conditions are obviously $n = 1$, $n_e = r_1 = r_2 = 0$. In the system (1), Ω is the three photon Rabi frequency for the g-e transition, Γ is the ionization rate of the excited state and Δ is the dynamical detuning, i.e. the difference between the three photon energy $3\omega_1$ and the atomic frequency ω_{eg} shifted by dynamical Stark effect. The terms involving Δ and Ω produce the laser-induced Rabi nutation, Γ represent the laser-induced ionization and the last term contains collective radiation effects, which are characterized by the time constant T_{SR} already introduced in superradiance theory[10].

$$T_{SR} = \left(\frac{2\pi}{hc} \rho \, L\omega_3 \left| p_{eg} \right|^2 \right)^{-1} = (N_o \, \mu \, \gamma_{rad})^{-1} \quad (2)$$

ρ is the atomic density, L the interaction length, ω_3 is the emitted frequency and p_{eg} the electric dipole of the g-e transition. T_{SR} is equal, leaving out the geometrical factor μ^{-1}, to $(N_o \, \gamma_{rad})^{-1}$, N_o being the initial number of neutral atoms and γ_{rad} the radiative decay rate for a single atom in the e state. The "collective character" of the rate $\frac{1}{T_{SR}}$ is obvious on this expression. Let us define atomic parameters in units of T_{SR}^{-1}: $\omega = \Omega \, T_{SR}$, $\gamma = \Gamma T_{SR}$ and $\delta = \Delta \, T_{SR}$. These parameters plus the interaction time in units of T_{SR} determine the collective behaviour of the system.

RESULTS OF THE MEAN-FIELD MODEL FOR A RESONANT EXCITATION

Due to the lack of an analytical solution, the system (1a) to (1d) has been numerically integrated, assuming a spatially uniform excitation and a square laser pulse. Sample results are shown in Fig. 1 to 3. The populations n_g, n_e and the ionization probability $n_i = 1-n_g-n_e$ are plotted versus time expressed in units of T_{SR}. On separate figures we have drawn the instantaneous third-harmonic intensity and its time-integrated value, which is the number of emitted photons per atom. The present figures correspond to a resonant excitation, but off-resonance results have been published too[9]. Since a strong ionization rate is unfavourable to harmonic generation, this study is restricted to the case $\Omega > \Gamma$, namely $\Omega = 10\,\Gamma$ in Fig.1-3.

First, consider the case of low pressure, i.e. $\Omega T_{SR} > 1$. In Fig.1, the parameter ω is 10 and γ is 1. One observes the well known damped Rabi oscillations of the atomic populations. The third-harmonic energy, characterized by the total number of generated photons Nph remains weak. The ionization rate is close to its zero-pressure value $\frac{\Gamma}{2}$. Third harmonic emission, due to quadratic terms in system (1), has a time dependence which is exponential with a time constant Γ^{-1} if we average it over a Rabi period. Assuming an interaction time long enough to reach saturation, the total number of ions is the initial number of neutral atoms :

$$N_i = N_o \tag{3}$$

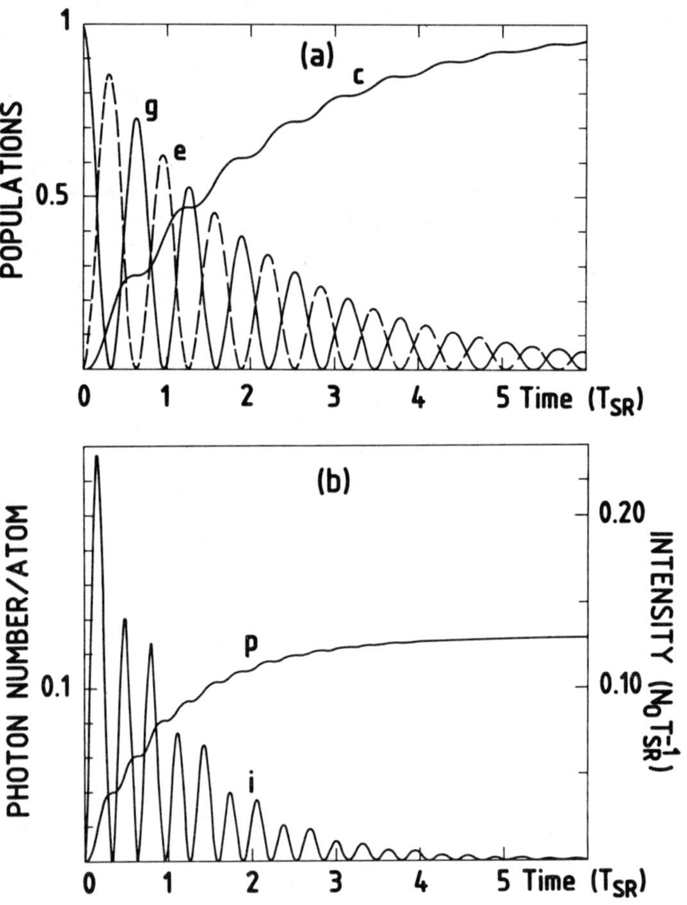

Figure 1 - Time evolution of the atom (Fig. 1(a)) and the radiated field (Fig.1(b)) during the interaction. The reduced parameters, defined in the text, are ω=10, ν=1, δ = 0. Abscissas : time τ = t/T_{SR}. Ordinates - Fig.1(a) : g ground state population; e excited state population; c ion number per atom - Fig.1(b) : p photon number per atom the corresponding scale being on the left; i radiated intensity, with the scale on the right, in units $N_0^2 \mu \nu_{rad}$.

and the number of generated photons is given by

$$N_{ph} = N_0^2 \frac{\mu \, \gamma_{rad}}{8\Gamma} \tag{4}$$

This quadratic dependence versus pressure is the mark of a collective behaviour.

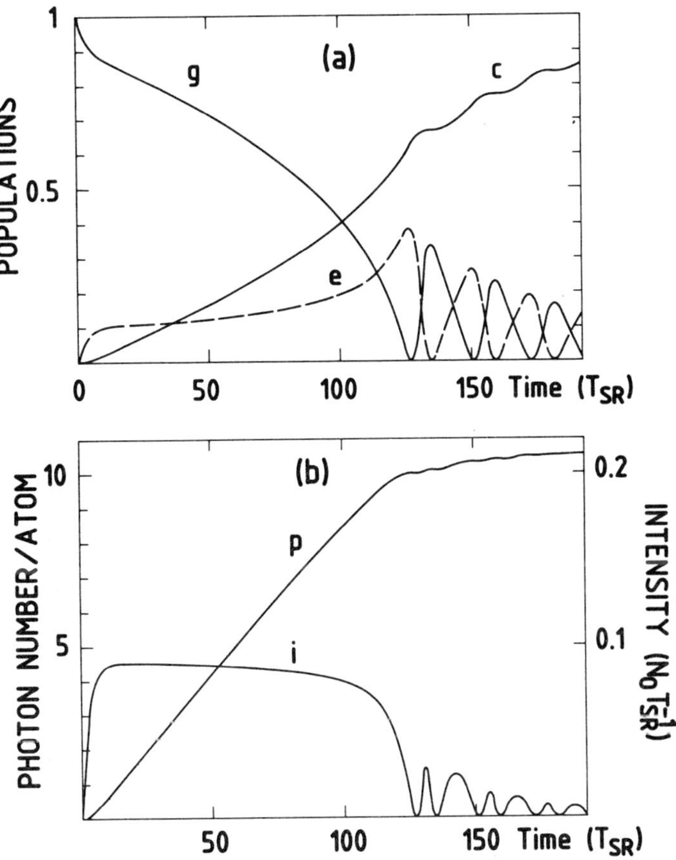

FIGURE 2. Time evolution of the atom and the radiated field. Same situation as in Fig.1, except for the pressure which is 32 times higher. Therefore $\omega = 3.1 \times 10^{-1}$, $\gamma = 3.1 \times 10^{-2}$, $\delta = 0$. The abscissa scale is such that equal lengths in Fig.1 and 2 corresponds to equal interaction times, and reduced times τ in a ratio of 32.

On the other hand, Fig.2 illustrates high pressure situation. The laser intensity is the same as in Fig.1, and the pressure is 32 times higher. A striking difference is the disappearance of Rabi oscillations during a first stage. This stationary behaviour can be understood as a balance between the laser excitation and the radiative decay when a suitable inversion of population has been reached. A very simple argument leads to predictions which are in good agreement with more involved calculations in the case $\gamma < \omega \ll 1$ and before saturation ($n_i < 1$). Since at dynamical resonance the atomic coherence is purely imaginary ($r_1 = 0$), a stationary solution of the system (1) is given by :

$$r_2 \simeq \omega \tag{5}$$

and thus, using a general property of the system (1)

$$n_e n_g = r_1^2 + r_2^2 \tag{6}$$

and the assumption that n_g is close to unity, we get

$$n_e \simeq \omega^2 \tag{7}$$

The ionization rate, given by Γn_e, is in the present case $\Gamma \omega^2$ and the total ion number is given by

$$N_i = N_o \Gamma \omega^2 t = \frac{1}{N_o} \frac{\Gamma \Omega^2}{\mu^2 \gamma^2_{rad}} t \tag{8}$$

The ionization rate is multiplied by $2\omega^2$, which is inverse quadratic with pressure, and the total ion number is inversely proportional to pressure. The emitted energy calculated in the same way :

$$N_{ph} = N_0 T_{SR}^{-1} \int_o^t r_2^2 (s)\, ds = N_0 T_{SR} \Omega^2 t = \frac{\Omega^2}{\mu \gamma_{rad}} t \tag{9}$$

is pressure independent. At longer interaction time, the depletion due to ionization reduces the atomic populations and the stationarity condition (5) no longer holds. The effect of collective radiation becomes unsignificant and the Rabi oscillations can reappear, as seen in Fig.2. A crude estimate of the time characterizing the transition from the non-oscillating regime to the oscillating regime is given by :

$$t_{cr} \simeq \frac{1}{2\omega^2 \Gamma} \tag{10}$$

Neglecting the harmonic generation posterior to t_{cr}, we we obtain for the saturation value of N_{ph}, using eq(9) and eq (10):

$$N_{ph}\,(t>t_{cr}) \simeq N_o^2 \frac{\mu\, \gamma_{rad}}{2\Gamma} = \frac{N_o}{2\gamma} \qquad (11)$$

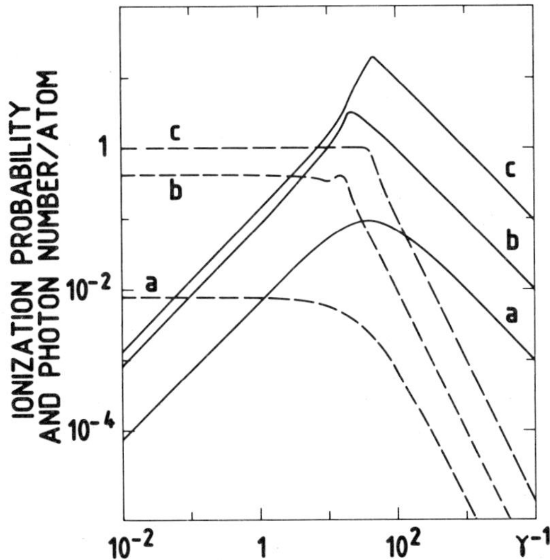

FIGURE 3. Pressure dependence of the ionization probability (broken lines) and photon number per atom (solid lines) for a constant laser intensity. If the laser intensity is 10^{10} W.cm^{-2}, a typical order of magnitude is $\gamma^{-1}=1$ when $p \simeq 4 \times 10^{-6}$ torr. The various curves correspond to various interaction times. a : $t = 1/\Omega$; b: $t = 10/\Omega$; c : $t = 100/\Omega$.

We have distinguished between two ranges of pressure. The pressure dependence of the ionization probability N_i/N_o, and of the photon number per atom N_{ph}/N_o is shown in Fig.3 for various interaction times. One can show[9] that the transition range between the two regimes is characterized by the condition:

$$\omega = \frac{\Omega}{N_o\, \mu\, \gamma_{rad}} \simeq \frac{1}{\sqrt{8}} \qquad (12)$$

With a laser intensity of 10^{10} W.cm^{-2}, the pressure defined by (12) is about 3×10^{-3} torr, in reasonable agreement with the experimental value for which the ion signal saturates. The parameter ω characterizes the pressure regime, while emitted energy at long interaction times depends on γ.

Lastly, the third harmonic energy depends on the laser

intensity I. At low intensity, eq.(9) applies and shows that N_{ph} is cubic versus I. At high intensity, relation(4) holds and N_{ph} varies like Γ^{-1} which is I^{-2} if the resonant state is two-photon ionized. Therefore one can define an optimum laser intensity also given by (12).

If we go back to experimental results, we see that large pressure dependent shift on third-harmonic signal cannot be explained in this mean-field model. Phase-matching condition is important for non linear processes in a dense medium as emphasized by Y. PRIOR in his talk at this conference. In non linear optics, third-harmonic generation with focused beams needs a negatively dispersive medium(11) and is thus possible only in the blue wing. Nevertheless, the present model provides a simple dynamical approach. The decrease of the ion number, the I^3 dependence of the third harmonic signal and its saturation with pressure are correctly predicted. Furthermore, the excited state population is treated in a non perturbative way, which extends the validity of this theory to very high laser intensities.

OTHER THEORETICAL WORKS

JACKSON and WYNNE, in the above-mentioned letter [6] have derived the ionization rate from a simple argument based on the Fermi golden rule and Maxwell equations. They emphasize that the bound-bound g-e transition results from the interference between two amplitudes : absorption of three laser photons and absorption of one third harmonic photon. After some absorption lengths, the third harmonic field can be calculated as the forced solution of the Maxwell equation at frequency ω_3. Using a simple expression for the wave-vector mismatch near resonance, they show that the two above-mentioned amplitudes exactly cancel each other, whatever the laser intensity and the gas pressure. This fact is to compare to relation (5) which expresses the same cancellation in the high pressure limit of our model. Conversely such a cancellation is not possible with two counter propagating laser beams.

In a sensibly more general framework, PAYNE and GARRETT [12] have developped another semiclassical theory including propagation effects, laser-bandwith and beam focusing. The Schroedinger equation is solved by means of perturbation theory, the polarization at third-harmonic frequency is then calculated over all the interaction volume. The third-harmonic field is then obtained through the Green function solution of the Maxwell equation. Off resonance, they find that the phase-matching condition governs the efficiency of third-harmonic generation. The absorption of one third-harmonic photon is more efficient than the absorption of three laser photons in the discrete spectrum; thus ionization is due to one third-harmonic photon plus additional laser photons, and peaks on the blue wing, approximately like the phase-matching curve. Near resonance, the three photon Rabi frequency and the third harmonic Rabi frequency cancel each other, therefore at a sufficient pressure there is no net ion production.

CONCLUSION

The aim of this short review was to point out that collective radiation can be interpreted in several ways and thus give way to a rather large variety of models. Treatments derived from non linear optics and perturbation theory are useful to understand propagation and focusing effects for a moderate laser intensity or not on exact resonance. The mean-field approach exposed here fails for an optically thick medium, but remains valid for a high laser intensity. It is amazing to note that the cancellation of the Rabi frequency near resonance appears in each model, after some coherence lengths in the space dependent theory[6, 12], in the so-called high pressure regime in the present mean-field model. A same physical phenomenon can be approached through various models which are complementary and sometimes also convergent. New higher-order effects versus the laser intensity have been reported by NORMAND et al[7]. This emphasizes the need for a theory accounting for propagation, focusing and high laser intensity. Harmonic generation near multiphoton resonance is still an open domain.

REFERENCES

1. N.B. Delone, Sov. Phys. - Usp. (English translation). 18,169 (1976);
   P. Lambropoulos, Advances in Atomic and Molecular Physics 12, 87 (1976);
   J.H. Eberly, P. Lambropoulos, Multiphoton Processes, Proceedings of an International Conference at the University of Rochester, Rochester N.Y. June 6-9, 1977, John Wiley and Sons (1978); J. Morellec, D. Normand, G. Petite Advances in Atomic and Molecular Physics 18, 97-164(1982).
2. C. Grey Morgan, Rep. Prog. Phys. 38, 621-665 (1975).
3. G.H.C. New and J.F. Ward, Phys. Rev. Lett. 19, 556 (1967); J.F. Ward and G.H.C. New, Phys. Rev. 185, 57 (1969).
4. J.C. Miller, R.N. Compton, M.G. Payne , W.R. Garrett, Phys. Rev. Lett. 45, 114 (1980); J.C. Miller and R.N. Compton Phys. Rev. A25 , 2056 (1982).
5. J.H. Glownia, R.K. Sander, Appl. Phys. Lett. 40, 648 (1982); Phys. Rev. Lett. 49, 21 (1982).
6. D.J. Jackson, J.J. Wynne, Phys. Rev. Lett. 49, 543 (1982)
7. D. Normand, J. Morellec, J. Reif, J. Phys. B 16, L227 (1983).
8. K. Aron, P.M. Johnson, J. Chem. Phys. 67, 5099 (1977).
9. M. Poirier, Phys. Rev. A 27, 934 (1983).
10. R. Bonifacio, P. Schwendimann, F. Haake, Phys. Rev. A 4; 302 (1971); 854 (1971).
11. D.C. Hanna, M.A. Yuratich, D. Cotter, Non linear optics of free atoms and molecules, Springer Verlag series in optical sciences vol. 17 (1979).
12. M.G. Payne, W.R. Garrett, H.C. Baker, Chem. Phys. Lett. 75, 468 (1980); M.G. Payne, W.R. Garrett, Phys. Rev. A, 26, 356 (1982); Phys. Rev. A (in press).

MULTIPHOTON SELECTIVE EXCITATION: ON THE ROLE OF CHAOTIC DYNAMICS WITH MANY BASINS OF ATTRACTION (+)

FORTUNATO TITO ARECCHI
Istituto Nazionale di Ottica and Phys. Dept., University, Firenze, Italy

Abstract Considering the multiphoton excitation of a large molecule as a classical nonlinear dynamics, some speculations are offered on the possible implications of a chaotic dynamics with many basins of attraction.

The problem of selective laser excitation of complex molecules by infrared multiphoton absorption has received a large attention, because of its possible applications to chemical engineering on one side, and of its implications in statistical mechanics on the other side. I wish to point the attention on the problem of relaxations in unimolecular reactions.

Since many reviews are available on this subject [1,2] I shall assume that the current conjectures on the loss of selectivity after a few-photon absorption are well known. To be precise, molecular activation is based on RRKM theory (uniform heating of a molecule even for resonant excitation of one vibrational mode) only for very rapid intramolecular relaxation[3]. The theory breaks down if the energy is supplied to one degree of freedom within a time shorter than the time

(+) Work partly supported by Contract CNR-INO

scale of intramolecular vibrational transfer to other modes, thus allowing for selective bond chemistry.

The non-RRKM chemistry is modeled after the spin-lattice relaxation theory in NMR [4] or the electric dipole-bath relaxation in quantum optics [5], that is, it is based on a coherent interaction between an IR field and a resonant vibrational mode, disturbed by diagonal and off-diagonal relaxation terms toward a thermal bath made up by all the other degrees of freedom of the molecule. This model implies a single time scale

$$t \ll T_1, T_2$$

where T_1, T_2 are the relaxation times toward a structureless bath having the usual maximum entropy properties.

It has been recently shown [6,7] that chaotic phenomena, occurring on a time scale corresponding to a high frequency spectrum, can be also characterized by a second, much longer, time scale (corresponding to a low frequency component) if the following circumstances are verified: i) the system has at least two attractors; ii) the attractors are near to be destabilized or they have just become unstable; and iii) the system is "open" to external fluctuations, i.e., the presence of white noise is essential to yield jumps between different basins of attraction. The long time scale yields a diverging low frequency spectral power going as $f^{-\alpha}$. In our spectra [6] the slopes should all be multiplied by a factor 2, because of a misleading use of the power calibration in the spectrum analyzer. Hence most slopes cluster around

$\alpha = 1.2$. The above conditions show that we are in presence of a phenomenon which occurs beyond the usual approach to chaos by either one of the current scenarios [8]. A simple jump between two attractors is not sufficient to explain the phenomenology. In fact experiments [6] show that, when leaving an attractor, the representative point in phase space has a long erratic motion before landing onto another attractor. Usually our spectra extend over less than two decades, at frequencies much less than the driving frequency. For the CO_2 laser [7], we have a $f^{-0.6}$ spectrum between 1 and i00 Hz, while the driving frequency is around 60 KHz.

To understand the above features we have investigated numerically [9] the dynamical equations of the systems of Ref. 6 and 7. However the 1/f region is so narrow that this procedure gives a poor insight. We have preferred to build a recursive map allowing for two independent attractors, namely

$$x_{n+1} = (a-1)x_n + a x_n^3$$

on the interval (-1,1), disturbed by white noise of variance between 10^{-7} and 10^{-5}. a is set around 3.980006 in order to have two period-three attractors which have become strange and extended over intervals of the order of 10^{-3}. As in the experiments [6,7], we observe $f^{-\alpha}$ regions with $\alpha = 1$ for intermediate noise levels, and $\alpha \to 0$ for high noise and $\alpha \to 2$ for low noise. These 1/f regions extend over 2 decades with slope changes at the two extrema, thus appearing as a suitable superposition of Lorentzians. As in equilibrium systems a 1/f slope is the result of a sum over a large number of Lorentzian curves suitably weighted [10], similarly we formulate the conjecture that in a nonlinear system we have to sum over many stability valleys to extend the 1/f law over many decades.

A speculative application of the above considerations to selective photochemistry is as follows. In multiphoton dissociation or isomerization reactions, consider the molecular potential as the anharmonic potential of a classical dynamical system. The nonlinear evolution under a driving radiation field near to resonance with the linearized eigen frequency (corresponding to the harmonic part of the potential) may allow for two or more independent attractors, as in Ref. 6,7,9.

How "classical" is the motion of a many-atom molecule is still an open question [11]. Let me remind however two powerful theorems [12,13] which state that, when coupling two systems (e.g. one an electromagnetic field and the other a molecular system) if one is classical, the other evolves from the ground state to a coherent state, which is very different from an energy eigenstate. This coherent state (of the field, if the driving source is a classical current [12]; of the atom if the source is a classical e.m. field [13]) is a minimum uncertainty state. Hence the natural quantum state of a system driven by a classical source is very well approximated by a classical description. We should expect that a molecule driven by a high intensity IR field reaches a coherent state with a specific algebra depending on the symmetry of the problem. Coherent states for harmonic oscillators were introduced in Ref. 12, for N 2-level atoms in Ref. 13, and N-levels atoms in Ref. 14. These References may offer hints on how to work coherent states for particular molecular potentials. What here matters, however, is the fact the current pictures on classical chaos may be relevant to describe a multiphoton unimolecular process.

Thus, it should be easy to adjust the control parameters (frequency and intensity of the IR field) in order to have a many-attractor dynamics, and hence a structured thermal bath made of two, or more, basins of attraction, with individual relaxation times toward each of them plus much longer jump times. If these longer times occur much above the picosecond range, it should then be easy to observe phenomena which are secular with respect to the current picosecond intramolecular relaxations. For instance, it should be possible the excitations of one, over two, local groups of bonds even with rather long pulses. Such a possibility was already shown by Lin et al.[15] for a non-centro-symmetric molecule as sucrose solid in KBr.

In this short outlook of "chaotic photo-chemistry" I have used terms which may not be familiar to the audience. I find thus convenient to add a tutorial Appendix on very recent investigations of these phenomena.

Appendix

Nonlinear dynamical systems and turbulence.

Let us consider a dynamical system, ruled by the equation

$$\dot{X} = F(X;\mu) \tag{1}$$

where X is an n-dimensional vector, F a nonlinear function and μ an m-dimensional control parameter.

We study the equilibrium solutions

$$F(X;\mu) = 0 \tag{2}$$

for different μ. For a critical value of μ, a stationary solution may switch from stable to unstable (bifurcation). An example is the laser threshold, where the branch $x_1=0$ becomes unstable, and it appears a new stable branch $x_2 \neq 0$.

We are not interested in the general treatment of bifurcations, but just how a sequence of them may lead a system to turbulence, or chaos. We call turbulence the appearance of a continuous power spectrum, which corresponds to unpredictable features (a line power spectrum would correspond to a periodic correlation function).

Before 1963, the Landau-Hopf model [15], based on mode-mode coupling in a fluid due to the nonlinear hydrodynamic equations, hypothized the generation of a large amount of uncommensurate frequencies, which eventually accumulated into a continuum. The hydrodynamic equations, being field equations correspond to an infinite number of coupled equations. In 1963 Lorenz [16] showed that three coupled nonlinear equations were enough to reach chaos. Now, three equations can mimic field equations with drastic cutoffs due to the boundary conditions.

If in eq. (1) we take $X = (x,y,z)$, we have a three-dimensional phase space. After a transient, the trajectories becomes closed loops (periodic or stable attractors) corresponding to power spectra made of discrete lines. But for crucial values of the control parameters $\vec{\mu}$, one has a strange attractor, that is, an intricate loop which never appears to close on itself, and correspondingly a continuous spectrum.

The Lorenz equations, with parameter values suitable to give a strange attractor, are given by

$$\begin{aligned} \dot{x} &= -10\,x + 10\,y \\ \dot{y} &= -x\,z + 28\,x - y \\ \dot{z} &= xy - 8/3\,z \end{aligned} \tag{3}$$

Notice that Lorenz chaos is a deterministic one, because we do not have noise sources.

Equivalent to the Lorenz model is a system of two 1st order equations (or 2nd order equation) plus an external modulation. An example is the driven Duffing oscillator[17,6] ruled by

$$\ddot{x} + \gamma \dot{x} + \omega_0^2 x - \beta x^3 = A \cos \omega t \tag{4}$$

Equation (4) is equivalent to 3 coupled equations

$$\begin{aligned} \dot{x} &= y \\ \dot{y} &= -\gamma y - \omega_0^2 x + \beta x^3 + A \cos z \\ \dot{z} &= \omega\,. \end{aligned} \tag{4)'$$

The potential corresponds to a single minimum. For different control parameters $\vec{\mu}$ (either modulation amplitude A or frequency ω) it may give a sequence of subharmonic bifurca-

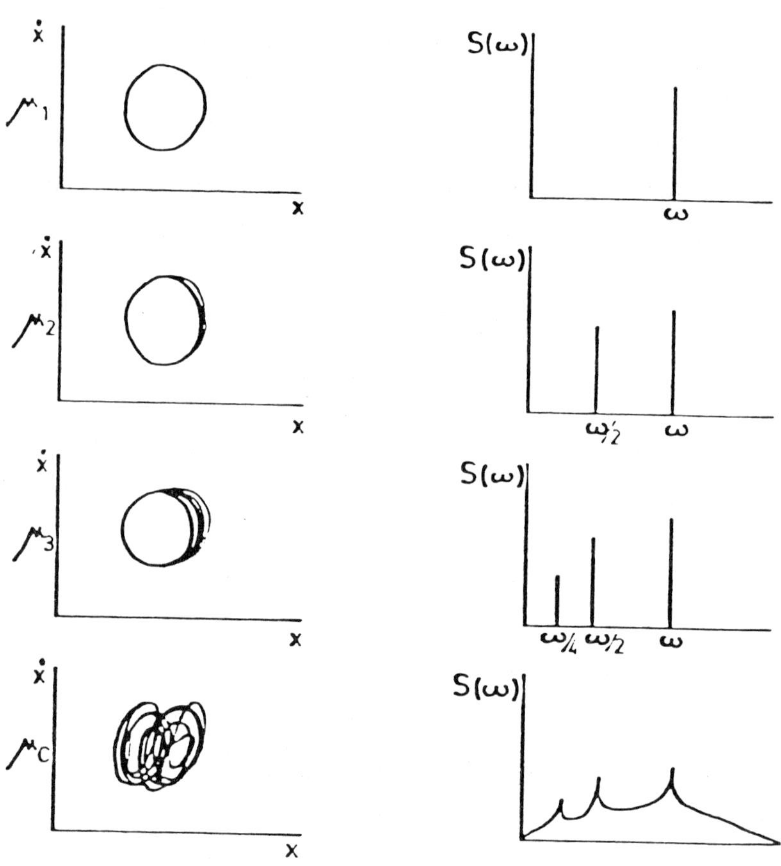

FIGURE 1. Phase space plots $(\dot{x}, x)$ and power spectra $S(\omega)$ for different control parameters.

tions leading eventually to chaos as shown in Figure 1.

The sequence of subharmonic bifurcations corresponds to the successive appearance of period $T = 2\pi/\omega$, 2 T, 4 T... $2^n T$ in the output. If we call μ_n the value of the parameter at which the period $2^n T$ appears then the following relation is verified

$$\delta = \frac{\mu_{n+1} - \mu_n}{\mu_{n+2} - \mu_{n+1}} \xrightarrow[n \gg 1]{} 4.669\ldots$$

This number has been shown by Feigenbaum [18] to be universal.

If, in the space $(\dot{x}, x, t)$ one considers the discrete transformation from the point $(\dot{x}, x)$ at time t to the point $(\dot{x}, x)$ at time t + T after integration of equation (4) over that interval one has the discrete mapping

$$\vec{x}_{t+T} = f(x_t) \tag{5}$$

Such a correspondence is illustrated in Figure 2. It is called stroboscopic, or Poincaré, map. The difference equation (5) is fully equivalent to the differential equation (4). Of course, having performed a time integral, one has reduced the three-dimensional recurrence (5). In many cases of physical interest the points of the Poincaré map cluster over an almost one-dimensional manifold.
Since the interesting bifurcations are associated with a change of slope, the one-dimensional map can be studied around a maximum. Thus a quadratic map as

$$x_{n+1} = \lambda x_n (1 - x_n) \tag{6}$$

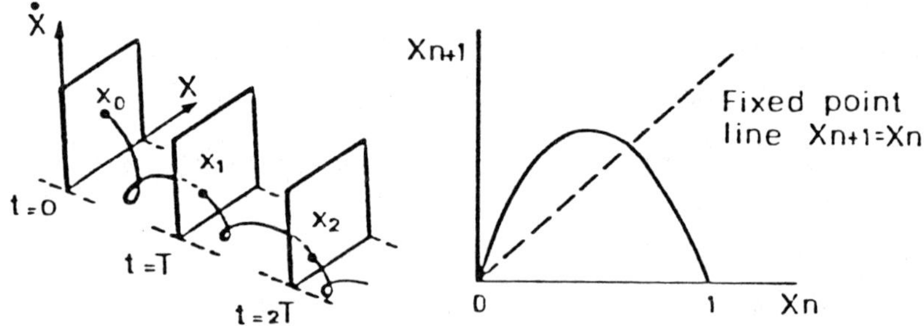

FIGURE 2. Trajectory in 3 D phase space (solid line) and direct mapping $x_t \to x_{t+T}$ (dashed line). Example of quadratic map.

is sufficient to display most of the features of equation (3) and the role of μ is here taken by λ . One can develop a straightforward set of transformations using discrete maps [19]. For instance the second iterate is

$$x_{t+2T} = f(x_{t+T}) = f(f(x_t)) = f^{(2)}(x_t) \quad ,$$

and a fixed point is ruled by the equation

$$x_{t+T} = x_t = f(x_t) \quad .$$

Of course, a fixed point in the map does not mean a single equilibrium point, like in the differential equation, but a limit cycle of period T, whereby the position in phase space goes onto itself at each T. A period 2T would appear as a solution of the equation

$$x_{t+2T} = x_t = f^{(2)}(x_t)$$

and so on. Feigenbaum has evaluated his δ value by showing that in such cascades of period doublings, the local stracture of the attractor is reproduced at a rescaled size in successive bifurcations, with the rescaling parameter being a universal constant.

So far we have sketched how, by varying the control parameter, a system ruled by equation (4) or (5) undergoes a sequence of bifurcations eventually leading to chaos.

Noise is not essential (deterministic chaos), but if we add it, the number of subharmonic bifurcations before chaos become smaller and smaller. This can be put in terms of a scaling law where the variance of the external noise appears somewhat as a modification of the control parameter. Let us now change the sign of the potential, getting two stable valleys (Figure 3). This is equivalent to the new Duffing equation:

$$\ddot{x} + \gamma \dot{x} - \omega_0^2 x + \beta x^3 = A \cos \omega t \qquad (4)''$$

Depending on the initial conditions, we have two independent attractors. Let us increase μ until both attractors become strange. Now, addition of a random noise may trigger jumps from one to the other. These jumps give a low frequency divergence in the power spectrum [6]. These jumps may be considered as a superchaos insofar as they couple two strange attractors, otherwise independent. Here presence of random noise is essential.

Figure 4b shows the appearance of the low frequency, or 1/f divergence. A similar effect was observed in a Q-modulated CO_2 laser.

Figure 4 shows bistability, that is, the simultaneous coexistence of two attractors corresponding respectively to f/4 and f/3 subharmonic. Increasing the modulation depth m, the attractors become strange and correspondingly one observes the spectral divergence as in Figure 4c. Thus associating low frequency divergence with jumps between two attractors (multistable situations) seems a successful conjecture.

REFERENCES

1. N. Bloembergen and E. Yablonovitch, Physics Today, May 1978, page 83.
2. A. H. Zewail, Physics Today, November 1980, page 27
   W. Fuss and K. L. Kompa, Progr.Quant.El. 7,117 (1981).
3. P. J. Robinson, K. A. Holbrook, Unimolecular Reactions, Wiley-Interscience, 1972.
4. A. Abragam, Nuclear Magnetism, Oxford 1961.
5. H. Haken,in: "Laser Handbook", ed. by F. T. Arecchi and E. Schulz Du Bois, North Holland 1972, vol. 1.
6. F. T. Arecchi and F. Lisi, Phys. Rev. Lett. 49,94 (1982).
7. F. T. Arecchi, R. Meucci, G. Puccioni and J. Tredicce, Phys.Rev.Lett. 49,1217 (1982).
8. J. P. Eckman, Rev.Mod.Phys. 53,643 (1981).
9. F. T. Arecchi, R. Badii and A. Politi (to be published).
10. R. F. Voss and J. Clarke, Phys.Rev. B13,556 (1976).
11. W. E. Lamb, in "Laser Spectroscopy IV", ed. by H. Walther and K. W. Rothe, Springer 1979.
12. F. Bloch and A. Nordsieck, Phys.Rev. 52,54 (1937)
    R. J. Glauber, Phys.Rev. 131,2766 (1963).
13. F. T. Arecchi, E. Courtens, R. Gilmore and H. Thomas, Phys.Rev. A6,2211 (1972).

14. F. T. Arecchi, R. Gilmore and D. M. Kim, Lett.N.Cimento 6,219 (1973).
15. L. Landau and E. Lifshitz, "Statistical Physics", Pergamon Press, 1958.
16. E. Lorenz, Jour.Atmos.Sc., 20,130-141 (1963).
17. J. P. Crutchfield and B. A. Huberman, Phys.Rev.Lett. 43, 1743.
18. M. J. Feigenbaum, J.Stat.Phys. 19,25 (1978); 21,669(1979)
19. P. Collet and J. P. Eckmann, "Iterated Maps, Birkhäuser (1980).
20. C. T. Lin, J. B. Valim and C. A. Bertran, in: "Lasers and Applications", ed. by W. O. N. Guimares et al., Springer Series in Optical Sciences, vol. 26, 1981.

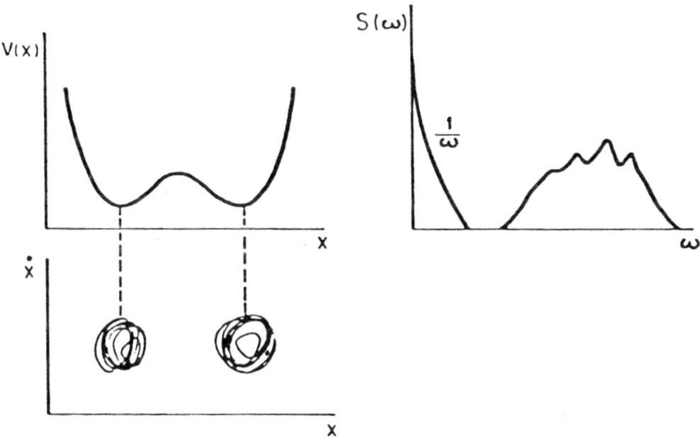

FIGURE 3. a) Bistable potential V(x) (change of sign in an-harmonic force of eq.(4)).Coexistence of two strange attractors in $(\dot{x},x)$with possible mutual jumps induced by external noise (line with arrows)

b) Birth of $1/\omega$ branch in power spectrum associated with the above jumps.

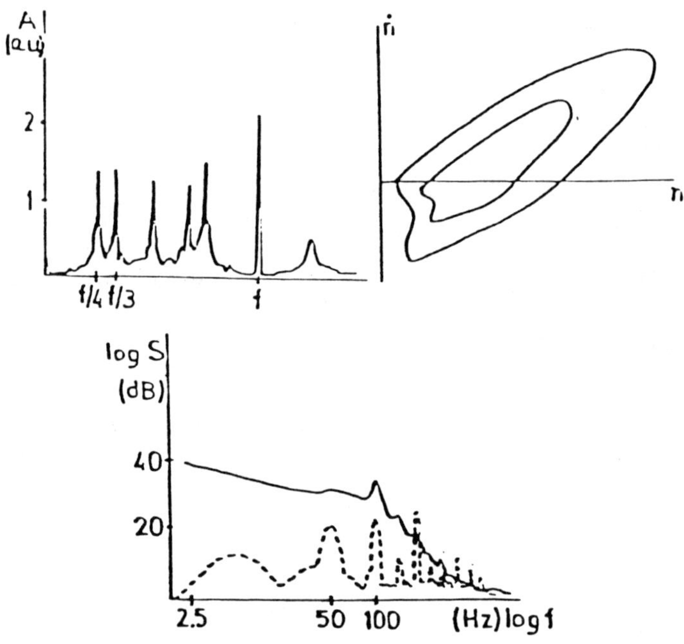

FIGURE 4. Bistability and 1/f noise in a CO_2 laser with loss modulation

a-b) coexistence of two attractors (period 3 and 4 respectively). The two superposed spectra correspond to two starts with different initial conditions

c) comparison between the low frequency cut-off (dashed line) when the two attractors are stable and the low frequency divergence (solid line, slope α = 0.6) when the two attractors are strange.

PHOTODISSOCIATION AND RAMAN SCATTERING OF DIATOMIC MOLECULES IN INTENSE FIELDS.

ANDRE D. BANDRAUK and GILLES TURCOTTE
Département de chimie, Faculté des sciences,
Université de Sherbrooke, Sherbrooke, Qué., Canada
J1K 2R1

Abstract Photodissociation and Resonance Raman Scattering of a diatomic molecule are studied as a function of field strength via a coupled equations approach. It will be shown that new bound states induced by the laser field can strongly affect these and other multiphoton processes.

INTRODUCTION

Molecules in intense electromagnetic fields are much more complex systems than atoms, since one must now consider the nuclear degrees of freedom. Thus in addition to electronic levels, one must also include rotational and vibrational levels in any meaningful description of molecules[1]. This of course implies that molecules are intrinsically multi-level systems so that two level descriptions such as used in the dressed atom picture[2,3] are no longer applicable. It is the basic aim of the present article to examine a tractable molecular problem where one can calculate with reasonable accuracy the effect of a strong laser field on a molecular process. The simplest such process which has been examined theoretically and experimentally in the

Fermi-Golden rule limit (weak field = classical photochemistry and physics) is direct photodissociation[4]. We will show in the presentation to follow how such a simple molecular process is modified by an intense laser field. In particular we will focus our attention on the Ar_2^+ molecule for which good ab initio potentials now exist and the important electronic transition moments have also been calculated[5,6]. We will examine the effect of a strong field on the direct photodissociation (see fig. 1) in this molecule ans show how new laser induced bound states modify the angular distribution as compared to the well known traditional (weak field) results used currently in photochemistry[4]. Finally since one expects a molecule imbedded in an intense laser field to fluoresce strongly, the corresponding Resonance Raman (RRS) spectrum should contain much spectroscopic information about the dressing of the molecule as in the atomic case[7]. It is with this in mind that we have developped numerical methods of calculating such a spectrum for the Ar_2^+ molecule. Our numerical calculations cover thus both weak fields (classical spectroscopy) and strong fields and are expected to be quite accurate since for Ar_2^+ multiphoton excitations should be quite negligible compared to the first absorption which leads to a very strong photodissociation due to the large ${}^2\Sigma_u^+ \leftrightarrow {}^2\Sigma_g^+$ transition moment

(2 a.u.)[5-6]. All other multiphoton processes are at least three photons away from the excited $^2\Sigma_g^+$ continuum and must pass through virtual (nonresonant states), see fig. 1. Finally at the fields we will be considering, $I \sim 10^{10}$ W/cm^2, it is known that Ar is quite stable since breakdown occurs at intensities beyond 10^{12} W/cm^2 for Ar gas at 1 atm pressure[8]. Thus the theoretical and numerical results presented here, we hope will be an inducement for experimentalists to examine the "dressing" of molecules by intense laser fields.

PHOTODISSOCIATION AS PREDISSOCIATION

Multiphoton processes are nothing but redistribution of energy between the molecular and field states. Thus, as illustrated in fig. 1, the ground electronic-vibrational (v) - rotational (J) - n photon bound state $|^2\Sigma_u^+, v, J, n\rangle$ of total energy $E = E_v + E_J + n\hbar\omega$ is degenerate with some (n-1) photon continuum state $|^2\Sigma_g^+, C, J_c, n-1\rangle$ of total energy $E = E_c + E_{J_c} + (n-1)\hbar\omega$. On a total energy scale, the field electronic potentials $[V(^2\Sigma_u^+) + n\hbar\omega]$ and $[V(^2\Sigma_g^+) + (n-1)\hbar\omega]$ cross at some distance R_c (crossing point) and are coupled via the radiative coupling $\vec{\mu}\cdot\vec{E}$ where μ is the electronic transition moment $\langle ^2\Sigma_u^+|\vec{r}|^2\Sigma_g^+\rangle$ and E is the electric field amplitude. It is clear then that photodissociation is the analogue of predissociation[9] where now the bound-continuum coupling is radiative[10-16]. We thus have the classical case of a discrete state embedded in a continuum, the time evolution of which can be studied by the method of Green's functions[17-19].

If originally the initial bound state was in the eigenstate $|a\rangle$ of some Hamiltonian H_o, then through some perturbation V (radiative in our case), at time t it will have evolved according to the expression (we set $\hbar = 1$)

$$\psi(t) = {}^{-iHt}|a\rangle \quad , \tag{1}$$

where $H = H_o + V$ and $H_o|a\rangle = E_a^o|a\rangle$. The decay amplitude, $I_a(t)$, i.e., the probability amplitude that the system starting at $t = 0$ in $|a\rangle$ remains in this state after a time t is defined by

$$I_a(t) = \langle a|\psi(t)\rangle \quad \langle a|e^{-iHt}|a\rangle \quad , \tag{2}$$

which may be expressed in terms of the Green's function as[17-19]

$$I_a(t) = \frac{1}{2\pi i}\int_{-\infty}^{\infty} dE e^{-iEt} G_{aa}(E) \quad . \tag{3}$$

The transition amplitude between the state $|a\rangle$ and the continuum $|c\rangle$ is given by

$$I_{ca}(t) = \frac{1}{2\pi i}\int_{-\infty}^{\infty} dE e^{-iEt} G_{ca}(E) \quad . \tag{4}$$

Let $G(E) = (E-H_o-V)^{-1}$ and $G^o(E) = (E-H_o)^{-1}$ be the perturbed and unperturbed operators. Then by iterating the equation for G[18],

$$G(E) = G^o(E) + G^o(E)VG(E) \quad , \tag{5}$$

we get,

$$G(E) = G^o(E) + G^o(E)VG^o(E) + G^o(E)VG^o(E)VG(E) \quad , \tag{6}$$

from which it follows by evaluating matrix elements,

$$G_{aa}(E) = (E-E_a-i\Gamma_a/2)^{-1} \quad , \tag{7}$$

where $E_a = E_a^o + \Delta E_a$. ΔE_a is an energy shift given by a principal part integral or the real part of the following integral, the imaginary part of which is also the linewidth,

$$\Delta E_a - i\Gamma_a/2 = \lim_{\varepsilon \to o} \int \frac{|V_{ca}|^2 dE_c}{E+i\varepsilon-E_c} = \int \frac{|V_{ca}|^2 dE_c}{E-E_c} - i\pi |V_{ca}|^2 . \tag{6}$$

Examples of models for which this integral can be solved analytically are reported in references[20-22]. The photodissociation Green's function, G_{ca} can be obtained from (5),

$$G_{ca}(E) = G_c^o V_{ca} G_{aa} = \frac{V_{ca}/(E-E_c)}{(E-E_a + i\Gamma_a/2)} \quad . \tag{9}$$

The transition amplitudes are found next by performing contour integrations in the lower half energy plane in equations (2) and (3). Thus from the simple singularity in G_{aa} (equation 7) we obtain

$$I_a(t) = e^{-iE_a t} e^{-\Gamma_a t/2} \quad , \tag{10}$$

from which it follows that the population of state $|a>$ is

$$P_a(t) = |I_a(t)|^2 = e^{-\Gamma_a t} \quad . \tag{11}$$

The photodissociation amplitude is obtained similarly by

considering the two poles in the lower half energy plane in equation (9), i.e., at $E_c - i\varepsilon$ and $E_a - i\Gamma_a/2$. The resulting integration gives

$$I_{ca}(t) = \frac{V_{ca}}{E_c - E_a + i\Gamma_a/2}\left[e^{-iE_c t} - e^{-iE_a t} e^{-\Gamma_a t/2}\right]. \qquad (12)$$

Integrating the square of this expression over the continuum energy E_c to get the total photodissociation probability P_c from state $|a\rangle$ (we assume that V_{ca} and therefore ΔE_a and Γ_a are independent of energy, thus avoiding extra poles from new bound states created by energy dispersion of these parameters[20]), we obtain,

$$P_c(t) = \int |I_{ca}(t)|^2 dE_c = \frac{2\pi |V_{ca}|^2}{\Gamma_a}(1 - e^{-\Gamma_a t}) \quad . \qquad (13)$$

Since $\Gamma_a = 2\pi |V_{ca}|^2$, equation (8), we obtain finally

$$P_c(t) = 1 - e^{-\Gamma_a t} \quad ; \quad P_a + P_c = 1 \quad . \qquad (14)$$

The above results are <u>exact</u> for a model of <u>one</u> bound state $|a\rangle$ coupled to <u>one</u> continuum $|c\rangle$. Molecular systems are <u>multilevel</u> systems, so these expressions must be generalized. More general expressions for the Green's functions can be obtained for multilevel systems[16,19,23]. We will show in the next section how this can be also done using coupled equations of scattering theory. We must however emphasize that in any given state to state experiment or calculation, one must take into account the lifetime and shifting, i.e., the <u>dressing of the initial state</u>, as seen in equation (12). Furthermore at weak fields and short times so that the condition $\Gamma_a t$ or rather $\Gamma_a t/\hbar << 1$, we readily obtain the Fermi-Golden rule expression[24]: $P_c(t) = \Gamma_a t/\hbar = 2\pi|V_{ca}|^2/\hbar$. Irradiation for long times with a laser beam will result in violation of this condition, so that we shall be considering conditions for which $\Gamma_a t|\hbar > 1$. For such cases, the Fermi-Golden ruel expression for Γ_a, equation (8), may still be valid at weak fields. However, for strong fields, renormalization of the spectrum via multiphoton processes will invalidate this expression and new bound states induced by the laser will be of importance for a valid description[10,14-16].

COUPLED EQUATIONS AND SEMICLASSICAL THEORY

For a direct photodissociation as illustrated in fig. 1a, the effect of the laser field on some initial rotational vibrational state in the ground electronic potential ${}^2\Sigma_u^+$ which is coupled via the radiative perturbation $V = \vec{\mu}\cdot\vec{E}$ to the continuum electronic potential ${}^2\Sigma_g^+$ can be studied as a predissociation problem[14,15] in the electron-field or dressed molecule representation, fig. 1b. Sharp bound states will appear as narrow resonances[25,26] in the elastic scattering matrix element S_{CC} for the continuum potential

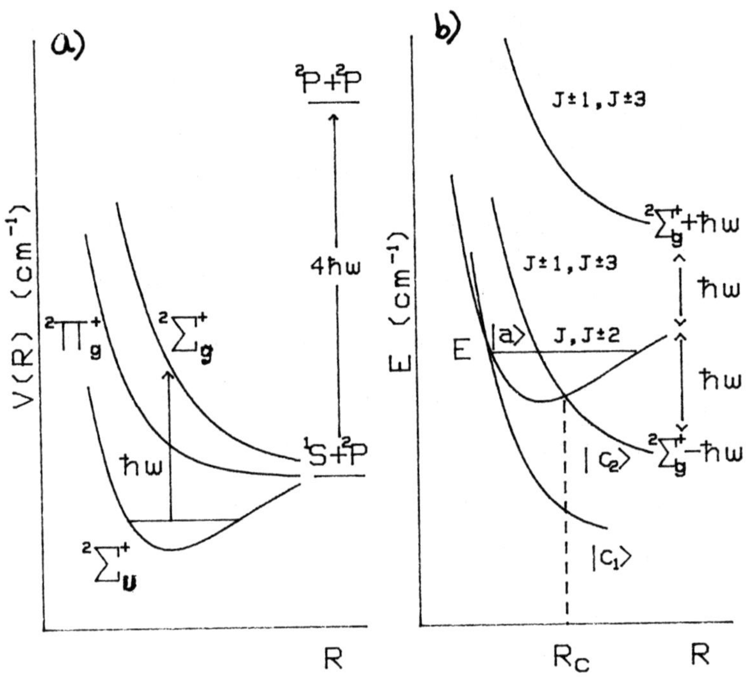

FIGURE 1. a) Electronic potentials and asymptotic states for Ar_2^+; b) Electron field potentials of a) - ω = 28328 cm^{-1} (XeF laser).

$V_c(R)$ coupled to the molecular field bound states of $V_a(R)$. These new sharp states induced by the laser field occur whenever a diabatic (unperturbed) bound state of $V_a(R)$ is in resonance in the dressed picture (molecule + field) with the new adiabatic bound states of the new adiabatic potential $V_+(R)$ formed by radiatively induced avoided crossings[9,14-15], i.e.,

$$V_+(R)=\frac{1}{2}[V_a(R)+\hbar\omega+V_c(R)]+\frac{1}{2}\{[V_c(R)-V_a(R)-\hbar\omega]^2+4\vec{\mu}\cdot\vec{E}\}^{\frac{1}{2}} \quad (15)$$

One can also obtain these resonances as poles in the inelastic scattering matrix element $S_{C_1C_2}$ where $|C_1>$ is the $^2\Pi_g$ continuum (fig. 1a) and $|C_2>$ is the $^2\Sigma_g^+$ continuum (fig. 1b). Thus $|C_1>$ can serve as a probing incoming channel since it is only weakly coupled radiatively to the original $^2\Sigma_u^+$ bound states[5]. This inelastic S matrix element is given in the semiclassical limit by[14,27]:

$$S_{C_1C_2} = 2i(U_1P_2U_2)^{\frac{1}{2}} \cos\phi_2 \sin(\beta-\phi_1)/\Delta \quad , \qquad (16a)$$

$$\Delta=\cos\beta+U_1\cos\phi_1 \exp[i(\phi_1-\beta_1)]+P_1U_2\cos\phi_2 \exp[i(\phi_2-\beta)] \qquad (16b)$$

The phase β is the semiclassical action due to the diabatic potential $V_a(R)$, i.e.,

$$\beta = \int_\alpha^\beta k_a(R)dR \ ; \quad k_a(R) = \hbar^{-1}\{2M[E-V_a(R)]\}^{\frac{1}{2}} \ , \qquad (17)$$

where α and β are left and right turning points on the potential $V_a(R)$. The phase integrals $\phi_2(\phi_1)$ are the new adiabatic semiclassical actions for the newly created potentials $V_+(R)$ by the continua $|C_2>$ ($|C_1>$) in interaction with $V_a(R)$ as follows from equation (15). Thus for the case we are interested in, i.e., strong interaction between V_a and $V_{C_2} = V(^2\Sigma_g^+)$, then

$$\phi_2 = \int_{\alpha_2}^\beta k_2^+(R)dR + \chi(\varepsilon) \quad , \qquad (18)$$

where $k_2^+(R) = \hbar^{-1}\{2M[E-V_+(R)]\}^{\frac{1}{2}}$ is the semiclassical wave vector of the new adiabatic potential $V_+(R)$. $\chi(\varepsilon)$ is a phase correction between 0 and $-\pi/4$[9]. P is related to U by

$$P_i = U_i + 1 = \exp(2\pi\varepsilon_i) \quad ,$$
$$\varepsilon_i = (\vec{\mu}\cdot\vec{E})^2/\hbar v \, [V'_a(R) - V'_{C_i}(R)]_{R=R_{C_i}} \quad . \tag{19}$$

Thus ε_i is a Landau Zener parameter at the ith crossing point R_{C_i} and $V'(R) = dVdR$.

If all phases, diabatic (β) and adiabatic (ϕ), correspond to quasibound states, then one can expect the quantization rule[27],

$$\beta = (n+\tfrac{1}{2})\pi + (E-E_\beta)\alpha_\beta \ , \ \phi_i = (m+\tfrac{1}{2})\pi + (E-E_i)\alpha_i ,$$
$$E = E_r - i\Gamma_r/2 \ , \quad \alpha = \partial\phi/\partial E = \pi/\hbar\omega_\phi \quad . \tag{20}$$

The poles of (16a) corresponding to the zeros of (16b) are obtained by applying the relations (20) to the zeros of (16b). One obtains by separating real and imaginary parts of $\Delta = 0$, the resonance energies E_r and widths Γ_r,

$$E_r = \frac{E_\beta + x_2 E_{\phi_2}}{1 + x_2} \ , \ \frac{\Gamma_r}{2} = \frac{\pi x_2 (1+U_2)(E_\beta - E_{\phi_2})^2}{\hbar\omega_{\phi_2}(1+x_2)^3} \ . \tag{21}$$

We readily see from equation (21) that the dressed states are mixtures of the <u>diabatic</u> (E_β) and <u>adiabatic</u> (E_ϕ) laser induced states. The mixing coefficient $x_i = U_i\omega_\beta/\omega_\phi$, where $\omega_{\beta(\phi)}$ are the diabatic (adiabatic) vibrational frequencies, is governed by the expression (19) for U_i. Thus for very strong fields, $x_i \to \infty$, and $E_r \to E_{\phi_2}$, i.e., only new laser induced (adiabatic) states appear. Finally, for any coupling strength, whenever $E_\beta = E_{\phi_2}$, i.e., for coincidence between diabatic (zero field) molecular states and adiabatic (field diagonalized) states, (equation (15)), then very narrow resonances are to be expected. This creates also zeroes in the inelastic S-matrix element $S_{C_1C_2}$, equation (16a). These criteria were utilized to locate and correlate new laser induced stable bound states in two channel calculations (one bound + one continuum). The stability and energy convergence of these resonances was verified by increasing the number of channels, especially for high field strengths where one expects considerable interaction, i.e., mixing of channels via the radiative coupling $\vec{\mu}\cdot\vec{E}$ with the corresponding selection rules $\Delta J = 0, \pm 1$.

We now describe the computational procedure. Assuming z-incident polarization, the general radiative matrix element becomes[28]

$$\langle JM\Omega|V_a^r|J''M''\Omega\rangle = \left(\frac{2J+1}{2J''+1}\right)^{\frac{1}{2}}\langle J1M0|J''M''\rangle\langle J1\Omega 0|J''\Omega\rangle D, \quad (22)$$

where z is the direction of the transition motion μ for $^2\Sigma \to {}^2\Sigma$ transitions, $\Omega = \frac{1}{2}$, $D = \langle\Omega|\mu_z|\Omega\rangle E$. Thus for computational purposes, the coupled equations depend on three types of coupling according to the selection ruels $\Delta J = 0, \pm 1$:

$$V^r(J,J+1) \quad \frac{D}{2(J+1)} [(J+1+M) \; (J+1-M)]^{\frac{1}{2}} ,$$

$$V^r(J,J) = \frac{D\Omega M}{J(J+1)} , \; V(J,J-1) = \frac{D}{2J} [(J-M) \; (J+M)]^{\frac{1}{2}} . \qquad (23)$$

Using these radiative couplings, the multichannel coupled equations for the radial nuclear functions $F_i(R)$ become,

$$[T_N - E + W_i(R)]F_i(R) + \sum_{i \neq j} V_{ij}(R)F_j(R) \equiv 0 \; . \qquad (24)$$

These differential equations are integrated numerically by a Fox-Goodwin integrator method[29]. In equation (23), T_N is the nuclear kinetic energy operator; $W_i(R)=V_i(R)+n_i\hbar\omega_i-\hbar^2\hat{N}^2/2MR^2$ are the molecular field potentials; N is the nuclear angular momentum; $V_{ij}(R)$ are the <u>radiative</u> couplings of equation (23). For bound states, $F_j(R = \infty) = 0$, and for continua, the asymptotic forms are

$$F_j(R)=k_j^{-\frac{1}{2}} \sin (k_jR-N\pi/2)+k_i^{-\frac{1}{2}}R_{ij} \cos (k_iR-N\pi/2). \qquad (25)$$

From these aymptotic forms, one obtains the quantities R_{ij} from which one can calculate the general S-matrix elements by the relation $S = (1-iR)^{-1} (1 + iR)$.

To calculate photodissociation and light scattering cross sections from an initial bound state $|a\rangle$ of energy E_a^o, we make use of the artificial channel method of Shapiro[30] which we have generalised to predissociation and RRS[22,31]. This method is implicit in equation (16a) where the entrance channel $|C_1\rangle$ is the weakly coupled continuum

$^2\Pi_g$ (fig. 1). One could also use an infra-red laser to probe with this channel the upper dressed levels of the $^2\Sigma_u^+$ ground state[15]. Thus as illustrated in fig. 2, the entrance (artificial) channel $|C_1\rangle$ is coupled to the initial free molecular state $|0\rangle$ which is then coupled to $|a,J,n,0\rangle$ which is dressed by the continua $|C,J\pm1,n-1,0\rangle$, where n is the initial number of photons ω_1, 0 is the initial number of photons ω_2. At zero (weak) intensities, the overlap $\langle o|a\rangle = 1$ but at higher intensities, the $|a\rangle$ state will mix with other bound states and continua, so that the quantity $|\langle o|a\rangle|^2$ will represent the occupation of the initial (free molecule) state in the new dressed states. We emphasize here that since we are dealing with a multilevel system, the present calculational method involves many dressed states as opposed to the one dressed state approximation of equation (12). Furthermore since the intensity I_1 (incident photon ω_1) >> I_2 (scattered photon ω_2), radiative corrections from spontaneous emission will be negligible compared to laser-induced effects, so these small corrections are neglected. Finally the photodissociation amplitude will be proportional to the total exit amplitude T_{oc} to continuum $|c\rangle$ whereas the RRS amplitude will depend on the transition amplitude T_{od} between the initial free molecular state $|0\rangle = |a,J,n,0\rangle$ and the final continuum $|d\rangle$. The amplitudes

T_{oc} and T_{od} can be obtained[22,31] from the numerical coupled equations S-matrix elements S_{1c} and S_{1d},

$$T_{oc} = \frac{iS_{1c}(E-E_o+i\Gamma_o/2)}{2\pi V_{1o}}, \quad T_{od} = \frac{iS_{1d}(E-E_o+i\Gamma_o/2)}{2\pi V_{1o}}. \tag{26}$$

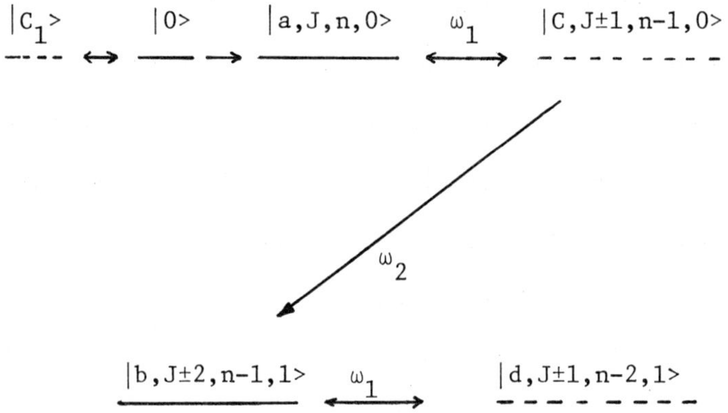

$|C_1\rangle$ ↔ $|0\rangle$ → $|a,J,n,0\rangle$ ω_1 ↔ $|C,J\pm1,n-1,0\rangle$

ω_2

$|b,J\pm2,n-1,1\rangle$ ω_1 ↔ $|d,J\pm1,n-2,1\rangle$

FIGURE 2. Emission of photon ω_2 in dressed representation of incident photon ω_1. $|C_1\rangle$ - entrance continuum; $|o\rangle$ - initial state; $|a\rangle$, $|b\rangle$, bound states $({}^2\Sigma_u^+)$; $|c\rangle$, $|d\rangle$ - continuum states $({}^2\Sigma_g^+)$.

The quantities on the right, S_{1c}, S_{1d}, Γ_o, V_{1o} are amenable to numerical calculations by the integrator method described above for coupled equations such as equation (24). Thus one obtains exact numerical results for the photodissociation amplitude T_{oc} and the RRS amplitude T_{od}. These results are described in the next section.

NUMERICAL RESULTS AND CONCLUSION

In fig. 3, we present a converged calculation of linewidths of the v = 8, J = 9.5, M = 0.5 level of Ar_2^+ in its ground electronic state ${}^2\Sigma_u^+$ coupled radiatively to the ${}^2\Sigma_g^+$ continuum. For this state, the $\Delta J = \pm 1$ couplings at 10^{10} W/cm^2 are about 300 cm^{-1} for $\mu \simeq 2$ a.u.[15]. Since the vibrational frequency ω_β = 300 cm^{-1}, we thus see we are no longer in the realm of perturbation theory. We note in fig. 3 that <u>saturation</u> occurs at about 5 x 10^9 W/cm^2 (maximum in Γ). However as the field intensity is increased, desaturation occurs with a minimum at 1.5 x 10^{10} W/cm^2. The Γ vs I oscillates with saturations occuring at various values of I. Thus multilevel systems have many saturation intensities as opposed to the two level case. Furthermore, at the desaturation intensity of 1.5 x 10^{10} W/cm^2, it was found that eleven channels (five bound states, J,J±2, J±4 and six continua J±1, J±3, J±5) were needed to stabilize the resonance.

As a result of this saturation and desaturation behaviour of linewidths, the calculated photodissociation angular distribution can become anisotropic and will deviate considerably from the expected classical distribution proportional to $\cos^2\Theta$[4]. This is clearly illustrated in fig. 4 where the solid line is the traditional weak field $\cos^2\Theta$

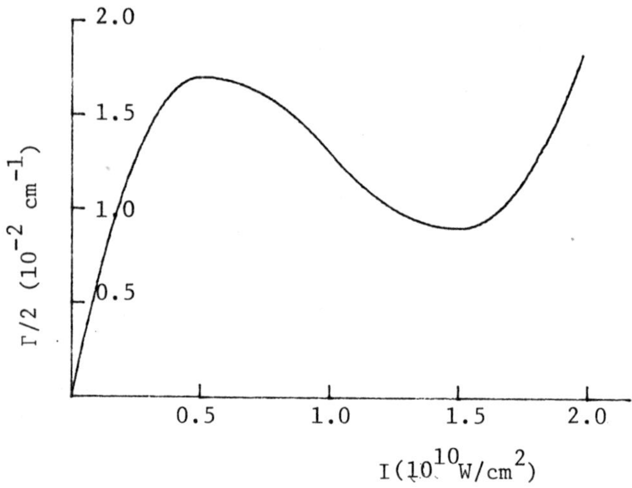

FIGURE 3. Behaviour of Γ as a function of intensity I for the $^2\Sigma_u^+$, v = 8, J = 9.5, M = 0.5 level of Ar_2^+ in a XeF laser beam (ω = 28328 cm^{-1})

dependence. Thus at 1.5 x 10^{10} W/cm^2, where the M = 0.5 component of the J = 9.5 level desaturates (see fig. 3), the angular distribution peaks at about 60° (for the dotted curve, ΔJ = 0 transitions were neglected). This is explainable in terms of the minimum of Γ at that intensity for M = ½. For this component, a delay in dissociation occurs due to the presence of new adiabatic (laser induced) levels predicted in equation (16a), i.e., $\phi_2 \simeq (m + \frac{1}{2})\pi$. Calculating the phase integral ϕ_2, equation (18) at the energy of the diabatic (unperturbed) level confirmed the quantization of this new level[15]. The semiclassical picture is thus seen to apply and confirms therefore the creation of new laser induced bound states and their drastic effect on the photodissociation angular distribution. It is to be noted, that the traditional $\cos^2\Theta$ amplitude is nearly statistical. The presence of the adiabatic states renders the distribution in fig. 4 highly <u>nonstatistical</u>.

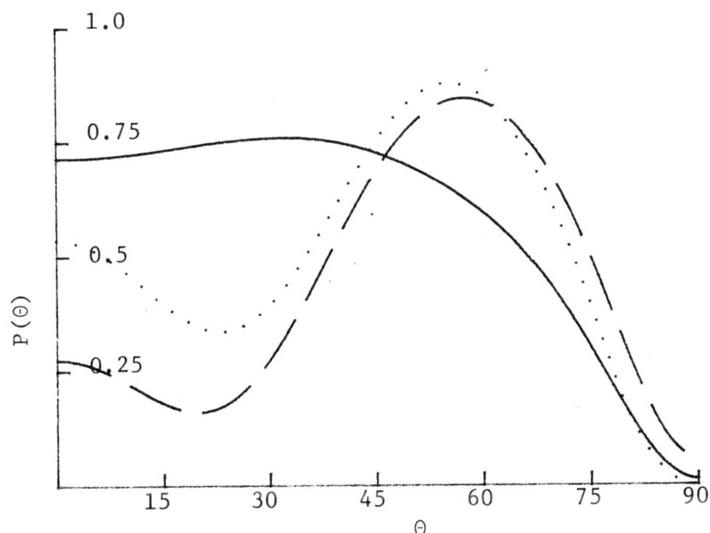

FIGURE 4. Photodissociation angular distribution: ——— weak field ($I < 10^8$ W/cm^2); .... $I = 1.5 \times 10^{10}$ W/cm^2 without $\Delta J = 0$ transitions; ------- $I = 1.5 \times 10^{10}$ W/cm^2 including $\Delta J = 0$ transitions.

Finally we present an example of RRS for the same initial state for which fig. 3 applies. As shown in fig. 2, photon ω_2 is emitted between the two bound states |a> and |b> in continuous interaction (dressing) with the continua |c> and |d>. Both initial |a> and final states |b> acquire laser induced linewidths Γ_a and Γ_b. We have already pointed out in the introduction (see also equation 12) that in the photodissociation at high intensities the initial state is considerably broadened by the laser field. In the RRS calculation, both initial and final states are now unstable. This is reflected in the broadened lines of the 0 branch ($\Delta J = -2$) emanating from the initial 9.5 and 10.5 J values taken as an example (fig. 5) for their relative sharpness.

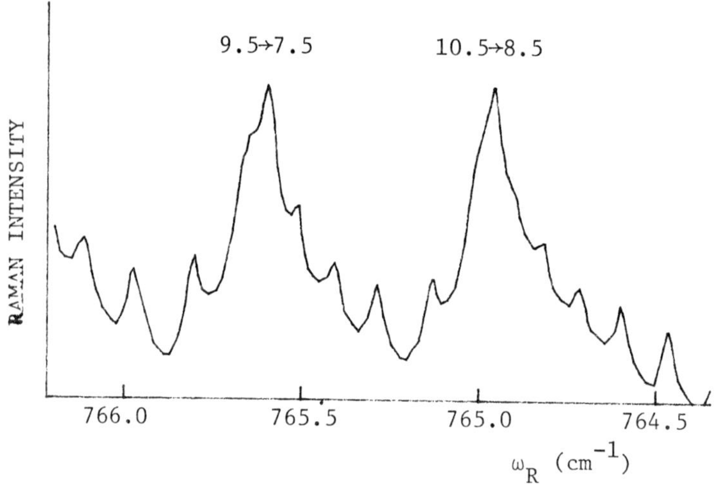

FIGURE 5. Raman spectrum of 0 branch ($\Delta J = -2$) for initial $J = 9.5$, 10.5 at intensity $I = 5 \times 10^9$ W/cm^2.

For the intensity of 5×10^9 W/cm^2, in fig. 3 we have seen that the $M = \frac{1}{2}$ component saturates, i.e., attains maximum dissociation. At such intensity it is found that the photodissociation angular distribution is highly statistical, since all components give maximum photodissociation. At this intensity, it was found that the depolarization ratio in RRS also approaches the statistical value of 1[32].

In conclusion we have shown that if the electronic potential and transition moments of a diatomic molecule are known, then coupled equations such as described in equation (24) must be used to properly describe intense field effects on molecules. In the case of bound continuum transitions, the presence of laser created (adiabatic) bound states may be first located by a semiclassical theory (equation 16). This was the starting point of all our calculations. The more elaborate coupled equations calculation introduces multilevel mixing via the large radiative couplings. In photodissociation, the broadening of the initial state results in different photodissociation rates for different M components. Highly anisotropic angular distributions are obtained (fig. 4) in the presence of new adiabatic (laser induced) levels. Similarly in RRS spectra, Stark splitting due to different energy shifts for the different components now appear (fig. 5). The RRS spectra are considerably broadened since both initial and final states are destabilized by the field. We have thus explicitly illustrated the effect of the dressing of molecular states in photodissociation and RRS.

REFERENCES

1. G. Herzberg, Electronic Spectra of Polyatomic Molecules (Van Nostrand, N.Y. 1966).
2. C. Cohen-Tannoudji, in Frontiers in Laser Spectroscopy, eds. R. Balian, S. Haroche (North Holland, Amsterdam 1975).
3. C.R. Stroud, Phys. Rev. A3, 1944 (1971).
4. R.N. Zare, Mol. Photochem. 4, 1 (1972).
5. W.J. Stevens, M. Gardner, A. Karo, P. Julienne, J. Chem. Phys. 67, 2860 (1977).
6. W.R. Wadt, J. Chem. Phys. 68, 402 (1978).
7. F.Y. Wu, R.E. Grove, S. Ezekiel, Phys. Rev. Lett. 35, 1426 (1975).
8. G.M. Weyl, D.I. Rosen, J. Wilson, W. Seka, Phys. Rev. A26, 1164 (1982).
9. A.D. Bandrauk, M.S. Child, Molec. Phys. 19, 95 (1970).
10. A.I. Voronin, A.A. Samokhin, Sov. Phys. - JETP 43, 4 (1976).
11. A.M.F. Lau, Phys. Rev. 13, 139 (1976).
12. T.F. George, I.H. Zimmerman, J.M. Yuan, J.R. Laing, P.L. Devries, Acc. Chem. Res. 10, 449 (1977) and references therein.
13. J.M. Yuan, T.F. George, J. Chem. Phys. 68, 3040 (1978).
14. A.D. Bandrauk, M.L. Sink, Chem. Phys. Lett. 57, 569 (1978).
15. A.D. Bandrauk, M.L. Sink, J. Chem. Phys. 74, 1110 (1981).
16. A.D. Bandrauk, G. Turcotte, J. Chem. Phys. 77, 3867 (1982).
17. M.L. Goldberger, K.M. Watson, Collision Theory (John Wiley & Sons, N.Y. 1964), Chap. 8.
18. K.M. Watson, J. Nuttall, Topics in Several Particle Dynamics (Holden-Day, San Francisco, 1967), Chap. 1.
19. L. Mower, Phys. Rev. 142, 799 (1966).
20. J.P. Laplante, A.D. Bandrauk, J. Chem. Phys. 65, 2592 (1976).
21. M.L. Sink, A.D. Bandrauk, Chem. Phys. 33, 205 (1978).
22. K. Kodama, A.D. Bandrauk, Chem. Phys. 57, 461 (1981).
23. A. Lami, N.K. Rahman, Phys. Rev. A26, 3360 (1982).
24. A.S. Davydov, Quantum Mechanics (Pergamon Press, 2nd edition, N.Y. 1976), page 431.
25. M.L. Sink, A.D. Bandrauk, J. Chem. Phys. 73, 4451 (1980).

26. A.U. Hazi, Phys. Rev. A19, 920 (1979).
27. M.L. Sink, A.D. Bandrauk, J. Chem. Phys. 66, 5313 (1977).
28. O. Atabek, R. Lefebvre, M. Jacon, J. Chem. Phys. 72, 2670 (1980).
29. D.W. Norcross, M.J. Seaton, J. Phys. B6, 614 (1973).
30. M. Shapiro, J. Chem. Phys. 56 2852 (1972).
31. A.D. Bandrauk, G. Turcotte, R. Lefebvre, J. Chem. Phys. 76, 225 (1982).
32. G. Turcotte, A.D. Bandrauk, Chem. Phys. Lett. 94, 175 (1983).

PREDISSOCIATION IN A STRONG ELECTROMAGNETIC FIELD: THEORY OF DOUBLE RESONANCE

ALESSANDRO LAMI[*] and NASEEM K. RAHMAN[*+]
*Istituto di Chimica Quantistica ed Energetica Molecolare, C.N.R., Pisa, Italy.
+Istituto di Chimica Fisica, Università di Pisa, Pisa, Italy.

Abstract The problem of predissociation in a strong e.m. field is investigated. The absorption spectrum, for a weak probe field is calculated in presence of an intense e.m. field coupling the predissociative state to another bound state resonantly. A general expression is given for the absorption cross section as a function of the intensity of the strong field and of the two frequencies.
The role of asymmetry parameters for the predissociative resonance as it is seen from both the bound states is also brought out. Some typical lineshapes are shown which exhibit the characteristic features of such double processes. Sharp enhancements of the absorption cross sections are seen at critical intensities, which may be attributed to the formation of field-induced bound states.
The applicability of the theory to autoionization as well is pointed out.

INTRODUCTION

The continuum part of the molecular spectrum has been investigated from the beginning of modern spectroscopy[1]. The main features of the bound continuum absorption spectra, which have been rationalized in terms of the Franck-Condon principle, appear to be the following:
i) the spectrum mimics the behaviour of the vibrational

wavefunction of the initial state, having the same number of nodes and maxima (the so called reflection principle[1,2]).

ii) The energy variation of the absorption cross-section is normally quite smooth (at least for the lowest vibrational states).

The point ii), which could be discussed more precisely and quantitatively[3], has the important consequence that "normally" one observes the exponential depletion of the initial state coupled to the continuum. "Normally" means here "with the light intensities currently used in this kind of experiments".[3] Far from the saturation region this also gives rise to a time independent photoabsorption rate (or photodissociation rate). In the energy domain one can say that the molecular continuum appears to be practically undistorted, for the almost whole range of light intensities used. When predissociation exists, however, the continuum becomes structured and this clearly appears in the absorption spectrum in the form of peaks (of more or less distorted Lorentzian form) superimposed on the smooth continuum background[1,4]. These peaks can be attributed to quasi-bound states embedded in the continuum. In fact, in the usual interpretation, they are bound states of a zero-order hamiltonian (Born-Oppenheimer) which becomes coupled to a dissociative continuum (and hence becomes part of the continuum spectrum) due to a perturbation term.

The field induced coupling of the initial state to the continuum can then be sharply peaked at some energies, corresponding to predissociative resonances. In this case it can be shown[3,5] that, by increasing the field strength, a non-exponential depletion of the initial state can be produced. This happens for a light intensity such that the photodissociation width of the initial bound state becomes comparable with the linewidth of the predissociative resonance. This non-exponential depletion can be interpreted as a signal that the field strength has become strong enough to distort in some way the continuum, i.e. to alter in a substantial manner the original resonance.

THE DOUBLE-RESONANCE SCHEME

The distortion of continuum can be, in principle, detected by measuring the absorption spectrum. However one should be able to measure the transient spectrum since at the intensity required, a rapid saturation occurr, which destroys any structure. One can also analyze the energy distribution of fragments with high resolution in a beam experiment. The most direct way however, seems to us the spectroscopic one. This is why we proposed[5,6] to study the phenomenon by a double resonance technique. The present paper contains a further theoretical investigation on the problem.

Let us first discuss the kind of scenario we envisage. This is depicted in Figure 1 where the curve-crossings resulting in the (electronic) predissociative resonances are illustrated (left of Figure 1a and 1b) together with the levels considered in the theoretical treatment (right of Figure 1a and 1b).

One starts from the ground state $|1\rangle$ and excites by a frequency ω_1 the predissociative resonance. A second source ω_2 is then used to couple the same resonance to another vibro-rotational state |3> belonging (this is the case in Figure 1a) to the same electronic state. The case where the state |3> is in another electronic state is identical mathematically and what we describe below applies equally well for that case. (Figure 1b).

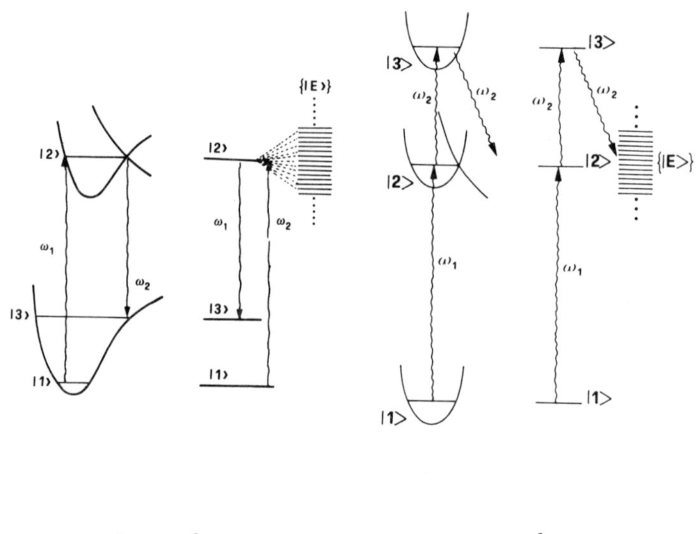

FIGURE 1a and 1b - Two possible schemes for a double resonance predissociation.

The predissociative resonance consists of the state |2> being coupled to the degenerate continuum $\{|E>\}$ by the intramolecular perturbation. Both state |2> and the continuum $\{|E>\}$ carry oscillator strength from the state |1> or the state |3>, giving rise in general to asymmetric (Fano) lineshapes[7] (exciting the resonance separately from state $|1\rangle$ or state $|3\rangle$). The source ω_2 is designed to have an intensity I_2 sufficient to distort the original resonance, whereas the intensity I_1 is weak: it acts as the probe for the "new continuum". Since the whole process begins with the absorption of a photon ω_1 (we assume that the frequencies ω_1 and ω_2 are far apart and consider only resonant or quasi-resonant transitions), the saturation problem is avoided in this way.

Notice that in principle other transitions (single and multiphoton) induced by the strong field ω_2 are possible, starting from the ground states or even from the excited states. Hence some care must be taken in designing the experiment. In particular one should avoid broad predissociative resonances since these will require a very strong field to exhibit some distortion.

THEORY

The theoretical grounds of the present approach have been established in Ref. 5 where we also deal with the general problem of two strong fields. Here we sketch only an essential outline of the theory and give a completely general formula (which has not been given previously) for the case of weak I_1 intensity.

The time dependent Schrödinger equation for the level and the coupling scheme in Fig. 1 is solved in the resonant approximation by an effective hamiltonian formalism. The resonant hamiltonian is first written in a spectral way

$$H = \varepsilon_2 |2\rangle\langle 2| + \varepsilon_3 |3\rangle\langle 3| + \int E\, |E\rangle\langle E|\, dE + \tag{1}$$
$$+ \Big\{ \int V^{P}_{2E}\, |2\rangle\langle E|\, dE + \int V^{R_2}_{3E}\, |3\rangle\langle E|\, dE +$$
$$+ \int V^{R_1}_{1E}\, |1\rangle\langle E|\, dE + h.c. \Big\}$$

where now

$$\begin{aligned} |1\rangle &\equiv |1\rangle \otimes |\omega_1, \omega_2\rangle \\ |2\rangle &\equiv |2\rangle \otimes |0, \omega_2\rangle \\ |3\rangle &\equiv |3\rangle \otimes |0, 2\omega_2\rangle \\ |E\rangle &\equiv |E\rangle \otimes |0, \omega_2\rangle \end{aligned} \tag{2}$$

In Eq. (1) V^{R_1} and V^{R_2} are the molecule-radiation interaction terms and V^{P} is the intramolecular (predissociation) interaction. The quantity we want to evaluate is the probability of leaving the initial state by absorption of a photon ω_1, i.e.

$$W_1(t) = 1 - |U_{11}(t)|^2 \equiv 1 - |\langle 1|\, e^{-iHt}\, |1\rangle|^2 \tag{3}$$

(the time t=0 corresponds to the switching on of the two fields).

The continuum $\{|E\rangle\}$ is projected out and one obtains a 3 x 3 effective hamiltonian which drives the time evolution in the space of states $|1\rangle$, $|2\rangle$ and $|3\rangle$. With respect to other equivalent ways of handling the problem, this procedure has the advantage that all the relevant quantities in the problem are collected together as matrix elements of a complex symmetric effective hamiltonian matrix. One has

$$H^{eff} \equiv \begin{vmatrix} -i\gamma_1+\delta_1+\omega_1+\omega_2 & -i\gamma_{12}+\delta_{12}+V_{12}^{R_1} & -i\gamma_{13}+\delta_{13} \\ -i\gamma_{21}+\delta_{21}+V_{21}^{R_1} & \varepsilon_2-i\gamma_2+\delta_2+\omega_2 & -i\gamma_{23}+\delta_{23}+V_{23}^{R_2} \\ -i\gamma_{31}+\delta_{31} & -i\gamma_{32}+\delta_{32}+V_{23}^{R_2} & \varepsilon_3-i\gamma_3+\delta_3+2\omega_2 \end{vmatrix} \tag{4}$$

where

$$-i\gamma_{ij}+\delta_{ij} = \int \frac{\langle i|V|E\rangle\langle E|V|j\rangle}{\varepsilon - E}\, d\,E \tag{5}$$

and the perturbations V are those necessary for realizing the effective coupling (i.e. via the continuum). For example (see Figure 1)

$$-i\gamma_{12}+\delta_{12} = \int \frac{\langle 1|V^{R_1}|E\rangle\langle E|V^{P}|2\rangle}{\varepsilon - E}\, d\,E \tag{6}$$

(the continuum states are energy normalized and $\gamma_{ii} \equiv \gamma_i$, $\delta_{ii} \equiv \delta_i$).

The γ's and δ's are in principle weakly energy dependent but this dependence may be neglected and therefore, they will be considered as fixed.

The problem of calculating transition amplitudes between states $|1\rangle$, $|2\rangle$, $|3\rangle$ can then be solved[5] by diagonalizing H^{eff}.

Since we are interested in the case of weak ω_1 field and the only amplitude we need is $U_{11}(t)$, a perturbative approach can be followed. One has

$$U_{11}(t) \cong e^{-iE_1 t} \tag{7}$$

where E_1 is the eigenvalue of H^{eff} which corresponds to the unperturbed complex energy of state $|1\rangle$, i.e. H^{eff}_{11} in Eq.(4). A perturbative solution is easily obtained. Let us first rewrite the effective hamiltonian Eq.(4) by redefining the energy scale and rearranging the extra-diagonal matrix elements

$$H^{eff} \equiv \begin{vmatrix} -i\gamma_1 & -\gamma_{12}(q_{12}+i) & -\gamma_{13}(q_{13}+i) \\ -\gamma_{12}(q_{12}+i) & -i\gamma_2+\Delta' & -\gamma_{23}(q_{23}+i) \\ -\gamma_{13}(q_{13}+i) & -\gamma_{23}(q_{23}+i) & -i\gamma_3+\Delta'' \end{vmatrix} \tag{8}$$

where

$$q_{12} = -\frac{V^{R_1}_{12} + \delta_{12}}{\gamma_{12}} \tag{9a}$$

$$q_{13} = -\frac{\delta_{13}}{\gamma_{13}} \tag{9b}$$

$$q_{23} = -\frac{V_{23}^{R_2} + \delta_{23}}{\gamma_{23}} \tag{9c}$$

$$\Delta' = \varepsilon_2 + (\delta_2 - \delta_1) - \omega_1 \tag{10a}$$

$$\Delta'' = \varepsilon_3 + (\delta_3 - \delta_1) + \omega_2 - \omega_1 \tag{10b}$$

The secular equation for H^{eff} is written as

$$E = -i\gamma_1 - \frac{\gamma_{12}^2(q_{12}+i)^2(-i\gamma_3+\Delta''-E) - \gamma_{13}^2(q_{13}+i)^2(-i\gamma_2+\Delta'-E)}{(-i\gamma_2+\Delta'-E)(-i\gamma_3+\Delta''-E) - \gamma_{23}^2(q_{23}+i)^2} \tag{11}$$

and the perturbative solution for $\widetilde{E}_1$ is obtained iterating one time. Putting on the right the zero order value $\widetilde{E}_1 \cong -i\gamma_1$ one has

$$\widetilde{E}_1 \cong -i\gamma_1 - \frac{\gamma_{12}^2(q_{12}+i)^2(-i\gamma_3'+\Delta'') - \gamma_{13}^2(q_{13}+i)^2(-i\gamma_2'+\Delta')}{(-i\gamma_2'+\Delta')(-i\gamma_3'+\Delta'') - \gamma_{23}^2(q_{23}+i)^2} \tag{12}$$

where now $\gamma_2' = \gamma_2 + \gamma_1$; $\gamma_3' = \gamma_3 + \gamma_1$. Since the perturbative approach is well justified only for $\gamma_1 << \gamma_2, \gamma_3$, in the following we take $\gamma_2' = \gamma_2$ and $\gamma_3' = \gamma_3$. The absorption probability is (see Eq. (3))

$$W_1(t) = 1 - e^{-2\Gamma_1 t} \tag{13}$$

where

$$\Gamma_1 = \text{Im}\,\tilde{E}_1 = \gamma_1 + \frac{BC - AD}{C^2 + D^2} \tag{14}$$

and

$$A = \gamma_{12}^2\left[(q_{12}^2-1)\Delta''+2\gamma_3 q_{12}\right] - \gamma_{13}^2\Delta'(q_{13}^2-1+2q_{13}) \tag{14a}$$

$$B = \gamma_{12}^2\left[2q_{12}\Delta''-\gamma_3(q_{12}^2-1)\right] - \gamma_{13}^2\left[2q_{13}^2\Delta'-\gamma_2(q_{13}^2-1)\right] \tag{14b}$$

$$C = \Delta'\Delta''-\gamma_2\gamma_3-\gamma_{23}^2(q_{23}^2-1) \tag{14c}$$

$$D = -(\gamma_2\Delta''+\Delta'\gamma_3+2\gamma_{23}^2\, q_{23}) \tag{14d}$$

For $\Gamma_1 t \ll 1$ one has a time independent absorption rate. The cross section is

$$\sigma_1 = (2\,\omega_1 / I_1)\Gamma_1 \tag{15}$$

Eqs. (14) and (15) gives the general formula for the absorption cross section. q_{12} and q_{23} have the meaning of Fano parameters[7] for the separate transitions to the predissociative resonance from state $|1\rangle$ and $|3\rangle$ respectively[6]. Putting $q_{12} = \infty$, for example, amounts to considering that the line-shape for the absorption from state $|1\rangle$ is Lorentzian. q_{13} is a kind of mixed asymmetry parameter.

Eqs. (14) and (15) cover a large number of possibilities. The limiting case $q_{12} = \infty$ can be obtained, for example, by replacing everywhere $q_{12}\gamma_{12}$ by $-V_{12}^{R_1}$ and putting otherwise $\gamma_{12} = 0$.

The quantities γ_{ij} and γ_i, γ_j are related . (For a general discussion see Ref.5). Here we consider, for simplicity, that by the combined effect of selection rules for electronic predissociation and one-photon transitions (linear polarization), the partial wave of the continuum to which one can acced from ($|1\rangle$, $|2\rangle$ and $|3\rangle$) is the same. In this case one has simply[5]

$$\gamma_{ij} = \pm(\gamma_i \gamma_j)^{\frac{1}{2}} \tag{16}$$

This corresponds physically to the fact that both direct and indirect (i.e. via predissociation) photodissociation from state $|1\rangle$ or $|3\rangle$ can only produce fragments with the same angular momentum.

DISCUSSION

Previously[5,6] we calculated some absorption spectra for particular values of q's and γ's. For example we showed that in the case $q_{23} = \infty$ the field ω_2 modifies the spectrum in a different manner depending on if $q_{12} > 1$ or < 1, for the resonant case[6] (Eq.(10)):

$$\Delta' = \Delta'' = \Delta \tag{17}$$

Here we report, some new results for the case $q_{12} = \infty$ (i.e. one cannot acceed directly to the continuum from state $|1\rangle$, and the absorption profile is lorentzian, when ω_2 is switched off). The frequency ω_2 is fixed at resonance (i.e. Eq.(17) holds). The predissociation width γ_2

is assumed as energy unit; q_{23} and I_2 are varied.

The intensity I_2 does not appear directly but it is incorporated in γ_3, which is proportional to I_2. For example, the curve a in Figure 2 is obtained for $\gamma_3/\gamma_2 = 1$. This means that we are utilizing a field I_2 such that the photodissociative width from state $|3\rangle$ is equal to the undisturbed predissociation width. This may corresponds to a strong intensity if γ_2 is large, or to a moderately weak intensity if it is small. Assuming, for example, as a typical case $\gamma_2 \sim 1\ \mathrm{cm}^{-1}$ (predissociation lifetime $\sim 5.10^{-12}$ sec) one can roughly estimate[5] that $\gamma_3 \sim \gamma_2$ for $I_2 \sim 10^{10}\ \mathrm{W/cm}^2$.

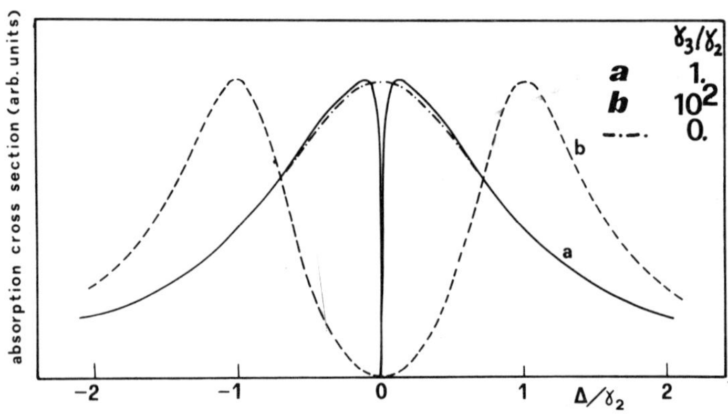

FIGURE 2 - Absorption cross-sections as a function of the detuning of the probe field for the case $q_{12} = \infty$, $q_{23} = \infty$, at various intensities of the strong field. For further explanation see the text.

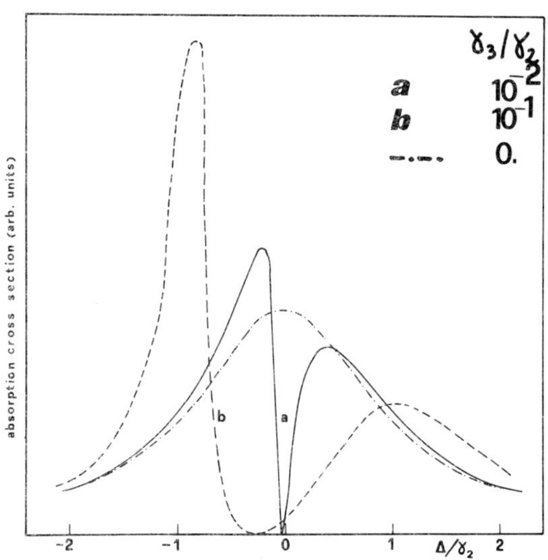

FIGURE 3 - Absorption cross-sections as a function of the detuning of the proble field for the case $q_{12} = \infty$, q_{23} = 3 and for various intensities of the strong field. For further explanation see the text.

The Figure 2 refers to the case $q_{12} = \infty$ and $q_{23} = \infty$ and exhibits the characteristic dip of the coherent trapping[6,8], which combines with a dynamical Stark shift.

Figures 3 and 4 show what happens for q_{23} = 3 and q_{23} = 1. (in all the cases $q_{12} = \infty$) respectively. Here the symmetry is broken and a sharp peak appear which moves towards the left and becomes a "delta" for a critical intensity. Increasing further γ_3 (i.e. I_2) the peak continues to move and becomes broader and broader. This is illustrated only in Figure 4, for q_{23} = 1, but it is a common feature of all the cases with $q_{12} = \infty$. It can be attributed[5] to the creation of a bound state embedded in a continuum, as a result of diagonalization of the two "strong" perturbations V^{R_2} and V^P. From Eq. (14), after a little algebra, one obtains that this "delta" peak appear at $\Delta = -q_{23}$ for an I_2 such that $\gamma_3/\gamma_2 = 1$. The above described behaviour is similar to that derived in Ref. 9 for discussing the photoelectron spectrum in a single, strong e.m. field, tuned near an atomic autoionization resonance.

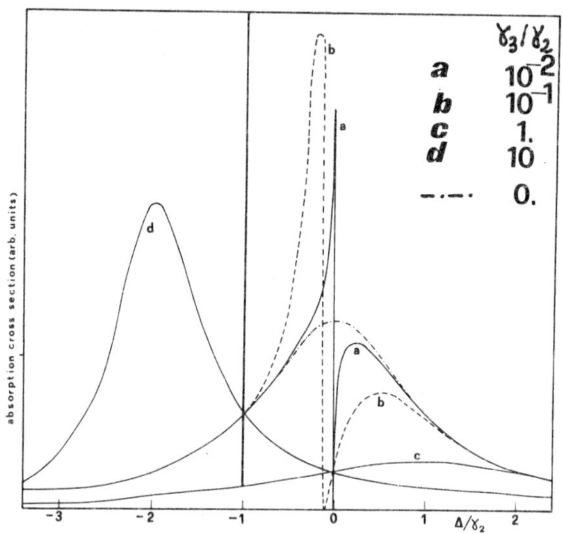

FIGURE 4 - Absorption cross-sections as a function of the detuning of the probe field for the case $q_{12} = \infty$, $q_{23} = 1$ and for various intensities of the strong field. For a further explanation see the text.

The Ref. 9 just mentioned serves also to establish a close contact with the formally analogous problem of autoionization in a strong e.m. field[10]. The whole discussion in this paper can in fact be translated, mutatis mutandis, to the case of autoionization.

The double resonance scheme discussed here can also be utilized to investigate the possibility of increasing the predissociation yield (a photopredissociation yield) by a second field ω_2. The process has been called[11] photon-catalysis since photons ω_2 are not consumed. Eqs. (14), (15) are however not sufficient for that, since we calculate the population that leaves the ground state (and gives rise to the absorption signal) but we do not distinguish what part goes to the state $|3\rangle$. The problem can be tackled by computing[5] the total population of the continuum; i.e.

$$P_C(t) = 1- |U_{11}(t)|^2 - |U_{31}(t)|^2$$

The present approach to the problem of field induced modification of molecular processes is not unique. One can proceed considering diabatic field + molecule electronic surfaces and try to solve the problem of nuclear motion or scattering (see two other papers in this volume and references therein). In this way are taken into account aumatically the role of all the vibrorotational manifold supported by the electronic surfaces considered. The two approaches are, in our opinion, complementary. If one has to do with resonant or near-resonant transitions, as in the present case, the isolation of a small number of states is justified and our way of handling the problem is the simplest one. If virtual transitions play an important role, the number of states that one must include becomes larger and larger and the "global" dressed surface approach presents some advantages.

Eq. (14) has been derived assuming that the two fields ω_1 and ω_2 could be described by Fock states. If they are coeherent fields, Eq. (14) does not work, in principle. This corresponds to a situation were the phase of the two fields will play a crucial role (in general it is well known that double resonance experiment can be very sensitive to photon statistics[12]). However, if the two fields are uncorrelated and can be described by diagonal density matrices (as in the case of coherent fields with random phase[13]), Eq. (14) can be considered as the starting point for averaging over the right photon distribution. If the latter is sharply peaked Eq.(14) itself can be considered as correct (this avoids unnecessary complications due to statistical properties of the two fields).

CONCLUSIONS

We end with the following final considerations:

(i) In this article we have shown that the absorption cross-section for a double resonance experiment involving a predissociating state can be explicitly written down in the form given in Eq. (14). The γ's and q's appearing in that expression may be viewed in various manner. Some of them are obtainable from ab initio calculations and in fact if all of them are calculated with precision, an experiment will serve to confirm the theory. Realistically, this may not be possible and one may utilize some of the parameters as given from usual experimental data and then fit the result of a double resonance experiment to obtain the remaining parameters. These latter can also, as stated earlier, be calculated theoretically and checked against measured values. In a nutshell, one sees a fairly wide ranging interplay between theory and experiments to unravel some of novelties of a double-resonant experiment involving predissociation.

(ii) A fairly substational amount of literature has been accumulating in recent years on studying the effect of strong field on an autoionizing state. There is a strong affinity of some of this and what we have studied for the predissociation. Since the research on autoionization is somewhat separated from predissociation even when the strong similarities between the two phenomena cannot be ignored, it is our belief that the theoretical technique utilized in this and our previous papers on this subject should be considered as a fruitful and rather simple approach worth pursuing for the treatment of autoionization.

(iii) In a previous volume and elsewhere, the process of photon catalysis has been analyzed. What has been studied here is essentially the same phenomenon albeit from a more realistic point of view since we consider the molecule to be initially in the ground state. The effectiveness of photon catalysis in general cannot be guaranted from what we have deduced so far. What can be said, however, is that this study fills a gap on a possible scenario of photon catalysis.

(iv) Double resonance experiments and theoretical treatments of the same has shown the rather fascinating possibility of coherence trapping[8]. The phenomenon of coherence trapping with one state in the molecular continuum has not been pursued experimentally, an understandable fact due to the lack of theoretical considerations. (Ideas on population trapping in atoms have been discussed in some recent works[14]). Work reported here and elsewhere is rather encouraging regarding such a prospective experimental study.

(v) Last, but surely not the least is the following: the phenomenon of predissociation is a rich and well known molecular curiosity. It has been studied more carefully in recent years since it reveals the various facets of a molecular state and the subtleties of intramolecular dynamics. The experimental tools of the most recent kind have been necessary to study the details of the predissociating process. Our investigation has been based on the belief that once the basic predissociating phenomena has been studied with a good deal of accuracy, one may prospect rather challenging newer kind of experiments with a clear and methodical theory that will enable one to learn much more about dynamics with molecules involving intense e.m. field than have been hitherto attempted. Since the theory of these experiments can now be said to be on a firm foundation, only the future experimental feasability of these will show how much of our considerations have been useful as well as timely.

REFERENCES

1. See, e.g., G.Herzberg, Molecular Spectra and Molecular Structure (Van Nostrand, New York, 1950) Vol.I, II,III and references therein.

2. E.A.Gislason, J.Chem.Phys. 58, 3702 (1973); J.P.Laplante and A.D.Bandrauk, Chem.Phys.Letters 42, 184 (1976); J.Vigué, Ann.Phys.Fr. 3, 155 (1982).
3. A.Lami and N.K.Rahman, Il Nuovo Cimento 63B, 407 (1981).
4. See, e.g., J.A.Beswick and J.Durup, in Proceedings of the Summer School on Chemical Photophysics, Les Houches 1979, edited by C.N.R.S.
5. A.Lami and N.K.Rahman, Phys.Rev.A. 26, 3360 (1982).
6. A.Lami and N.K.Rahman, Opt.Comm. 43, 383 (1982); J.Mol.Structure 93, 295 (1983).
7. U.Fano, Il Nuovo Cimento, 12, 256 (1935); Phys.Rev. 124, 1866 (1961).
8. E.Arimondo and G.Orriols, Lett.Nuovo Cimento 17, 333 (1976); H.R.Gray, R.M.Whitley and C.R.Stroud Jr., Optics Lett. 3, 218 (1978).
9. K.Rzazewski and J.H.Eberly, Phys.Rev.Lett. 47, 408 (1981); Phys.Rev.A 27, 2026 (1983).
10. P.Lambropoulos and P.Zoller, Phys.Rev.A 24, 379 (1981); Y.S.Kim, P.Lambropoulos, Phys.Rev.Lett. 49, 1698 (1982); M.Crance and L.Armstrong, J.Phys.B: At.Mol.Phys. 15, 3199 (1982), ibid. 15, 4637 (1982); J.S.Agarwal, S.L.Hann, K.Burnett and J.Cooper, Phys. Rev.Lett. 48, 1164 (1982); P.T.Greenland, J.Phys.B: At.Mol.Phys. 15, 3191 (1982).
11. A.M.F.Lau and C.K.Rhodes, Phys.Rev.A 15, 1570 (1977), ibid. 16, 2392 (1977); A.M.F.Lau in Dynamics of the Excited State edited by K.P.Lawley, (John Wiley & Sons, New York, 1982).
12. P.Zoller, Phys.Rev.A 19, 1151 (1979), ibid 20, 1019 (1979); A.T.Georges and P.Lambropoulos, Phys.Rev.A. 18, 587 (1978), ibid 20, 991 (1979).
13. W.H.Louisell, Quantum Statistical Properties of Radiation (John Wiley & Sons, New,York 1973), Chap.3.
14. P.E.Coleman and P.Knight, J.Phys.B: At.Mol.Phys. 15, 3191 (1982).

COHERENT EXCITATION OF POLYATOMIC MOLECULES UP TO HIGHLY EXCITED VIBRATIONAL STATES BY LASER LIGHT.

A.GIARDINI-GUIDONI[(*)], E.BORSELLA[(*)], R.FANTONI[(*)], C.D.CANTRELL[(o)].
(*) - ENEA, Dip. TIB, CRE Frascati (00044 Italy).
(o) - University of Texas at Dallas, Richardson (U.S.A.)

Abstract Coherent interaction of polyatomic molecules with intense IR laser field is discussed. The occurrence of multiphoton resonances strongly sensitive to molecular vibrational states is peculiar of this interaction. Experimental evidences of multiphoton resonances are shown in case of absorption and dissociation measurements performed by one or two IR lasers.

INTRODUCTION

The past decade has witnessed an extraordinary interest on the part of the chemical physics community in the phenomena of infrared multiple-photon excitation (MPE) and dissociation (MPD) of polyatomic molecules.[1] From the perpective of collision theory, MPE can be considered a half-collision between a molecule and a laser beam in which only the alternation of the state of the molecule is of interest. The ability of molecules containing four or more atoms to absorb enough photons to raise the energy of excitation above the dissociation threshold was initially surprising, in view of the anharmonicity expected of any molecular vibration. The splitting of degenerate excited vibrational states in symmetric molecules[2] and the existence of several modes of vibration with similar frequencies in asymmetric molecules[3] have been shown to provide pathways of excitation in which the steps are nearly equally spaced, thus effectively compensating the decrease in the spacing of successively higher energy levels that one might expect for a simple anharmonic oscillator. Many questions of understanding and of the proper interpretation of experimental data remain, however, chiefly as the result of a protracted debate over the nature of highly excited vibrational states in polyatomic molecules and the processes by which these states are pumped by laser light. The conventional picture of MPE and MPD divides the vibrational states of a molecule undergoing MPE into three distinct regions of successively higher energy[4], corresponding respectively to a sparse spectrum at low energies, a dense spectrum of discrete molecular eigenstates below the dissociation threshold E_D (The Quasicontinuum), and a continuum above E_D. MPE in all three regions can in principle be described by the Schroedinger equation as long as the time between collisions is sufficiently long for the molecules to be effectively isolated from one another. In practice MPE in the quasicontinuum has been described most often by rate equations for the populations of the vibrational states. While the fundamental validity of rate equations in the quasicontinuum remains in doubt, this approach has been somewhat successful in modeling the total energy absorbed during MPE.

DYNAMICS IN THE QUASICONTUUM

One of the major questions about MPE concerns the distribution of energy within molecules that have absorbed energy from the laser beam. The distribution of the velocities of the molecular fragments formed in MPD has been found[5] to be consistent with the equipartition of energy among all the degrees of freedom of the molecule prior to dissociation. This does not prove that the energy absorbed from the laser is distributed thermally below the threshold of dissociation. Indeed, Lyman and co-workers[6] found that the infrared absorption spectrum of SF_6 undergoing MPE bore no resemblance to the spectrum of thermally heated SF_6, thus demonstrating that equipartition of energy does not occur in this case for molecules well below the dissociation threshold. It is obvious that if energy were equipartitioned during MPE, then it would be impossible to produce unimolecular or bimolecular reaction rates with laser excitation that differ substantially from those that can be obtained with thermal excitation. As with other aspects of MPE and MPD, an accurate experimental probe and a satisfactory theory of the intramolecular distribution of energy during MPE have yet to be found.

The equipartition of energy in MPE is often ascribed by those who believe in its existence to intramolecular relaxation (IMR). Some of the problems in understanding IMR during MPE have arisen previously in other contexts, while others are new. One of the most venerable models for the dynamics of a quasicontinuum[7-11] is that of a single state coupled to N closely spaced discrete states. We shall call such a system a (1,N) system. Essentially all scattering processes of a particle with no internal degrees of freedom involve a (1,N) system in which the single state is the incident wave and the N discrete levels are the other eigenstates in a finite volume with energies below a certain cutoff value. The Schroedinger equation in this case is

$$i\hbar d(\psi_K)/dt = \hbar^2 k^2/2m(\psi_K + \Sigma_K V_{K,K'} \psi_{K'}) \quad (1)$$

where the time-dependent wave function $\psi(r,t)$ is written

$$\psi(r,t) = \Sigma_K U_K(r) \psi_K(t) \quad (1b)$$

in terms of normal-mode eigenfunctions $U_K(r)$.

Correspondingly, the Schroedinger equation for a single lower level with amplitude C_O linked by dipole-allowed transitions with N upper levels with amplitudes C_1, C_N is, in the rotating-wave approximation (1b),

$$i\hbar d(C_O)/dt = - \sum_{1=M}^{N} E(t) \mu_{O,M} C_M \quad (2a)$$

$$i\hbar d(C_M)/dt = - \hbar \delta_M - E(t) \mu_{(O,M)} C_M \quad (2b)$$

where $\delta_M = (\omega - E_M)/\hbar$ is the detuning from resonance, ω being the laser frequency. We have assumed that the N upper levels are eigenstates of the Hamiltonian in the absence of the laser field, and therefore do not interact with one another.

Ordinarily in scattering problems one is concerned with solving Eq.(1) for an incident wave that is formed from a superposition of many eigenfunctions U_K, while in MPE a molecule can be assumed to be in a single initial state. Further in scattering theory one normally makes a continuum approximation, thereby avoiding the recurrence effects that result from the nonzero spacing of the energy levels. Under these circumstances it is appropriate to suppose, as Weisskopf and Wigner[8] did, that the amplitude of the lower state decays exponentially in time. The damping constant is

$$\gamma = \pi/\hbar \; (V^2_{N,O} \; G(\omega)) \tag{3}$$

According to their calculation, where $G(\omega)$ is the density of states. Correspondingly, over a time interval that is short compared with the recurrence time, the amplitude $C_O(t)$ determined by solving Eq.(2) (for a laser field that is suddenly switched from zero to a finite value E_O) decays exponentially[12]. The time in which the ground-state amplitude decays may be considered as the intramolecular relaxation time, more will be said on this point below. In this case the damping constant depends on the amplitude of the laser field, on the magnitude of the dipole matrix elements $\mu_{O,N}$, and on the density of states. If the width of the dipole distribution in frequency is large compared with the peak rabi frequency $\mu_{O,N}E_O/\hbar$, in fact, the damping constant is

$$\gamma = \pi/4\hbar \; \mu^2_{0,N} \; E^2_0 \; G(\omega) \qquad (4)$$

the dependence of the IMR time $2\pi/\gamma$ on the laser field strength is not surprising in view of the analogy between Eqs.(1) and (2), but surprises many who regard the IMR time as an intrinsic property of the molecule. The dependence of the IMR time on the conditions of excitation has been confirmed in theoretical calculations for pulses that are not switched on suddenly[13].

At this point it is logical to ask why we regard $2\pi/\gamma$ as the IMR time. The IMR time is usually defined as the time required for the overlap of the initial state $\psi(0)$ with the state at a later time $\psi(t)$ to decay towards its asymptotic, long-time value. A preferable definition, and one that emphasizes the role of coherence, is that the IMR time is the time required for the decay of a quantity such as the expectation value of the dipole moment, which will be nonzero only if the system is in a coherent superposition of upper and lower states. For the (1,N) system these definitions lead to essentially the same IMR time, which is evidently in both cases the time that is required for the ground-state amplitude to decay to zero.

Another important question concerns the re

lation of the IMR time to the distribution of energy within the molecule. While this question, which goes to the heart of the issue of the equipartitioning of energy versus selectivity in laser chemistry, is far from resolution at the present time, we offer a few comments. The IMR time we have calculated above does not depend on the specific nature of the N molecular eigenstates that constitute the upper "level". One state, for example, may be a superposition of many normal vibrational modes, while others may involve a few modes. Short IMR times, which accompany high laser intensities, result from the excitation of a broad spectrum of molecular eigenstates, which may involve the excitation of many different parts of the molecule. The state produced by such strong pumping may well be difficult to distinguish experimentally from a state in which the energy is equipartitioned. However, excitation with lower laser intensities will produce longer IMR times, will excite a narrower spectrum of states, and may possibly be manipulated (by adjusting the laser frequency and intensity) to produce a more selective excitation.

The vibrational states of the electronic ground state of real polyatomic molecules differ in major respects from the simple model (1,N) system.

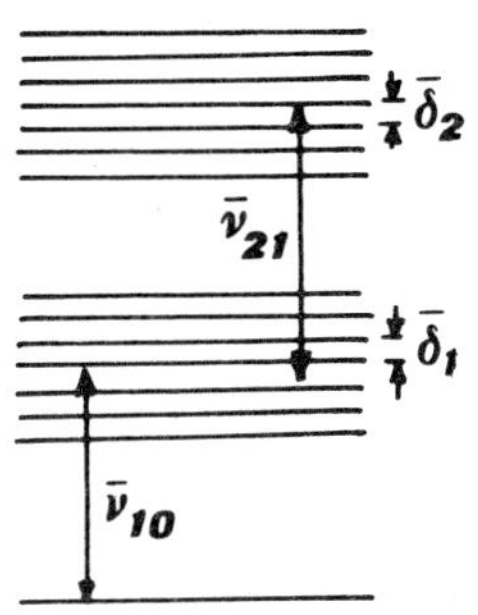

FIGURE 1. Excitation of a (1, N,N') system.

A model that more nearly represents in a qualitative way the spectrum of vibration-rotation states of a polyatomic molecule is the (1,N,N',...) system,which consists of a single lower level and successively higher bands of closely spaced levels (see Figure 1). The new feature in this type of system is that of multiphoton resonances, that is, direct transitions from one band to a non-neighboring band by the absorption of more than one laser photon. The experimental work we have carried out givies direct evidence for the existence of this phenomenon in MPE.

EXPERIMENTAL EVIDENCES OF MULTIPHOTON RESONANCES

Multiphoton resonances have already been observed in SF_6[14] spectra measured at low temperature and low exciting intensity. Their assignment[15] was accomplished taking into account the anharmonic splitting[2] of vibrational levels in the octahedral force field. In case of less symmetric molecules at room temperature contributions from rotational sublevels and vibrational hot bands can be relevant to compensate the anharmonicity. Excited states, coming from compounds vibrations and overtones of vibrational modes different from the pumped one, can also contribute to form the (1,N,N') system of polyatomic molecule levels.

Narrow resonances have been observed in the MPE spectra of CF_3Br and CF_3I excited by means of a continuosly tunable CO_2 laser. Results are reported respectively in Figure 2 and 3 as

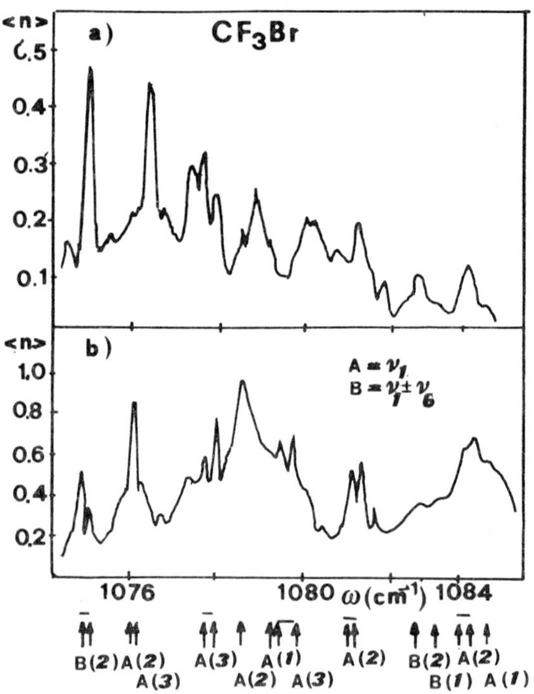

FIGURE 2. MPE spectrum of CF_3Br.(a) Measurements at $\emptyset$=0.35J/cm^2,p=0.5torr and T$\sim$300°K. The continuous curve joins experimental points separated by 0.09 cm^{-1}. (b) Calculations (see text).

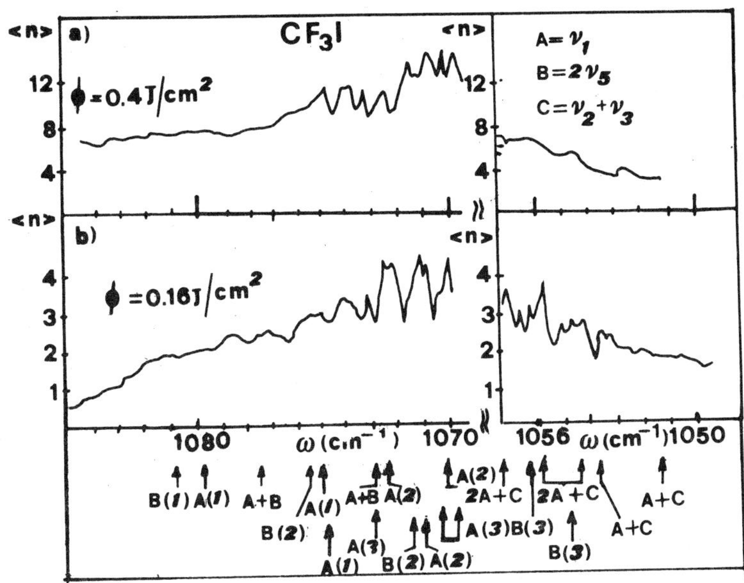

FIGURE 3. MPE spectrum of CF_3I measured at P=0.5 torr and T ~ 300°K. (a) Ø=0.4J/cm^2, (b) Ø=0.16J//cm^2. The continuous curve joins experimental points separated by 0.09 cm^{-1}.

measured by means of optoacustic detection[16]. One, two and three photon resonances in the ν_1 ladder and in the $\nu_1 \pm \nu_6$, $\nu_1 \pm 2\nu_6$, $\nu_1 \pm \nu_3$ have been identified in case of CF_3Br by using a simple (1,N) model[17,18] for its vibrorotational structure. Calculations and arrows on the bottom of Figure 2 mark the assignment in the investigated (1074-1085 cm^{-1}) frequency range. As far as CF_3I is concerned, in order to understand data in the region 1070-1085 cm^{-1} it is necessary to take into account the $2\nu_5$ overtone together with the ν_1

mode; while in the region 1050-1056 also the $\nu_2 + \nu_3$ compound vibration seems to play a role. In Figure 3 arrows on the bottom mark this assignment[18].

The structures, observed at rather low energy fluence, are kept in the MPE spectra also when the fluence increases. Two frequency MPE experiments performed at room temperature in the optoacoustic cell by means of two line tunable CO_2 lasers show clearly some resonances in CF_3Br (Figure 4). In case of CF_3I (Figure 5) a rather

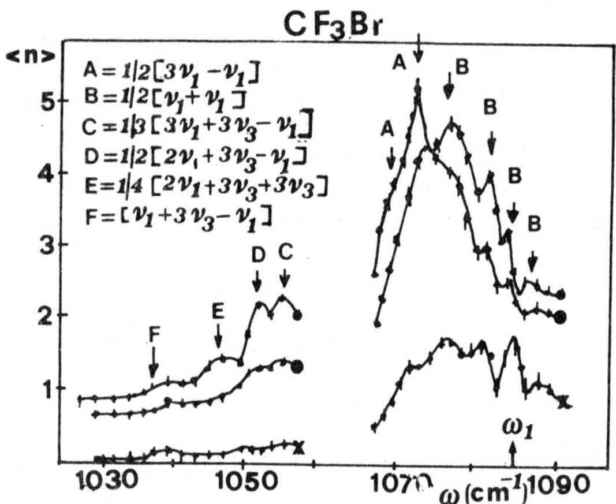

FIGURE 4. Two frequency MPE spectrum of CF_3Br measured at p = 0.5 torr and T ~ 300°K; ω_1 = = 1084.6 cm^{-1} $\emptyset_2$ = 1.2 J/cm^2 o $\emptyset_1$ = 0.4 J/cm^2,• $\emptyset_1$ = 0.8 J/cm^2, x single frequency $\emptyset$ = 1.2 J/cm^2.

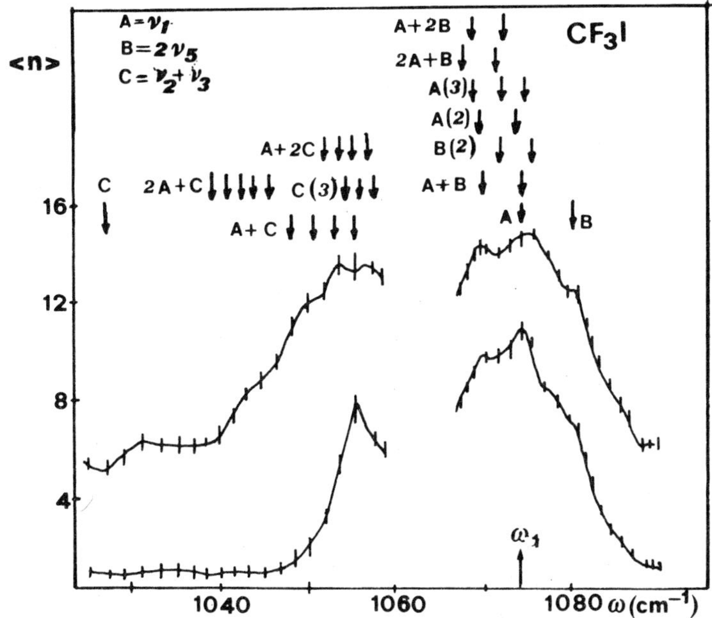

FIGURE 5. Two frequency MPE spectrum of CF_3I measured at p = 0.3 torr and T ~ 300°K; ω_1 = 1074.6 cm^{-1} $\emptyset_1$ = 0.15 J/cm^2 $\emptyset_2$ = 0.35 J/cm^2.

broad convolution of peaks is measured due to the much higher number of possible transition and to the rough frequency sampling. As the arrows in the red portion (1028 - 1057 cm^{-1}) of Figure 4 shows, the $3\nu_3$ overtone contributes to the CF_3Br MPE in this region. Data reported both in Figure 4 and Figure 5 show that resonances starting from the first ν_1 excited state noticeably increase when this state is selectively preheated by the first exciting laser tuned on the ν_1. This indicates that the excitation is not lost i.e. , the energy is not yet fully randomized, within times of the order of the laser pulses and synchronization (about 50 nsec).

Pure vibrational multiphoton resonances have been observed also in a molecular beam dissociation experiment (MPD) performed on expansion cooled molecules by means of one and two line tunable CO_2 laser. Results are reported in Figure 6.

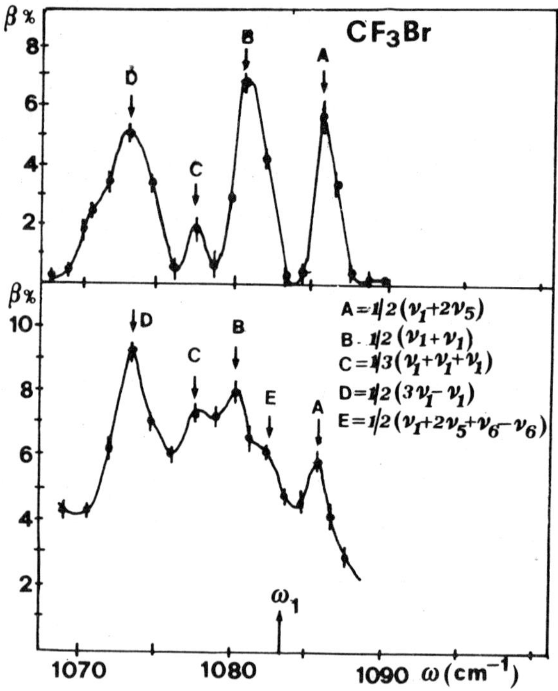

FIGURE 6. MPD spectrum of CF_3Br: (a) single frequency $\emptyset$=4.0 J/cm^2 1% CF_3Br in Ar; (b) two frequencies ω_1=1084.6cm^{-1}, $\emptyset_1$=1.5J/cm^2, $\emptyset_2$=3.5J/cm^2 10% CF_3Br in Ar.

A two photon resonance due to the presence of the $2\nu_5$ levels on the blue side of the ν_1 spectrum is evident. The single photon resonance lying in that region is absent[19] in MPD spectra because the anharmonic bottleneck prevents the excited molecules to reach E_D and to dissociate. Also one ν_6 hot band in the ν_1, $2\nu_5$ mixed ladder is observed in the spectrum measured at higher vibrational temperature (Figure 6a). The ν_1 excited state transition at 1074.6 cm^{-1} dominates the two frequency spectrum (Figure 6b), so demonstrating the relevance of coherent excitation also at high levels.

In case of the asymmetric molecule C_2F_5Cl the existence of several compound vibrations ($\nu_9+\nu_{14}$, $\nu_6+\nu_{10}$, $\nu_7+\nu_8$, $\nu_7+\nu_{15}$, $\nu_6+\nu_9$, $\nu_5+\nu_{16}$, $\nu_6+\nu_8$) near the ν_4 mode provides pathways of excitation leading to structures also in the two frequency dissociation spectrum reported in Figure 7. These peaks are evidence of strong spectroscopic structure in the quasi-continuum which might arise from intensity borrowing, due to anharmonic effects by combination bands that lie near the IR active ν_4 mode[3].

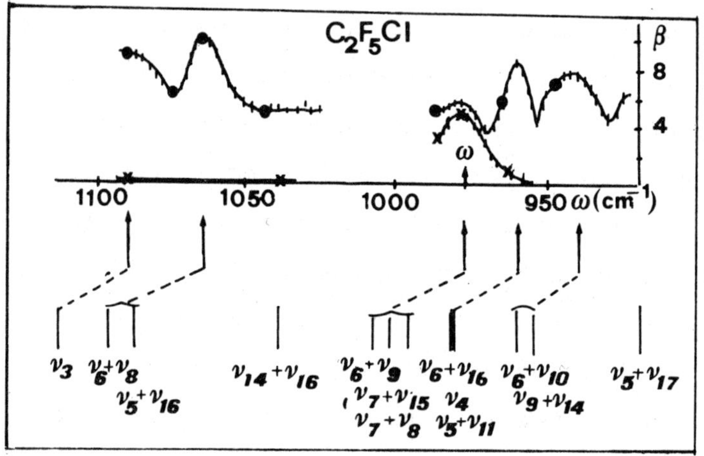

FIGURE 7. MPD spectrum of C_2F_5Cl, x single frequency $\emptyset = 2.2$ J/cm^2 • two frequencies ω_1=977.2 cm^{-1} $\emptyset_1 = 1.4$ J/cm^2 $\emptyset_2 = 2.2$ J/cm^2 pure C_2F_5Cl beam. Arrows mark assignment of multiphoton resonances.

ACKNOWLEDGEMENTS

Helpfull discussion with Prof.J.Reuss and Dr.M. Dilonardo are gratefully acknowledged, as well as the help of Dr.D.Adams, Dr.D.Masci, Dr.A.Palucci, U. Del Bello and A.Ferretti in the experimental work. Thanks are due to R.Belardinelli, P.Cardoni, I.Cenciarelli, M.Nardelli, S.Ribezzo and G.Schina for technical assistance.

REFERENCES AND FOOTNOTES

*. Research supported by the ENEA (Italy), CNR (Italy) and NATO (Grant N° 018.82).

1. For recent reviews, see (A) C.D. Cantrell, S.M. Freund and J.L.Lyman, in Laser Handbook, Vol.3, ed. M.L.Stitch (North-Holland, Amsterdam,1979), pp.485-576; (B) C.D.Cantrell, V.S.Letokhov and A.A.Makarov, in Coherent Nonlinear Optics: Recent Advances, Eds. M.S.Feld and V.S.Letokov (Berlin, Springer-Verlag, 1980), pp.165-269.
2. C.D.Cantrell and H.W.Galbraith, Optics Comm. 18, 513 (1976); Optics Comm.,21, 374 (1977).
3. E.Borsella, R.Fantoni, A.Giardini-Guidoni and C.D.Cantrell, Chem.Phys.Lett., 86, 284 (1982).
4. N.Bloenbergen, C.D.Cantrell and D.M.Larsen, in Tunable Lasers and Applications, Eds.A.Mooradian, T.Jaeger and P.Stokseth (Berlin, Springer--Verlag, 1976), pp. 162-176.
5. M.J.Coggiola, P.A.Schulz, Y.T.Lee and Y.R.Shen, Phys.Rev.Lett., 38, 17 (1977); J.D.Krajnovich, A.Giardini-Guidoni, AA.S.Sudbo, P.A.Schulz, Y.R. Shen and Y.T.Lee, in Laser-Induced Processes in Molecules, Eds. K.L.Kompa and S.D.Smith (Berlin, Springer-Verlag, 1979), p.176.
6. J.L.Lyman, L.J.Radziemski, Jr. and A.C.Nilsson, IEEE J.Quant.Electron., QE-16, 1174 (1980).
7. O.K.Rice, Phys.Rev., 33, 748 (1929); Phys.Rev., 34, 1451 (1929); Proc.Nat.Acad.Sci.(U.S.), 15, 459 (1929).
8. V.Weisskopf and E.Wigner, Zeits.F.Physik, 63, 54 (1930).
9. G.W.Robinson and R.P.Frosch, J.Chem.Phys., 37, 1962 (1962); J.Chem.Phys., 38, 1187 (1963).
10. M.Bixon and J.Jortner, J.Chem.Phys., 48, 715 (1968); J.Chem.Phys. 50, 3284 (1969); J.Chem. Phys., 50, 4061 (1969).
11. R.Lefebvre and J.Savolainen, J.Chem.Phys., 60, 2509 (1974).

12. A.A.Makarov, V.T.Platonenko and V.V.Tyakht, Zh.Eksp.Theor.Fiz., 75, 2075 (1978); Sov. Phys.JETP, 48, 1044 (1978).
13. R.S.Burkey and C.D.Cantrell, Optics Commun. 43, 64 (1982).
14. S.S.Alimpiev, N.V.Karlov, S.M.Nikiforov, B. G.Sartakov, E.M.Khokhlov and A.Shtarkov, Optics Commun.,31, 309 (1979).
15. M.Dilonardo, M.Capitelli and C.D.Cantrell, Applied Physics, B29, 181 (1982); D.P.Hodgkinson, A.J.Taylor, D.W.Wright and A.G.Robiette, Chem.Phys.Lett., 90, 230 (1982).
16. G.Sanna, M.Nardi and M.Bernardini, Proceedings of the International Conference on Lasers '81, (C.B.Collins, Ed.) (STS Press, 1982), p.83.
17. E.Borsella, D.Masci, M.Capitelli and M.Dilonardo, Chem.Phys. (in press, 1983).
18. E.Borsella, R.Fantoni, A.Giardini-Guidoni, D.R.Adams, C.D.Cantrell, Chem.Phys.Lett. (submitted).
19. E.Borsella, R.Fantoni, A.Giardini-Guidoni, D. Masci, A.Palucci and J.Reuss, Chem.Phys.Lett., 93, 523 (1982).

LASER-INDUCED BOUND STATES AT SURFACES: ION NEUTRALIZATION AND ADSORPTION

KAI-SHUE LAM, MICHAEL HUTCHINSON AND THOMAS F. GEORGE
Department of Chemistry, University of Rochester
Rochester, New York 14627 USA

Abstract A laser can be used to generate bound states, both electronic and vibrational, of a foreign atom on a solid surface, and is capable of enhancing processes like ion neutralization and adsorption.

INTRODUCTION

Two mechanisms for laser-generated bound states of a foreign species on a solid surface are discussed. The first, ion neutralization, leads to a bound electronic valence state of a projectile ion that is not degenerate with any electronic band states of the surface, while the second, radiative adsorption, gives a stable vibrational state of an adsorbed atom. The laser intensity plays the dominant role in the first process, whereas in the second a resonantly-tuned frequency is of greater importance.

ION NEUTRALIZATION

In many theories treating neutralization of ions scattered from solid surfaces,[1,2] resonance processes play a dominant role. This kind of resonance is between a discrete state and a continuum level. Thus a valence level of the projectile ion (the discrete state) is considered to have a position-dependent energy $\varepsilon_0(z)$ (see Fig. 1) which, at some region of small z, is resonant with a continuum of band levels of the electronic states in the solid surface; the position z represents some measure of the distance of the ion from the surface. The bound-continuum interaction is usually assumed to be significant only for small z, that is, when the ion is near the surface. Moreover, the strength of the interaction is assumed to be such that both the shift and width of the resonant state are small, and that the energy spectrum of the ion-surface system is the same as that of the band states of the solid surface. These limitations on the bound-continuum interaction, together with the particular nature of the model--one discrete state embedded in one continuum--lead to the situation where true bound states of the ion-surface system play no role at all in the discription of the mechanisms leading to charge transfer. If such mechanisms require true bound states, they have to be added to the model. For example, Auger neutralization may take place if a deep-lying unoccupied level of the incoming ion is available.[3]

Within the constraints of the model described above and a particular physical system, the bound-continuum interaction has a fixed strength, and thus true bound states either enter the picture or not at all. With the introduction of a laser, however, the situation is changed drastically. The fact of crucial importance is that both the laser frequency and the field strength are adjustable. Thus the same model, when it is understood that the bound-continuum interaction is due to a field coupling, can not only incorporate a variable $\varepsilon_0(z)$ (variable not only with respect to z but by amounts directly related to $\hbar\omega$) but also a variable coupling strength (directly related to the field strength). It is precisely these variable quantities which lead to the possible existence of true bound states, even when they are precluded in the absence of the field. The laser may then be used to enhance bound state mechanisms which are either unimportant or impossible in the field-free situation. With respect to the Auger neutralization process mentioned above, the laser may literally create a "deep-lying" valence state to act as receptor of an electron from a band level. Such a state may also interact resonantly with any core levels of the solid surface that happen to be approximately degenerate with it. In what follows we give a brief discussion of the theory behind the formation of the laser-induced bound state.

The schematic picture of the energy level structure in our model is given in Fig. 1. The Hamiltonian may be written as

$$H(z) = \sum_k \varepsilon_k n_k + (\varepsilon_0(z) + \hbar\omega)n_0 + \sum_k [V_k(z)c_0^\dagger c_k + \text{h.c.}], \quad (1)$$

where $V_k(z)$ is the bound-continuum interaction provided by the field coupling, and k is the band index. The emergence of possible bound states is most easily seen by focusing on the eigenvalue equation for H(z):

$$\varepsilon - (\varepsilon_0 + \hbar\omega) + g^2\int d\varepsilon' \frac{\rho(\varepsilon')|V(\varepsilon')|^2}{\varepsilon' - \varepsilon} = 0, \quad (2)$$

where we have introduced a coupling strength g into $V_k(z)$ such that

$$V_k(z) = gV(\varepsilon), \quad (3)$$

and the z dependence is not explicitly written on the RHS. For radiative coupling, g^2 is directly proportional to the field strength.

Under what conditions will a true bound state emerge? There will be bound states when Eq. (2) admits negative

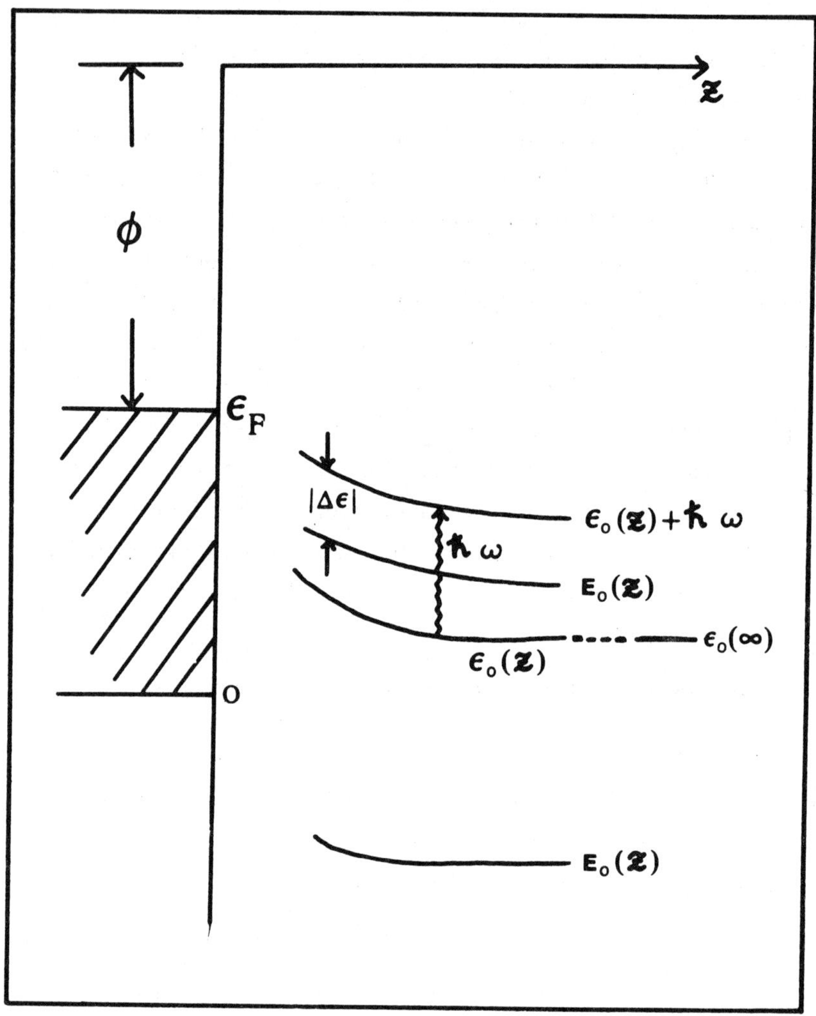

FIGURE 1. Energy level structure of the model describing ion neutralization at a solid surface. $\varepsilon_0(z)$ is the unoccupied valence level of the ion, ω is the laser frequency, and $\Delta\varepsilon$ is the level shift (this is shown to be negative in the picture). $E_0(z)$ is either a virtual level when it is degenerate with the conduction band, or a true bound state when it is outside the conduction band. ε_F is the Fermi energy, and ϕ is the work function.

energy solutions. Let this solution be $E_0 = -x$, where $x>0$. Eq. (2) then reads

$$x + \varepsilon_0 + \hbar\omega = g^2 \int d\varepsilon' \frac{\rho(\varepsilon')|V(\varepsilon')|^2}{\varepsilon' + x} . \tag{4}$$

Examination of the quantitative picture for a graphical solution quickly reveals that there will be a negative energy solution only when

$$g > g_{crit},$$

where

$$g_{crit} = \left[\frac{\varepsilon_0 + \hbar\omega}{\int d\varepsilon' \frac{\rho(\varepsilon')|V(\varepsilon')|^2}{\varepsilon'}} \right]^{1/2} . \tag{5}$$

Thus whenever the laser field strength is increased beyond a value specified by the critical coupling constant g_{crit}, a true bound state emerges. Furthermore, the model only admits one and only one such state.

For $g < g_{crit}$, however, the valence level of the projectile ion becomes an unstable (virtual) state with a shift in energy, $\Delta\varepsilon$, given by

$$\Delta\varepsilon = -g^2 P \int d\varepsilon' \frac{\rho(\varepsilon')|V(\varepsilon')|^2}{\varepsilon' - E_0} , \tag{6}$$

where P denotes the principal value of the integral. Eq. (6) implies that when

$$|\Delta\varepsilon| > \varepsilon_0 + \hbar\omega,$$

we have a true bound state. Looking at Fig. 1, then, the laser can be imagined to do the following thing: As the field strength is increased, the resonant valence level $\varepsilon_0 + \hbar\omega$ is pulled progressively down the conduction band. As long as $g < g_{crit}$, $E_0(z)$ stays within the conduction band, and at most we have an unstable state. But when

$g > g_{crit}$, $E_0(z)$ falls outside of the band, and a true bound state results.

RADIATION-ASSISTED ADSORPTION

We now consider an alternative mechanism for laser-generated bound states. This is the process of radiative adsorption, which is illustrated in Fig. 2. During the encounter with a surface, an adatom can undergo transition to a bound state of the adatom-surface potential by stimulated emission of a photon. If the surface were rigid, such a bound state would be unstable to the reverse process, photon absorption. However, by coupling the adatom motion to the phonon "bath" of the solid, there exists the possibility of a simultaneous decay to a lower-lying bound state by phonon creation. Such a state would be a true final state, except at temperatures sufficiently high that phonon annihilation (feedback) is important.

We shall now sketch the theory of this process for adsorption on a one-dimensional lattice (the extension to three dimensions is straightforward, but the notation for the one-dimensional problem is simpler). We may write the Hamiltonian as

$$H = H_0 + H^r_{int} + H^p_{int}, \tag{7}$$

where

$$H_0 = H^r + H^p + H^a, \tag{8}$$

and the superscripts r, p and a stand for the radiation, phonons and the adatom, respectively. The corresponding eigenvectors of these zeroth-order Hamiltonians are written

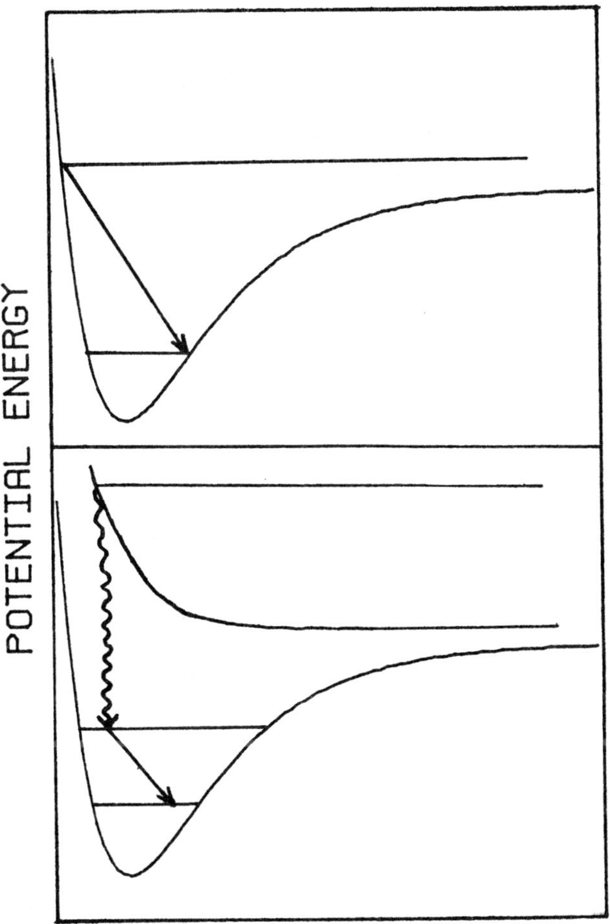

FIGURE 2. Adsorption events depicted in the space of the atom-surface electronic eigenstates. Top: adsorption from a continuum by the creation of a single phonon. Bottom: adsorption from a continuum by the "simultaneous" creation of a photon and a phonon. The phonon events are depicted by straight arrows.

as $|n^r_{k\sigma}>$, $|n^p_k>$ and $|E^+>$, $|f>$. The third and fourth eigenvectors (of H^a) are, respectively, the scattering state at energy E (with an outgoing wave boundary condition) and a final bound state f. The latter state thus represents the adatom bound to the surface. Using a projection operator Q to project out the bound intermediate state $|m>$ to which the radiative transition is made, it is now possible to write the adsorption rate R due to a plane-polarized laser as

$$R = |<n^p_k+1|<f|H^p_{int}|m>|n^p_k>|^2$$

$$\times |(G^+_{QQ})_{mm}<n^r_{k0}+1|<m|H^r_{int}|E^+>|n^r_{k0}>|^2$$

$$\times \delta((E-E_f) - \hbar\omega_p - \hbar\omega_r) , \tag{9}$$

where

$$G^+_{QQ} = \lim_{\varepsilon\to 0} (E + i\eta - H_{QQ} - H_{QP}G^+_0 H_{PQ})^{-1} , \tag{10}$$

$$P = 1-Q$$

and

$$O_{XY} = XOY .$$

G^+_0 is the Green's function for non-resonant scattering and is expanded in a product basis of surface plane waves and adatom scattering out-waves.

We now consider the contribution of a single-phonon transition to the rate by expanding the operator H^p_{int} in a Taylor's series in the lattice coordinate z_ℓ:

$$H^p_{int} = \sum_\ell d^\ell(z)\,(z_\ell - z^0_\ell), \tag{11}$$

where z is the corrdinate of the gas atom, and z^0_ℓ is the equilibrium (frozen lattice) coordinate of atom ℓ in the chain,

$$d^{\ell}(z) = \frac{\partial}{\partial z_{\ell}} V(z-z_{\ell})\big|_{z_{\ell}=z_{\ell}^{0}} \tag{12}$$

To arrive at the total averaged rate $\langle R\rangle$, it is necessary to average over initial and sum over final phonon states. Dropping for simplicity the Fock states of the radiation field, we are led by standard manipulations[4,5] to

$$\langle R\rangle = \sum_{\ell} \frac{|\langle f|\sum_{\ell} d^{\ell}(z)|m\rangle|^{2}}{NM} \frac{1}{\omega_{p}} (\bar{n}_{p} + 1)\, \rho(\omega_{p})$$

$$\times\ |(G_{QQ}^{+})_{mm} \langle m|H_{int}^{r}|E^{+}\rangle|^{2}, \tag{13}$$

where $\rho(\omega_p)$ is the phonon density of states as a function of the energy-conserving frequency, ω_p, $\bar{n}_p$ is the average occupation number of phonon mode p,

$$\bar{n}_p = (e^{\hbar\omega_p/kT}-1)^{-1},$$

and there are N atoms in the chain, each of mass M. An important feature of Eq. (13) is the natural separation of the gas-phase problem from that of the surface. The only point of contact is in $(G_{QQ}^{+})_{mm}$, which would contain width and level shift terms due to the phonons, and related terms which are due to the radiative interaction.

ACKNOWLEDGMENTS

This research was supported by the Air Force Office of Scientific Research (AFSC), United States Air Force, under Grant AFOSR-82-0046, the Office of Naval Research and the U.S. Army Research Office. The United States Government is authorized to reproduce and distribute reprints for governmental purposes notwithstanding any copyright notation hereon. TFG acknowledges the Camille and Henry Dreyfus

Foundation for a Teacher-Scholar Award (1975-84) and the John Simon Guggenheim Memorial Foundation for a Fellowship (1983-84).

REFERENCES

1. J. C. Tully, Phys. Rev. B, 16, 4324 (1977).
2. R. Brako and D. M. Newns, Surface Sci., 108, 253 (1981).
3. H. D. Hagstrum, Phys. Rev., 96, 336 (1954); 122, 83 (1980).
4. F. O. Goodman, Prog. Surf. Sci., 5, 261 (1974)
5. A. A. Maradudin, E. W. Montroll, G. H. Weiss and I. P. Ipatova, in Theory of Lattice Dynamics in the Harmonic Approximation (Academic Press, New York, 1971), pp. 261-375.